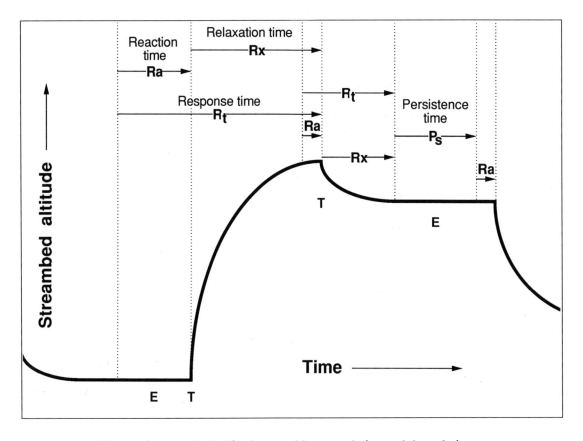

Changes in streambed altitude caused by aggradation and degradation are used as a reference plot to show the components of response time. $R_t$, response time, is the sum of the reaction time, Ra, and relaxation time, Rx. $P_s$ is the time of persistence of new equilibrium conditions, and T and E are threshold and equilibrium conditions, respectively.

Geomorphic Responses to Climatic Change

# Geomorphic Responses to Climatic Change

WILLIAM B. BULL

New York  Oxford  ☐  OXFORD UNIVERSITY PRESS  ☐  1991

Oxford University Press

Oxford   New York   Toronto
Delhi   Bombay   Calcutta   Madras   Karachi
Petaling Jaya   Singapore   Hong Kong   Tokyo
Nairobi   Dar es Salaam   Cape Town
Melbourne   Auckland

and associated companies in
Berlin   Ibadan

Published by Oxford University Press, Inc.,
200 Madison Avenue, New York, New York 10016

Oxford is a registered trademark of Oxford University Press

Library of Congress Cataloging-in-Publication Data
Bull, William B., 1930–
Geomorphic responses to climatic change / William B. Bull.
p. cm.   Includes bibliographical references
ISBN 0-19-505570-5
1. Climatic geomorphology.   I. Title.
GB447.B85   1991   551.4′1—dc20
90-32977

1 2 3 4 5 6 7 8 9

Printed in the United States of America
on acid-free paper

*To my family, whose support and love*
*were vital for the completion of this research,*
*and to Luna Leopold and Joe Poland*
*who instilled in me a desire for knowledge*
*and a dedication to do my best.*

# Preface

Landscapes are fascinating. Since about 1950, the emphasis on better understanding of the geomorphic processes of erosion and deposition, the development of dating methodologies that permit estimation of process rates, the rapid expansion and utilization of soils in landscape studies, and the application of many aspects of systems theory permit preliminary appraisals of important questions about changes in the landscapes around us. Earth scientists now address questions about active tectonics and climatic change that are important both for academic and applied reasons. Those curious about the world around them should be interested in such questions as, How sensitive to changes in climate are the processes of erosion and deposition in streams? On hills, how large must changes in temperature and precipitation be to influence vegetation, soil-profile formation, and sediment yield? How do past, present, or future climatic changes affect the productivity of the land? An investigator inquisitive about the origins of mountains may wonder, How does a pulse of uplift affect geomorphic processes and the shapes of hills and streams? Is the record of past mountain uplifts revealed in their landscapes? Are stream terraces synchronous or diachronous time lines that sweep through tectonically active landscapes? The answers to these and many other questions may be approached through multidisciplinary studies that involve geomorphology, paleoclimatology, structural geology, seismology, pedology, and Quaternary stratigraphy.

Process-oriented climatic fluvial geomorphology is the topic of this book; a future, companion volume, *Fluvial Tectonic Geomorphology*, will explore the impacts of tectonic uplift on fluvial systems. These two books thus discuss the most important controlling variables that change with time and greatly influence fluvial systems. Both books emphasize the importance of lithology (rock chemistry and fabric) and structure (fractures, joints, and shear planes). In areas where either climate or tectonic base level is the dominant changing variable, seemingly anomalous relations commonly are the result of local spatial variations of lithology and structure.

Vegetation and soil-profile development are key topics in all chapters because they greatly influence erosion and deposition. Fossil plants and relict soils provide data for paleoclimatic inferences and for dating of geomorphic responses to climatic change. Age estimates based on rock weathering and soil-profile development are especially useful for dating flights of stream terraces. Each chapter discusses the most useful pedogenic features for the climate and soil parent material of a different study area.

Much has been written about climatic geomorphology from the viewpoint of how landforms vary with worldwide climatic zones (Budel, 1981; Derbyshire, 1973,1976; Peltier, 1950; Stoddart, 1969; Tricart, 1974; Tricart & Cailleux, 1972). The emphasis is different in this book, which focuses on the subject of landscape change. For example, consider the truism that lithology is one of the most important factors influencing landscape morphology. Chapters by Douglas, Smith and Atkinson and by Thomas in Derbyshire's (1976) book evaluate the possible effects of five rock types on distinctive landforms in different climatic settings. By contrast, this volume deals with the question, How do climatic change, lithology, and uplift interact with the dependent variables of fluvial systems to determine the processes and landforms of a particular setting? Lithology, which varies in space, interacts with climate and geomorphic processes, which have changed during the late Quaternary. The approach used here is more akin to the concept of morphogenetic systems that result from climatic-process interactions (Wilson, 1968).

In this book I develop concepts through discussion of climate-induced changes in fluvial systems of four field areas. Studies in the Transverse and Coast Ranges of California, the southern Basin and Range Province of North America, Israel and the Sinai Peninsula of Egypt, and New Zealand provided data from markedly different climatic settings.

The basic topics of climate and paleoclimatology, vegetation, soils genesis, and geochronology are discussed in each chapter as essential background and to assess the responses of geomorphic processes to climatic change. Descriptions of present climates are compared with paleoclimatic inferences. Late Quaternary changes in temperature and precipitation were reflected quickly by changes in plant communities whose influence cascaded through hillslope, stream, and soil subsystems.

Most fluvial landscapes are polygenetic—they reflect the influences of both full-glacial and interglacial climates—because climate continues to change during the formation of mountains and piedmonts. Worldwide Pleistocene–Holocene climatic change and variations in climate during the Holocene profoundly affected fluvial systems in a variety of climatic and tectonic settings. Chapters 2, 3, 4, and 5 examine different aspects of climatic change on fluvial systems during the past 20,000 to 40,000 years in four study areas that presently receive less than 30 to more than 2000 mm mean annual precipitation. Geomorphic thresholds, feedback mechanisms, and response times to perturbations are a common theme in this book, so it is appropriate to discuss conceptual models for changing landscapes in the first chapter.

This book is written primarily for graduate students and colleagues. Because the concepts it presents are essential to many types of geomorphic analyses and instruction, it also may be useful as an advanced undergratuate text.

I am deeply grateful for the assistance and guidance of many persons. Students, faculty, visiting scholars, and staff of the Geosciences Department of the University of Arizona challenged concepts in seminar discussions, assisted in field work, gave editorial advice, and analyzed various drafts of this book. I am particularly indebted to Luna Leopold and Victor Baker for suggestions regarding stream power and resisting power. The Southern California Edison Company provided the initial financial support for the Vidal Valley study. Tom Freeman and Shingi Kuniyoshi of Woodward-Clyde Consultants helped in mapping the Vidal area and in collecting soil-carbonate samples. It was a pleasure to work with Teh-Lung Ku of the University of Southern California, who did the $^{230}TH/^{234}U$ dating. Chapter 3 would not have been possible without the discussions and field assistance of Asher Schick, Ran Gerson, Dan Yaalon, and Aaron Yair of the Hebrew University,

Jerusalem. The work for Chapter 4, and for parts of Chapters 2 and 3, could not have been done without financial support from the Office of Earthquake Studies of the U.S. Geological Survey. The Chapter 4 study was a team effort that involved two of my students. Leslie McFadden was in charge of the field descriptions and laboratory analyses of soil profiles. His expertise in soils gemorphology is reflected by his many fine contributions to Chapters 2 and 4. Christopher Menges was in charge of the Little Tujunga Canyon study and ably assisted in the shallow seismic work. The best study of diachronous stream terraces that I am aware of is discussed in a Ph.D. thesis by Ray Weldon. I greatly appreciate Ray's permission to use seven of his figures in Section 4.4.2. The opportunity to learn about landscape change in humid New Zealand was given by the U.S. National Science Foundation and the U.S. Geological Survey. Kelvin Berryman, Oliver Chadwick, Hugh Cowan, John Bradshaw, Jocelyn Campbell, Steven Forman, Peter Knuepfer, Matt McGlone, Jarg Pettinga, and Philip Tonkin assisted in the Charwell River study of Chapter 5.

I sincerely appreciate the substantial efforts and useful suggestions made by reviewers of book chapters: Pete Birkeland, Fred Cropp, Craig Kochel, Les McFadden, Phil Pearthree, and Kirk Vincent. Special thanks goes to Dorothy Merritts for many useful suggestions for improvement of the entire manuscript.

Authors have a deep sense of personal debt for those who work with them daily on what sometimes seems an endless job. My special thanks to Roger Bull for computer-aided drafting that was highly creative and efficient in conveying important visual information to the readers, and to Jo Ann Overs for precise and prompt word processing and transcription of reams of field notes that in part were patiently collected by my wife, Mary.

*Tucson, Arizona*                                        W.B.B.
*July*

# Symbols

Symbols used exclusive of mapping, metric, chemical, and soil-description symbols. Soil-description symbols are defined in Tables 2.5 and 2.6.

$\gamma$     Specific weight of sediment-water fluid

$\tau$     Shear stress exerted on streambed

$\omega$     Stream power per unit area of streambed

$\tau$     Total kinetic stream power

A     Cross-sectional area of streamflow or stream channel

Ad     Area of drainage basin

As     Area sampled

b     Empirical constant

d     Depth of streamflow

Df     Density of materials finer than gravel

Dg     Density of gravel clasts

E     Equilibrium condition

Gg     Percent change in gravel content

Gs     Suspended sediment transport rate

IS     Factors that favor invading species

ka     Thousands of years before present

ky     Thousands of years

L     Horizontal distance

Ma     Millions of years before present

Me     Mass eroded

Mr     Mass remaining

msl     Mean sea level

my     Millions of years

N     Number of measurements in a sample

n     Manning roughness coefficient

Pd     Average total precipitation for consecutive driest months

$P_s$     Time span of equilibrium conditions

Pw     Average total precipitation for consecutive wettest months

Q     Stream discharge

q     Stream discharge per unit width

R     Reaction time until effects of perturbation are observed

r     Stream channel hydraulic radius

$R_a$   Reaction time

RP   Resisting power

RS   Factors that favor resident species

$R_t$   Response time, the sum of reaction and relaxation times

$R_x$   Relaxation time needed to establish new equilibrium or threshold

S   Stream energy slope

SP   Stream power

Sp   Precipitation seasonality index

St   Temperature seasonality index

T   Threshold condition

Tc   Mean temperature of the coldest month

Th   Mean temperature of the hottest month

Tk   Thickness

v   Velocity of streamflow

Vd   Rock-varnish index

Vm   Modal P-wave velocity

Vp   Mean P-wave velocity

w   Width of streamflow

Wr   Thickness of weathering rind

Xmax   Maximum value of soil-horizon property

Xmin   Minimum value of soil-horizon property

Xp   Soil-horizon property

Xpn   Normalized soil-horizon property

# Contents

Geomorphic Responses to Climatic Change

# 1

# Conceptual Models for Changing Landscapes

Climatic change—past, present, and future—how suddenly it seems to occur, and how little we know about its impact! Climatic change, combined with actions of humans, may cause fragile farmlands to turn into deserts. Controversy abounds as to whether human activity has triggered a greenhouse effect by putting more carbon dioxide and methane gases into the atmosphere. Models suggest that doubling of atmospheric carbon dioxide may cause large temperature increases (Jung & Bach, 1985). Perhaps global warming has already begun! Most climatologists are now confident that measurements of warming during the past century are sufficiently real to rise above measurement uncertainties (Kerr, 1989,1990). Sea level rose $2.4 \pm 0.90$ mm/year between 1920 and 1970 (Peltier & Tushingham, 1989) and studies of a central Asian ice cap indicate that the past 60 years equaled the warmest part of the Holocene between 6 and 8 ka * (Thompson et al., 1989). How will natural or human-induced climatic change affect storm patterns, landslides, sediment yield, flood fre-

quency, stream channel stability, and agricultural productivity? Answers to these vital questions about the sensitivities of geomorphic processes to climatic change lie in studies of the effects of past climatic change on geomorphic processes (Brunsden & Thornes, 1979; Knox, 1983,1984; Solomon, 1987; Eybergen & Imeson, 1989).

The Quaternary is defined by climatic change strong enough to cause ice ages. The largest increase of global temperatures in the past 120 ky occurred during the recent transition from ice age to interglacial climate, and global temperatures continue to change (Mitchell, 1963; Jones et al., 1986). Full-glacial climatic conditions precipitated so much ice on the continents that sea level fell 130 m by 20 ka , but by 12 to 6 ka the warmest climate in the past 120 ky prevailed. The record of landscape change preserved in the hills, streams, alluvial deposits, and soils of humid and arid regions (see Table 2.1 for definitions of climatic terms) reveals fascinating changes in geomorphic processes caused by Pleistocene-Holocene climatic

---

*1 ky = 1000 years; 1 ka = 1 ky before present; 1 my = 1 million years; 1 Ma = 1 my before present (North American Commission on Stratigraphic Nomenclature, 1983).

change. The reasons for variations in sediment yield and runoff from hillslopes induced by climate change in the past, and for changing behavior of streams, provide invaluable lessons for those who ponder the effects of present and future climatic changes.

Climatic change greatly influences geomorphic processes on the hills and in the streams of fluvial landscapes, and its results may be recognized in altered landforms or processes. Changes in the amount and type of precipitation alter the magnitudes and rates of weathering, erosion, transportation, and deposition, thereby changing the shapes of the hills and streams that comprise a *fluvial system** (the drainage basin). Through feedback mechanisms, the changes in landscape cause still more changes in processes, and in arid regions more than 10 ky may be needed for complete adjustment to a *climatic perturbation*. On the other hand, climatic change commonly occurs with such frequency that fluvial systems are continuously adjusting. It is thus desirable to discuss in this first chapter appropriate conceptual frameworks for investigators interested in the impact of climatic change on landscapes over time spans of 0.1 to 100 ky. The short time spans of major floods also are important in terms of amount of work done and in the crossing of important thresholds in fluvial systems (Baker; 1988; Bull, 1988; Kochel et al., 1982; Kochel,1988).

This chapter concentrates on the conceptual foundation needed for better understanding of the response of fluvial watersheds to climatic and tectonic change. Thresholds and equilibrium conditions in open systems composed of both independent variables and variables affected by feedback mechanisms respond to climatic change in complex ways that need to be analyzed in a variety of ways.

## 1.1 Geomorphic Systems

Systems theory provides methods for simplifying the complexity of the real world by classifying forms and phenomena into logical structures. *Geomorphic systems* consist of interacting morphologic attributes or processes. Most geomorphologists consider landscapes as *open systems*—systems that involve the interaction of mass and energy in a physical environment within a defined space through time. An example is a *drainage basin* (a watershed) where mass is exposed to denudational processes by uplift of earth materials above the lowest altitude to which erosion can lower the landscape *(base level)*. Energy in the form of water is an external input—a function of climate—flowing through the system to do work. The potential energy of water from precipitation is determined in part by tectonic elevation of the landscape; it is converted to kinetic energy as the water flows downslope. Open systems are characterized by reversible processes, such as *aggradation* and *degradation* of valley floors by streams and the advance and retreat of glaciers. *Closed systems* are characterized by irreversible processes, such as the chemical reactions that cause concrete to set and the diagenetic expulsion of water from clayey sediments as they are converted to shale and slate.

In geomorphic systems analysis it is important to be aware of time and space frameworks (Schumm & Lichty, 1965). A geologist interested in denudation of mountain ranges is concerned with changes in large areas through time spans of millions of years. A forester or range-management specialist typically is interested in areas the size of watersheds and in time spans that cover the lifetimes of plants. A civil engineer working on the problem of scour around bridge piers is concerned mainly with the immediate physical environment of the bridge and with time spans of hours during flood discharges. All three workers may use an open-system model of thinking, but they are concerned with markedly different spaces and time spans. Time spans of 0.1 to 130 ky and drainage basin areas of $10^{-3}$ to $10^3$ km$^2$ are used in this book. A first step is to examine the fundamental components of such watersheds.

*See glossary at the end of the book for full definitions of geomorphology standard terms.

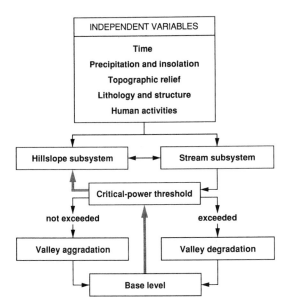

FIGURE 1.1 Basic elements of a fluvial system. Dashed arrows indicate feedback mechanisms.

TABLE 1.1 Partial list of independent and dependent variables of a fluvial system for a time span of more than 1000 years (1 ky).

| Independent Variables | Dependent Variables |
|---|---|
| Climate (source of kinetic energy) | Drainage basin area |
| | Hillslope morphology |
| Total relief (function of uplift; determines potential energy) | Drainage net |
| | Soil-profile development |
| | Vegetation |
| Base level at mouth of drainage basin | Fauna |
| Lithology | Water and sediment yield from hillslopes |
| Geologic structure (fractures, joints, shears, folds) | Stream-channel slope, patterns |
| Human activities | Water and sediment discharge from basin[a] |
| | Stream-channel scour and backfill[a] |
| | Human activities[a] |

[a] Applicable only for the largest flood events for the time span used in this table.

## 1.2 Variables of Fluvial Systems

The basic components of a fluvial open system are shown in Figure 1.1. Selected independent variables are listed at the top of the diagram. Fluvial systems may be divided into hillslope and stream subsystems, and even plant communities and soil-profile catenas may be considered subsystems. Stream subsystems have two disequilibrium *modes of operation,* aggradation and degradation, and one equilibrium mode (Section 1.3) operative during periods of no net change in variable interactions. An erosional-depositional threshold (the threshold of critical power) separates the modes of aggradation and degradation (Section 1.4). Changes in basic components of fluvial systems result in additional changes in the components through feedback mechanisms (Sections 1.2, 1.3).

A variable is an object or an attribute that varies in time, space, or both. Variables such as climate and lithology have little relation to other variables in a drainage-basin fluvial system and may be regarded as *independent variables* (Table 1.1). Independent variables exert primary control on fluvial systems, and may be thought of as external because they are partly or entirely the result of processes outside the fluvial system. Geologic structure is in part dependent on lithology, but for the purposes of landscape studies may be regarded as an independent variable. Most *dependent variables* may be thought of as internal because they are the result of process interactions within the system. Dependent variables—for example, vegetation—are controlled by both independent variables and other dependent variables. Hillslope morphology and drainage net may be considered independent variables for time spans on the order of 0.1 ky. Drainage basin area is an independent variable for time spans of less than 1 ky because it generally is constant during such short geologic time spans if resistant rocks underlie the basin.

Lithology and petrologic structures associated with rock formations undergo minimal changes with time, whereas climate and relief may change continually. Relief is a function both of uplift, which raises mountains, and of erosion, which lowers them.

A crude overall hierarchy of dependent variables may be made. Hillslope morphology, for example, is influenced mainly by independent variables but is also the result of dependent variables such as vegetation and drainage net. Slope morphology affects the soil development that controls plant growth, which in turn affects water and sediment yield. Meandering and braided stream-channel patterns are functions of the dependent variables listed above ''stream-channel slope and patterns'' in Table 1.1 as well as of the independent variables.

The human being may be regarded as both an independent and a dependent variable. Humans have the ability to control vegetation and other dependent variables of geomorphic systems—the ''control systems'' of Chorley and Kennedy (1971). Humans are an independent variable when they modify fluvial systems through such activities as grazing, clear-cutting of tropical rain forests, farming, construction, and damming and relocation of streams. Humans become a dependent variable when they find themselves at the mercy of their previous actions and of nature—damages suffered from floods, soil erosion, and landslides are examples.

Interactions between variables in a geomorphic system may result in *self-enhancing and self-arresting feedback mechanisms*. Progressive accumulation or erosion of hillslope colluvium (Fig. 3.1) are examples of self-enhancing mechanisms. Colluvium deposited on bare rock increases infiltration capacity, thereby providing both water and soil to support vegetation. The vegetation traps additional colluvial materials transported down the slope, thereby furthering the tendency for accumulation of colluvium. These interrelations may be reversed by change in one or several independent variables. For example, removal of vegetation is a *perturbation* (a change in one of the variables of the system) that may result in a stream-channel entrenchment that removes part of the colluvium from the footslopes. Channel entrenchment and removal of soil in either humid or arid regions

tends to reduce vegetation on adjacent hillslopes progressively and thereby to accelerate erosion of the hillslope sediment reservoir. For a reversal in the mode of operation to occur, a threshold must be passed that separates tendencies for progressive accumulation from those of progressive erosion of colluvium. Self-arresting feedback mechanisms promote equilibrium conditions.

## 1.3 Equilibrium in Fluvial Systems

*Equilibrium* is a condition of balance between processes operating in fluvial systems (Knox, 1976). Stream subsystems tend toward equilibrium even when variables are changing. Geomorphic equilibrium occurs when self-arresting mechanisms dampen the effects of perturbations, allowing adjustments among variables of a system, or part of a system, such that landscape morphology does not change with time. These internal adjustments may be able to accommodate minor changes in independent variables, but major changes tend to upset equilibrium. The adjustment to a ''graded (equilibrium) stream is one in which, over a period of years, slope, velocity, depth, width, roughness, pattern, and channel morphology delicately and mutually adjust to provide the power and efficiency necessary to transport the load supplied from the drainage basin without aggradation or degradation of the channels'' (Leopold & Bull, 1979, p. 195). Reaches of streams in equilibrium (as just defined) attain stable longitudinal profiles. For hillslope subsystems that become progressively lower through denudation, equilibrium is best defined as a time-independent (unchanging) configuration of the landscape (Hack, 1960). Thus, in general, hillslope morphology remains constant where the rate of hillslope denudation equals that of uplift. A similar approach can be used to define types of equilibrium in streams (Section 1.3.1.2).

Self-arresting feedback mechanisms tend to provide the regulation that is necessary for a reach to approach equilibrium. The adjustment of a stream to an increase of gravel from the hillslope is an example of self-regulation. Assuming no change in the discharge of water, the stream subsystem will change. In meandering stream channels, large increases of bedload cause alterations in all hydraulic variables, which in turn result in a decrease in channel sinuosity, so that more gravel can be transported. In braided channels, increases in bedload cause maximum alluviation in the drainage basin headwaters. This increase in streamflow gradient is the principal way stream power is increased. In both meandering and braided streams, adjustments of the hydraulic variables will continue as long as the stream has to transport the excess bedload. (Changes greater than those needed to transport the bedload will result in counterbalancing adjustments that will decrease the transporting capacity and competence of the stream.)

Many geomorphologists regard open systems in fluvial landscapes as approximations of steady-state conditions. According to Hack (1965a, p. 5) "the principle of dynamic equilibrium states that when in equilibrium a landscape may be considered a part of an open system in a steady state of balance in which every slope and every form is adjusted to every other." Thus, equilibrium of entire drainage basins occurs when all parts of a landscape downwaste at the same rate and local variations of landforms result from differences in variables such as lithology, types of erosional processes, and position of a given landscape element in the system. Many geomorphologists have used the model of *dynamic equilibrium (steady state)* to good advantage. Some workers prefer to emphasize the tendency toward steady state in geomorphic systems, while others emphasize attainment of steady state.

Leopold and Langbein (1962) pointed out that statistically, the energy distribution in geomorphic open systems is best considered a "most probable condition," that is, a compromise between the opposing tendencies of uniformly distributed rate of energy expenditure in space and minimum total work expended. They further postulated that "the most probable distribution of energy in certain geomorphic systems could be derived by consid-

ering the geomorphic system as an open system in steady state" (Langbein & Leopold, 1964, p. 784).

Different parts of fluvial systems vary greatly in the time needed to attain a new steady state after a change in one or more independent variables and in the length of time that equilibrium is maintained. Mean values of streamflow variables may remain constant during very short time spans of a few minutes. Changes in channel configuration and flow characteristics do not occur as sediment and water move through the channel. A meandering stream in a valley is an example of centuries-long steady state. The channel is not stationary—the meanders shift down and across the valley but retain similar geometry as long as the independent variables do not change. Millions of years may be needed to attain uniform rates of downwasting of hillslopes (Ahnert, 1970). Regardless of the time needed for attainment of steady state, "many natural systems tend to approach the new equilibrium at a rate proportional to their distance from the equilibrium value" (Howard, 1965). Thus, the rate of change decreases exponentially (asymptotically) as equilibrium is approached, although fluctuation of the independent variables may continue about certain values (Patton & Schumm, 1975).

### 1.3.1 EQUILIBRIUM IN STREAMS

#### 1.3.1.1 Classes of Stream Terraces
Before discussing equilibrium in streams, we need to define the classification of stream terraces used in this book. Two types of equilibrium, and the crossing of the erosional-depositional threshold, can be recognized by the presence of distinctive types of stream terraces. Former levels of streams may be classed as fill, fill-cut (alluvium is cut), and strath (rock is cut) terraces (Howard, 1959; Leopold & Miller, 1954). Each type of terrace may be paired or unpaired. *Unpaired terraces* (Davis, 1902) typically occur on the insides of meander bends of a stream that is steadily downcutting, but may even be isolated straths where a

stream impinges on bedrock hills. *Paired terraces* are remnants of a formerly continuous level of a stream. They commonly occur on both sides of a valley, but more importantly have continuity along the valley.

A *fill terrace* is formed by valley aggradation and subsequent channel incision into the alluvium, which leaves remnants of the former valley floor as the tread of a paired fill terrace. The tread represents a time of crossing of the critical power threshold (Section 1.5) as the stream operation changes from aggradation to degradation. Longitudinal profiles of fill terraces generally do not represent attainment of equilibrium, because crossing of the threshold tends to be abrupt and may occur at different times along the stream.

*Fill-cut terraces* form by lateral stream erosion into alluvium during brief periods when the stream stays at the same level (static equilibrium) followed by renewed stream-channel downcutting that isolates the terrace tread.

*Strath terraces* are common in tectonically active mountains. Erosional widening of valley floors in bedrock occurs during minor (static equilibrium) or major (type 1 dynamic equilibrium) strath formation. Minor strath terraces are genetically the same as fill-cut terraces: straths are surfaces beveled in bedrock and fill-cut surfaces are beveled in alluvium. Strath and fill-cut terraces differ from fill terraces in that only thin layers of stream gravel remain on the beveled surfaces: these may be regarded as lag deposits or cutting tools that would have been entrained by the next flood. Thick alluvial fills of the next climatically induced aggradation event may bury any strath surface.

### 1.3.1.2 Base Level of Erosion and Equilibrium
The concept of steady-state or equilibrium describes interactions between the variables of fluvial systems and is valuable for descriptions of minimal landscape change. This section discusses equilibrium of streams in terms of alterations in the altitude of a valley floor that is constant or is itself changing. It also considers other landforms indicative of the degree and type of adjustment between interacting variables in stream subsystems. Advances in landform dating made during the past few decades, together with a better understanding of rates of geomorphic processes, permit more complete definitions of equilibrium.

For some streams we can now describe rates of aggradation or degradation relative to concurrent uplift rates. A valuable concept that has been used for a century is that of base level of erosion. Base level of erosion involves all reaches of a fluvial drainage net. It is the equilibrium (graded) longitudinal profile below which a stream cannot degrade and at which neither net erosion nor deposition occurs (Powell, 1875; Barrell, 1917). A reach of stream at the base level of erosion has attained a time-independent configuration of its longitudinal profile that is maintained as long as the controlling variables do not change. This concept integrates the system, equilibrium, and base-level concepts. It considers the longitudinal profile spatially as being an infinite sequence of adjacent base levels (Gilbert, 1879), and temporally as being reestablished at multiple positions within a landscape. Temporal changes in any of the variables affecting stream power or resisting power may result in a new longitudinal profile for the base level of erosion. Similar base levels of erosion probably are associated with similar climatic conditions. Examples would include characteristic longitudinal profiles of streams during times of culmination of full-glacial or interglacial conditions.

The base level of erosion concept describes reaches of streams that have achieved one of two types of equilibrium-static equilibrium and type 1 dynamic equilibrium. Static equilibrium is characterized by a lack of either aggradation or degradation of the streambed (Leopold & Bull, 1979). By definition, downcutting streams are not in static equilibrium; they may not have attained the time-independent longitudinal profiles of the base level of erosion. Like degrading hillslopes, downcutting streams reflect orderly interactions between variables (Bull, 1975). Equations that describe longitudinal profiles of dynamic equilibrium may have different constants than those for longitudinal profiles of streams in static equilibrium.

The concept of dynamic equilibrium, which Hack (1960,1965a) applied to mountains, can be applied to streams. Two categories may be defined in terms of relative attainment of the base level of erosion. Type 1 dynamic equilibrium is present when the rate of tectonically induced downcutting equals the rate of uplift, allowing the longitudinal profile of the stream to attain a succession of base levels of erosion in a rising landscape. Diagnostic landforms include straths and a valley floor that is sufficiently wide for the formation and preservation of strath terraces. Type 2 dynamic equilibrium is present in streams with a strong tendency toward, but lack of attainment of, the base level of erosion. Diagnostic landforms include narrow valley floors whose longitudinal profiles plot as concave lines on arithmetic graphs and as straight lines on semilogarithmic graphs. Straths and strath terraces generally are not present. Disequilibrium is characteristic of degrading streams upstream from most type 2 reaches. Interactions between variables are distant from the base level of erosion in disequilibrium reaches. Diagnostic landforms include highly convex valley sideslopes in V-shaped canyons, and reaches whose longitudinal profiles plot as convex lines, even on logarithmic graphs. In this book the terms ''static equilibrium'' and ''dynamic equilibrium'' are used where appropriate for specific reaches of streams, and ''equilibrium'' is used in a more general sense to identify reaches that have attained the base level of erosion-either static or type 1 dynamic equilibrium.

Small ephemeral streams of arid regions that cross highly resistant rocks require more than 1 my to adjust their longitudinal profiles to tectonic deformation and achieve a new base level of erosion. Such sites are useful for evaluation of the long-term effects of tectonic deformation. Before the base level of erosion is attained, the orderly interactions between uplift and nontectonic variables can be described by the dynamic equilibrium conceptual model. Perennial streams that flow across soft materials in humid regions are better suited to studies of stream responses to uplift and changing climate during the past 40 ky. One such stream is the Charwell River of New Zealand (see Chapter 5) where all three types of equilibrium are present.

The piedmont reach of the Charwell River is an example of type 1 dynamic equilibrium. Changes in streambed altitude of this reach, which is downstream from the range-bounding Hope fault, have varied as a function of regional uplift (estimated to be about 0.5 to 1.3 m/ky) and of episodes of climate-change-induced aggradation. Tectonically induced downcutting by the river cannot proceed during times of aggradation characterized by full-glacial climates. Only during times of interglacial climates does the river have sufficient power, relative to the erodibility of the streambed materials, to reestablish type 1 dynamic equilibrium. Flights of major strath terraces record times of attainment of the base level of erosion in rivers that are being tectonically elevated. Rapid stream-channel downcutting of the Charwell ended about 4 ka. During the past 4 ky, uplift has raised the stream, which has downcut at a rate equal to the long-term regional uplift rate. Most fluvial erosion has occurred as horizontal cutting of soft bedrock to bevel a major strath. Major strath terraces, such as those estimated to have formed at roughly 40 and 30 ka, tend to be well preserved because they are buried under the deposits of the succeeding aggradation event. Major straths are frequent along the piedmont reach of the Charwell River and seem to be representative of fluvial system response to interstadial climatic conditions such as the present climate. Uplift rates are slow in this reach and ample stream power remains after the required amount of tectonically induced degradation has taken place.

The Charwell also provides an example of how self-arresting feedback mechanisms can bring about temporary static equilibrium in a stream that is entrenching bouldery valley fill. Times of aggradation, and degradation that exceeds uplift rates, are by definition periods of nonequilibrium. Degradation of a valley fill deposited during an aggradation event occurs when stream power exceeds that needed to transport bedload and the streambed is above the base level of erosion. Selective entrainment and transport of bedload by the Charwell caused winnowing of the valley fill and progressive accumulation of boulders on the streambed. This bouldery lag deposit armored and protected the streambed from further degradation by increasing the shear stresses needed to set the bed in motion and by increasing the hydraulic roughness. Streams with armored streambeds stop downcutting, and lateral erosion predominates where the stream impinges on the valley sides.

Brief episodes of static equilibrium may end abruptly when a perturbation such as a 1-ky flood, moves and breaks the boulders, reducing hydraulic roughness and shear stresses needed to initiate transport of streambed boulders. Renewed degradation can then form a lag gravel in a new streambed at a lower altitude. Remnants of the former streambed are preserved as treads of fill-cut or of minor strath terraces.

Thus without invoking either secular climatic or tectonic perturbations, a self-arresting feedback mechanism can occur repeatedly to form a flight of degradation (fill-cut and minor strath) terraces. Each terrace remnant has a well-sorted gravel cap that is more bouldery than the massive silty gravels of the preceding aggradation event. The treads of fill-cut or minor strath terraces record pauses in valley-floor degradation-static equilibrium-when adjustments between all variables interacted to transport the bedload with neither aggradation nor degradation of the streambed (Leopold & Bull, 1979).

In the drainage basin reach upstream from the Hope fault, uplift rates are greater (estimated at $3.8 \pm 0.2$ m/ky) and the fractured greywacke sandstone is more resistant to erosion. Small headwaters streams of the Charwell River are unable to degrade as rapidly as the mountains are rising (disequilibrium), but at present most reaches approximate a type 2 dynamic equilibrium condition.

Many possible interactions between variables can produce equilibrium. For a given reach of a stream, slope changes may be important in achieving equilibrium after vertical displacement caused by an earthquake. Changes in hydraulic roughness may be the dominant controlling variable after a tropical storm causes a landslide of bouldery materials into a stream. Streams are able to maintain equilibrium within a certain range of adjustments that encompasses interactions within limited bands of magnitudes and rates of the processes involved.

This introduction to the fundamental subject of equilibrium in streams utilizes the powerful concept of the base level of erosion. Now we turn our attention to a discussion of thresholds separating different geomorphic processes.

## 1.4 Geomorphic Thresholds, Response Times, and Threshold Ratios

Conceptual frameworks that emphasize prolonged equilibrium may be unsatisfactory for studies of the consequences of change in independent variables. *Geomorphic thresholds* provide a less restrictive framework. A threshold may be regarded as a balance between opposing tendencies-a boundary between separate system states. A geomorphic threshold in an open system separates different reversible processes that tend to change part of the system. A threshold in a closed system separates irreversible processes, for example, the change from stable to unstable conditions that is initiated by inception of landslide movement. The threshold conceptual framework includes within it the concept of equilibrium, for example when equilibrium occurs between times of aggradation and degradation (Section 5.4.4.1). Although the time of occurrence of the threshold and the magnitudes of the interacting variables cannot be forecast accurately beforehand, the change of mode of operation associated with a threshold usually is readily identifiable in the field. Recognition of the threshold phenomenon is an acknowledgement of the nonlinearity of fluvial geomorphic processes and their sensitivity to seemingly minor perturbations.

Differences between the threshold and equilib-

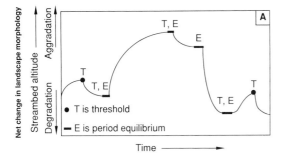

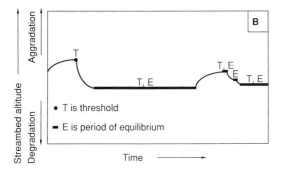

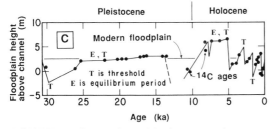

FIGURE 1.2 Streambed altitude versus time.
A. Hypothetical situation of frequent thresholds
and brief periods of equilibrium. Rates of change
in streambed altitude range from extremely rapid
to extremely slow. This system is characterized by
almost constant change. B. Hypothetical situation
of occasional thresholds and long periods of equilibrium. This system is characterized by attainment of equilibrium. C. Thresholds and equilibrium
periods for the Pomme de Terre River, Missouri,
during the past 30 ky (from Brakenridge, 1981).
The latest Pleistocene was characterized by 13 ky
of equilibrium, and the Holocene was characterized by numerous thresholds.

rium concepts are illustrated in Figure 1.2. Points
or periods in time that separate reversals of modes
of operation are thresholds, but they are not equilibrium conditions unless a landform has developed that is time independent. Periods of equilibrium are thresholds when they separate aggradational
and degradational modes of operation of the system. Figure 1.2A depicts a hypothetical situation
in which changes in the variables of a fluvial
system either are frequent or require long adjustment periods before a new equilibrium condition
is attained. Thresholds are common; periods of
equilibrium are infrequent and brief. Stream systems that rarely attain equilibrium for extended
time spans (1) may be affected by frequent changes
in independent variables (climate, uplift, and human impacts) so that self-arresting feedback mechanisms cannot operate, or (2) are inherently unstable, such as the common discontinuous ephemeral
streams of semiarid regions. Figure 1.2B depicts
the opposite situation, in which a stream subsystem maintains equilibrium much of the time. Equilibrium may be more readily attained in fluvial
systems of humid than of arid regions when situations of similar rainfall intensity/runoff ratios and
hillslope yields of bedload sediment are compared.
Perennial streams have a greater capacity to transport bedload because larger discharges per square
kilometer of watershed area provide more stream
power to do work. Because of their large annual
stream power, perennial streams may adjust quickly
to perturbations external to the fluvial system, such
as waterfalls created by fault displacements.

The field data needed to draw threshold-
equilibrium plots for streams include detailed information about stratigraphy and soils and many
age determinations. The summary threshold-equilibrium plot presented by Brakenridge (1981) for
the valley of the Pomme de Terre River of Missouri (Fig. 1.2C) has characteristics of both Figures 1.2A and 1.2B. Equilibrium prevailed during
the last period of full-glacial climate. In contrast,
during the Holocene the floodplain level changed
almost continuously, probably as a result of variable vegetation cover and amounts and types of

precipitation. Periods of aggradation appear to have been gradual compared with the brief episodes of channel downcutting and floodplain degradation.

One should resist the temptation to label as thresholds abrupt changes in degradation or aggradation rates, such as the equilibrium (E) parts of the plots of Figures 1.2A and 1.2B. The term "geomorphic threshold" purposely includes change of mode of operation and excludes acceleration of aggradation or degradation, even where equilibrium periods occur. A temporary variation in aggradation or degradation rate should not be considered a threshold because (1) from a practical viewpoint such a change may be difficult to identify in the field and (2) how abrupt the change must be to qualify as a threshold would require quantitative definition.

Use of the threshold conceptual framework offers several advantages. Thresholds may be abrupt or gradational (Begin & Schumm, 1984), and threshold concepts can be used in studies involving time spans that range from minutes to millions of years and spaces of equally great contrast. The threshold approach tends to focus attention on those variables and complex responses (Section 1.7) that are likely to cause change in the mode of system operation. Studies of both self-enhancing and self-arresting feedback mechanisms are encouraged by the threshold approach, whereas studies only of self-arresting feedback mechanisms are encouraged by the equilibrium approach. The threshold conceptual framework is particularly well suited for studies involving the interactions of humans with their environment, because it provides emphasis on how far removed a system is from stable conditions.

The concept of response time (Allen, 1974; Thornes & Brunsden, 1977; Brunsden & Thornes, 1979; Brunsden, 1980) following a change in an independent variable or in the internal operation of a system is introduced with the threshold-equilibrium plot of Figure 1.3. Climatically induced changes in streambed altitude do not occur instantaneously; there is a reaction time between perturbation and beginning of aggradation or degradation. The time needed to establish new equilibrium is the relaxation time, and the time this equilibrium endures is the persistence time. Response time is the sum of reaction and relaxation times. These terms are used primarily in equilibrium models because thresholds may have zero persistence times and relaxation time has previously been defined in terms of equilibrium. Relaxation and response times

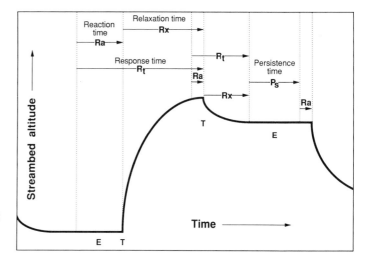

FIGURE 1.3 Threshold–equilibrium plot showing the components of response time, $R_t$, which is the sum of reaction time, $R_a$, and relaxation time, $R_x$. $P_s$ is the time of persistence of new equilibrium conditions, and T and E represent threshold and equilibrium, respectively.

can also be considered in terms of time spans needed to return to the mode or rate of process, or both, that prevailed before a perturbation (Fig. 4.28; Wolman & Gerson, 1978). The response time after a perturbation to the time of a new threshold or to equilibrium is a highly useful concept because it is a measure of the sensitivity of the system to change (Hicken, 1983; Wright, 1984).

Ratios are a convenient way of describing thresholds if the numerator and denominator describe opposing processes (Bull, 1980). The part of the system under consideration is at a threshold or equilibrium when the ratio is 1.0.

The ratio format has several advantages. The components of the threshold are identified and compared. The numerical index defines the relative condition that must be met to cross the threshold and change the mode of system operation. The variables included in the ratio may be simple or complex depending on the degree of complication of the subject being studied or on the desire of the scientist or engineer to simplify the complexity of the real world.

## 1.5 Threshold of Critical Power in Streams

The previous section touched on the great variety and types of thresholds in geomorphic systems. Some are simple and others complex; some are obvious and others obscure. This section discusses an obvious threshold in detail as a basis for subsequent discussions about the influences of tectonic and climatic perturbations. This important threshold-the threshold of critical power-separates the modes of aggradation (net deposition) and degradation (net erosion) in stream subsystems.

### 1.5.1 COMPONENTS OF THE THRESHOLD

The highly versatile *threshold of critical power* is defined as the ratio of power available (stream power) to power needed (resisting power) for entrainment and transport of bedload (Bull, 1979):

$$\frac{\text{Stream power (driving factors)}}{\text{Resisting power (resisting factors)}} = 1.0 \qquad (1.1)$$

All the variables affecting the numerator and denominator of Equation 1.1 interact to determine the capacity and competence of a stream to transport bedload.

#### 1.5.1.1 Stream Power

*Stream power* is the power available to transport bedload and consists of those variables that, if increased, favor bedload transport. Bedload transport is highly sensitive to changes in streamflow amount and gradient (for example, see Baker, 1973, Fig. 54). The importance of discharge on stream power is dramatically revealed by the marked increase in capacity for transport of bedload and suspended load that occurs with increasing discharge at a stream gauge station (Bagnold, 1977). For example, suspended sediment transport rate (Gs) increases by the large exponential factor of about 2.5 with increase in discharge (Q) (b is a constant) (Leopold et al., 1964):

$$Gs = bQ^{2.5} \qquad (1.2)$$

Graphs of bedload transport rates also have power-function exponents larger than 1.0 (Knighton, 1984, Fig. 3.9B).

Streams may be regarded as bedload-transporting machines and can be analyzed in terms of their stream power (Bagnold, 1973, 1977). Stream power is dissipated chiefly in maintaining fluid flow against flow resistance and moving the saltating and rolling bedload. Degradation of the streambed will occur when stream power is more than sufficient to transport the imposed bedload and overcome flow resistance. Part of the bedload will stop and the streambed will aggrade when stream power is insufficient. Bagnold described the total kinetic power, W, along a stream channel as

$$\Omega = \gamma QS \qquad (1.3)$$

or, in terms of power per unit area of streambed w,

$$\omega = \frac{\gamma QS}{w} = \frac{\gamma(wdv)S}{w} = \gamma dSv = \tau v \qquad (1.4)$$

where $\gamma$ is the specific weight of the sediment-water fluid, Q is stream discharge, S is the energy slope, w is streambed width, d is streamflow depth, v is mean flow velocity, and $\tau$ is the shear stress exerted on the streambed (Baker & Costa, 1987).

Uplift along range-bounding fault zones is a tectonic perturbation that increases stream power as increase in relief migrates upstream from mountain fronts. This tectonic input tends to increase stream power through increases in slope and orographically induced precipitation, but such increases occur only during long response times of 10 to 1000 ky. Such tectonic base-level changes undergo exponential spatial decay with increasing distance from active fault zones. In marked contrast, perturbations caused by late Quaternary climatic change affect entire watersheds (Fig. 1.8) with response times of 0.1 to 10 ky, and commonly change both stream power and resisting power.

### 1.5.1.2 Resisting Power
*Resisting power* is the power needed to entrain and transport the bedload supplied to the reach of a stream. It consists of those variables that, if increased, favor deposition of bedload. Resisting power increases with increases in hydraulic roughness (n) and size and amount of bedload. Thus the threshold is determined by many variables such as width, depth, and velocity that are included in the continuity equation

$$Q = wdv \qquad (1.5)$$

and the Manning equation (in metric units)

$$v = \frac{r^{2/3} \, S^{1/2}}{n} \qquad (1.6)$$

where r is stream-channel hydraulic radius.

Variations in resisting power are primarily the result of changes in the amount and size of bedload discharged from the hillslope subsystem and of hydraulic roughness for alluvium- and bedrock-floored streams. Variations on the order of 1 to 100 years result mainly from destruction of vege-

tation by fires, construction, grazing, and timbering—perhaps combined with extended droughts—and subsequent reestablishment of vegetation cover. Variations on the order of 1 to 10 ky result mainly from climatic changes that affect resisting power through the variables of vegetation, soil development, and bedload transport rate.

The components of the threshold defined by Equation 1.1 differ in their ease of measurement. Stream power may be estimated by measurements of discharge and stream gradient; for some sediment-laden flows specific weight also should be evaluated. Energy grade lines roughly parallel the longitudinal profiles of the water surface in most alluvial reaches (Leopold et al., 1964). Resisting power includes hydraulic roughness and, like this useful concept, it cannot be measured directly in the field. The power needed to transport bedload also is difficult to estimate, because amount and size of bedload are difficult to measure at peak discharges. Detailed studies by Emmett (1974, 1976), by Leopold and Emmett (1976), and by Andrews (1979,1983) are useful initial investigations of this important aspect of sediment transport. Bedload not only provides the cutting tools for streams in flood but also is deposited as alluvium during aggradation events.

E. W. Lane's idea that degradation and aggradation in streams may be considered as a delicate balance between opposing variables is illustrated in Figure 1.4. Magnitudes of bedload and stream discharge tend to drive stream subsystems to opposite modes of operation. Their influences are modulated by their leverage position on the balance beam through the variables of sediment size, hydraulic roughness, and stream slope. Interactions among these and other variables determine whether the system is close to threshold or equilibrium-a value of 1.0-or whether it departs greatly from them in either time or space.

A variety of field evidence may indicate that a given reach of a stream is close to the critical-power threshold-that the balance between aggradation and degradation is close to 1.0. The presence of stream-channel alluvium in amounts ex-

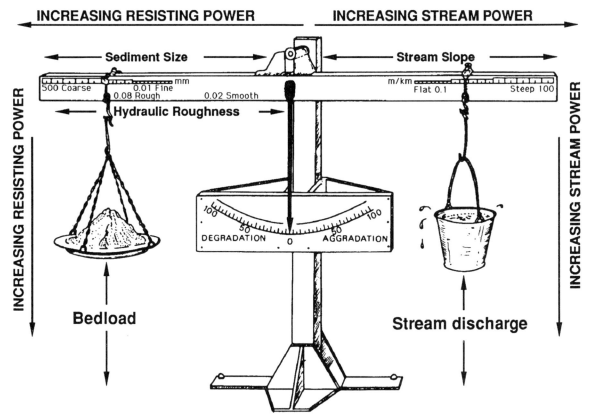

**FIGURE 1.4** Schematic balance between modes of aggradation and degradation in streams; zero is the threshold of critical power. Increases or decreases of one or more of the important variables may cause the mode of stream operation to depart markedly from the threshold condition. (Originally from notes of E. W. Lane; modified from Chorley et al., 1984.)

ceeding that scoured by large discharges suggests that net vertical erosion is minimal. At peak discharge stream width in downcutting reaches equals valley-floor width. When lateral cutting predominates over downcutting, valley-floor width exceeds channel width. Static equilibrium is present when neither net erosion nor net deposition is occurring. Measurements of erosion and deposition can be made in the field. For long time spans, dating of stratigraphy may be used (Fig. 1.2C). Evidence that the threshold has been passed might be a channel entrenchment that truncates strata or a backfilled stream channel. Parallel treads of longitudinal profiles of paired stream terraces are common; they suggest a return to a similar threshold or to equilibrium (type 1 dynamic equilibrium or static equilibrium) after adjustments to climatic perturbations. The evidence that many depositional settings fluctuated across the threshold is found in stratigraphies that contain numerous hiatuses. Although stratigraphic sections are indicative of net accumulation of deposits, planes of erosional truncation record times of partial removal of the section.

Before predicting future stream behavior one should estimate how far removed from equilibrium or threshold a given reach is. Reaches of streams at critical-power threshold or at equilibrium may be highly susceptible to accelerated downcutting or alluviation resulting from climatic, tectonic, or human-use perturbations that affect discharges of water and bedload. For such reaches, a moderate increase or decrease in resisting power may initiate an episode of aggradation or degradation. Larger perturbations are needed to reverse modes of operation in streams that are strongly aggrading or degrading.

### 1.5.2 TEMPORAL AND SPATIAL CHANGES

The concept of the threshold of critical power in streams is equally useful in humid or arid fluvial systems. The following subsections discuss aspects of aggradation or degradation in streams where flow typically increases or decreases downstream.

#### 1.5.2.1 Perennial Streams

Both stream and resisting power change with time. Changes in stream power during short time spans generally are the result of changes in discharge, because slope changes slowly. Changes in slope occur when stream-channel pattern changes or when nonuniform aggradation or degradation occurs along a reach of a stream. For example, slope will be doubled when a meandering stream with a sinuosity of 2.2 changes to a braided stream with a sinuosity of 1.1. Increase in streamflow width/depth ratios results in more efficient transport of bedload for a given stream power (Bagnold, 1977). Channel-pattern adjustments help achieve maximum efficiency quickly. Aggradation or degradation change valley-floor slope at a slower rate.

Variability of stream discharge causes major short-term variations in stream power. For this reason, definitions of aggradational and degradational modes of operation generally represent averages over a period of years.

Stream power is much more than is needed to transport bedload and overcome flow resistance in headwaters streams of mountainous regions; the ratio of the critical-power threshold is greater than 1.0. Even millions of years after cessation of the uplift that created the mountains, headwaters reaches will be far removed from equilibrium. Cross-valley morphologies of such disequilibrium reaches characteristically are V-shaped because of the persistent tendency of streams to cut down into bedrock and approach the base level of erosion.

Where lithology is spatially uniform, floods will degrade valley floors in downstream reaches of large streams to the base level of erosion much sooner than in upstream reaches. The reason for this important spatial variation in the time needed to attain the base level of erosion is that large streams have exponentially greater capacity (Eq. 1.13) and competence than small streams. Basin contributing areas, and therefore magnitudes of flood events generated by basin-wide rainfalls, increase markedly in the downstream direction. However, floods may have opposite effects in different parts of the same fluvial system, for example, rapid degradation in mountains and rapid aggradation in depositional basins.

Types of valley-floor materials are an important lithologic control in determining the effects of flood discharges. In both the Seaward Kaikoura Range of New Zealand (Mackay, 1981,1984) and in the Appalachian Mountains of the United States (Moss & Kochel, 1978), abundant cobbles and boulders greatly increase resisting power but also provide the cutting tools needed for efficient stream-channel incision in strongly degrading reaches.

Processes of bedrock erosion depend on how far removed a reach of a stream is from threshold or equilibrium conditions. Degradational reaches undergo maximum net incision into bedrock during peak streamflows. Longitudinal valley-floor slopes generally are decreased, but valley-floor widths remain narrow. Equilibrium reaches undergo erosion into bedrock less frequently because stream power exceeds resisting power only during large streamflows. Lateral erosion predominates in a stream close to static or type 1 dynamic equilibrium. Straths and floodplains indicate the permanent nature of the lateral erosion in both braided and meandering streams at the base level of erosion. Valley-floor width is increased but slope remains about the same. Temporary downcutting or deposition may occur but commonly is offset by the presence of the opposite process within the same reach, as in a meandering reach with point bars. The ratio of vertical to lateral erosion of bedrock during floods is highest in degrading reaches, is zero in aggrading reaches, and intermediate in type 1 dynamic equilibrium reaches.

### 1.5.2.2 Ephemeral Streams

Many arid mountains are sites of local intense rainfall that generates streamflow that quickly sinks into dry streambeds downvalley (*ephemeral streams*). Summer thunderstorms in the rugged, bare granitic mountains of southwestern Arizona

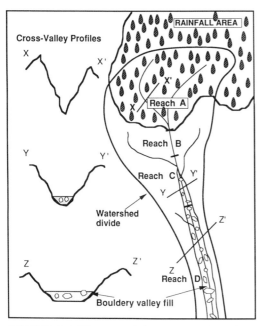

FIGURE 1.5 Diagram of drainage basin and cross-valley profiles for an arid rocky drainage basin (Fig. 2.12B) where 22 mm of rain falls in 30 minutes in the headwaters. See text for discussion. See Figure 1.6 for changes in hydraulic characteristics in the downstream direction.

(Fig. 2.12B) generate infrequent pulses of streamflow. Rainfall that equals or exceeds 25 mm in one day occurs about every 5 years. Flashy ephemeral streamflows degrade upstream reaches, create large dunes of grussy bedforms in intermediate reaches, and make smooth streambeds and aggrade downstream reaches. The intermediate reaches have a variety of bedforms and appear to have undergone minimal recent degradation or aggradation. Width/depth ratios and channel braiding increase in the downstream reaches. Data from aggrading reaches are shown in Figure 2.14.

Both stream power and resisting power change rapidly for such ephemeral streamflows. Figures 1.5 and 1.6 portray the effects of local thunderstorm rainfall of 22 mm in 30 minutes on bare granitic hillslopes in the headwaters of a large drainage basin. Power functions have been used to compare changes in the main components affecting the threshold of critical power; they are sketched for headwaters, intermediate, and downstream reaches (Fig. 1.5). The channel and water-surface slope approximates the longitudinal profile of the stream, whose slope decreases progressively downstream. Stream discharge increases with increase in source area downstream to the edge of the area of hillslope runoff. Then it loses water by infiltration into the dry, permeable, and progressively wider streambed. Bedload discharge is assumed to increase more rapidly than water discharge in reach A (like Eq. 1.2). Bedload transport rate continues to increase, but at a lesser rate in reach B because stream power is greater than resisting power. Bedload transport rate remains constant in reach C; neither aggradation nor degradation occurs as internal adjustments in the flow permit transport of bedload despite decreases in discharge and slope. In reach D, internal adjustments are insufficient to maintain bedload transport, and selective deposition of coarser particles occurs. Hydraulic roughness decreases downstream in flows of low to moderate suspended sediment concentration (Leopold et al., 1964, p. 244). Flow resistance decreases in reach A mainly because of increase in discharge and decrease in

FIGURE 1.6 Changes in the downstream direction of total sediment load, channel fall, stream discharge, and hydraulic roughness. This hypothetical ephemeral streamflow event results from a local rainfall of 22 mm in 30 minutes in the headwaters reach (A) of an arid rocky drainage basin. Equilibrium prevails in reach C, and the entire streamflow infiltrates into the streambed at the channel length of the right edge of the plot. p, j, c, f, b, m, r, and y are constants in power-function equations.

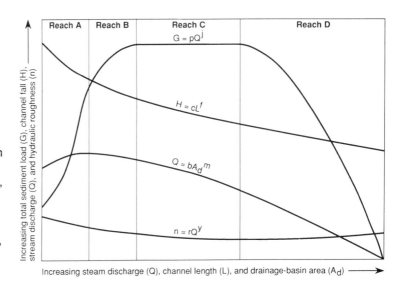

height of channel roughness elements relative to flow depth. The decrease in hydraulic roughness in reaches B and C is one type of internal adjustment-dune bedforms and highly turbulent flow of reach A give way to less turbulent flow and planar bedforms of reach C. The increase in hydraulic roughness in reach D is the result of decreasing flow depth to the point where the remaining streamflow disappears entirely into the dry streambed.

The interrelations between these and other variables directly affect stream power and resisting power and thus the degree to which the stream approximates threshold or equilibrium conditions. The spatial distribution of reaches A through D differ with each rainfall-runoff event and will be totally different for perennial streams where discharge continues to increase downstream. The example from the arid realm is better suited to illustrate obvious interactions, with rates of discharge of water and bedload transport decreasing to zero.

Stream power and resisting power are equal but changing in reach C (Fig. 1.5), which is in static equilibrium. Stream power is decreasing because of decreases in discharge of water and slope.

Bedload transport rate is constant or may even increase, but by definition it cannot decrease until the beginning of aggrading reach D. Hydraulic adjustments act as a self-arresting feedback mechanism to maintain equilibrium in reach C despite a decreases in discharge. The dune bedforms and highly turbulent flow that characterize reaches A and B give way to planar streambeds in reach C. Resisting power decreases with the decrease in hydraulic roughness-an example where hydraulic adjustments are sufficient to maintain equilibrium streamflow despite concurrent changes in several variables. Internal hydraulic adjustments are insufficient to allow attainment of equilibrium streamflow in reaches A, B, and D.

Changes in resisting power are a major cause of crossing the threshold of critical power in streams. Such crossing results either in valley-floor aggradation or in the formation of stream terraces. Because reach C is in static equilibrium, a moderate increase in resisting power may result in alluviation and a moderate decrease may initiate channel downcutting. The situation is different for reach A in which a large increase in resisting power will not stop the stream from downcutting but will lower the rate. For reach D, changes in resisting power may (1) accelerate aggradation, (2) return the mode of operation to static equilibrium, or (3) initiate channel incisement into the valley fill and thereby cause the threshold to be crossed.

Three possible interrelations between stream power and resisting power for the hypothetical arid fluvial system are summarized in Figure 1.7. This graph may be regarded as depicting variations that characteristically occur either with *stream order* or in the downstream direction. Stream power decreases with increasing distance from the headwaters because of downstream decreases of slope and discharge. Resisting power increases in reach A with increase in the power needed to transport bedload delivered from hillslopes and entrained from valley floors. Resisting power decreases in reach C because of decrease in hydraulic roughness; it decreases in reach D because of decrease in sediment load.

Changes in power for ephemeral and *perennial streams* (streams that flow continuously) can be compared by using the average exponents of the downstream hydraulic geometry equations (Leopold et al., 1964). For ephemeral streams of southwestern Arizona (Fig. 2.12B) as illustrated in Figures 1.5 through 1.7; $w \propto Q^{-0.5}$, $d \propto Q^{-0.3}$, $v \propto Q^{-0.2}$, and $S \propto Q^{-0.8}$.

Total stream power, $\Omega$, decreases markedly:

$$\Omega \propto wdvS \tag{1.7}$$

$$\Omega \propto Q^{(-0.5-0.3-0.2-0.8)} \tag{1.8}$$

$$\Omega \propto Q^{-1.8} \tag{1.9}$$

and stream power per unit width, w ,also decreases

$$\omega \propto dvS \tag{1.10}$$

$$\omega \propto Q^{-1.3} \tag{1.11}$$

Discharge increases downstream in perennial streams, and $w \propto Q^{+0.5}$, $d \propto Q^{+0.4}$, $v \propto Q^{+0.1}$, and $S \propto Q^{-0.8}$. Total stream power increases

$$\Omega \propto Q^{(+0.5+0.4+0.1-0.8)} \tag{1.12}$$

$$\Omega \propto Q^{+0.2} \tag{1.13}$$

and stream power per unit width decreases, but only slightly compared with the ephemeral stream case (Eq. 1.11).

$$\omega \propto Q^{-0.3} \tag{1.14}$$

### 1.5.2.3 Threshold-Intersection Points

Relative strengths of stream power and resisting power change in both time and space in most fluvial systems (Graf, 1982a,1983a,b), and thereby affect where alluvium is deposited or eroded. Threshold-intersection points occur along longitudinal profiles of valleys where the stream changes its' mode of operation. In Figure 1.8, aggradation of a reach that formerly was downcutting into bedrock moved the threshold-intersection point progressively upstream to location A at the time of maximum aggradation. Subsequent downcutting through the valley and piedmont alluvium then moved the point downstream to B.

Threshold-intersection points may shift over a great variety of time scales. More than 5 ky may

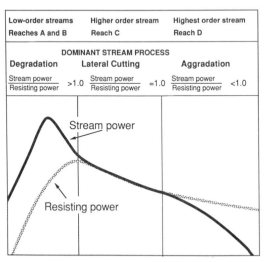

FIGURE 1.7 Diagrammatic graphs of stream power and resisting power for reaches in an arid rocky drainage basin where local rainfall of 22 mm falls in 30 minutes in the headwaters. See Figure 1.5 for map and topographic profiles of the basin.

be needed for the hypothetical shifts illustrated in figure 1.8. The bedload transport rate commonly peaks before water discharge peaks during floods. An upstream shift of the threshold-intersection point is followed by a downstream shift of the point during the later stages of the same flow event: resisting power decreases more rapidly than does stream power because of the diachronous nature of the water and sediment hydrographs.

A much different situation is depicted in the aggradational reach of Figure 1.8, where deposition has been continuous because resisting power exceeds stream power. Variations in the relative strengths of stream power and resisting power result merely in varying rates of aggradation. Such reaches are particularly obvious in basins of internal drainage where accumulation of playa and associated alluvial-fan deposits constitute a base-level rise that gradually tends to decrease slope and thereby stream power. An example is the large area of lacustrine deposition in the tectonically stable interior of Australia.

Locally steep and gentle reaches may occur within degrading or aggrading reaches. In Figure 1.9, temporary deposition has occurred in a bedrock stream channel. Its upstream and downstream ends are threshold-intersection points. Reach X is aggrading. Increases in vegetation density, flow width, and infiltration capacity all act as self-enhancing feedback mechanisms that promote aggradation. Stream power in reach X tends to equal resisting power through the process of selective sedimentation. This process decreases bedload and thereby reduces resisting power, tending to establish a condition of static equilibrium. For such equilibrium to be achieved, however, the decrease in sediment load must be sufficient to compensate for the concurrent decrease in stream power caused by alluviation on progressively gentle slopes. It is unlikely that aggrading reaches in streams of either humid or arid regions can attain static equilibrium, but the tendency toward equilibrium may result in roughly similar values of stream and resisting power.

The local deposition of alluvium illustrated in Figure 1.9 also results in the formation of reach

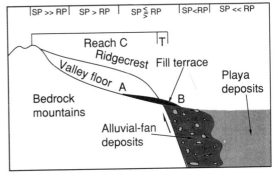

FIGURE 1.8 Relative strengths of stream power (SP) and resisting power (RP) along a hypothetical fluvial system in an arid closed basin. Tectonic perturbations are initiated in reach T, and climatic perturbations are initiated in reach C. The mode of operation is most likely to change in the reach at the mountain front as a consequence of climatically induced changes in hillslope water and sediment yield. A and B mark locations of threshold-intersection points.

Y, which is inherently unstable because its slope is comparatively steep. The local excess of stream power tends to cause channel entrenchment into the alluvium. This in turn tends to concentrate flow, and thus initiates a self-enhancing feedback mechanism that destroys protective vegetation and redistributes the alluvium.

Thus, local aggradation in either bedrock or alluvial reaches may result in short alluvial reaches where the relative rates of change of processes and landforms are dependent on the two offsetting

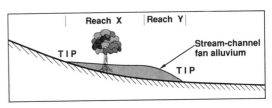

FIGURE 1.9 Diagram of a stream profile showing adjacent alluvial reaches that are more gentle (X) and steeper (Y) than the preexisting bedrock channel. Threshold-intersection points shown by TIP.

self-enhancing feedback mechanisms just noted. Alluviation will be temporary and threshold-intersection points will shift in a bedrock channel such as that illustrated in Figure 1.9. Where streams debouch onto permanent depositional areas, such as deltas and alluvial fans, local aggradation may be considered a small-scale analogue of the much larger deposit.

The concept of the threshold of critical power is used extensively in this book. This versatile concept allows evaluation of the impact of local perturbations such as uplift of a mountain front, regional perturbations such as climatic change that affects entire fluvial systems, and human actions. Thresholds are useful for describing the behavior of systems that tend toward equilibrium, such as streams. They also provide a useful format for studying interactions between variables that do not tend toward an equilibrium.

## 1.6 Non-Equilibrium Landforms

### 1.6.1 EVIDENCE FOR LACK OF EQUILIBRIUM

Spatial variations in strength of soil-profile development show that rates of erosion have not been uniform in many hillslope subsystems, and that aggradation and degradation have been more common than equilibrium (static or type 1) in many

TABLE 1.2 Assigned ages of Quaternary temporal terms, in thousands of years before present (ka).

| Age | Ka |
| --- | --- |
| Holocene | |
| Late | 0–4 |
| Mid | 4–8 |
| Early | 8–10 |
| Pleistocene | |
| Latest | 10–20 |
| Late | 10–125 |
| Middle | 125–790 |
| Early | 790–1650 |

stream subsystems. Soil-profile development becomes stronger with the passage of time, as is indicated by increases in strength of soil structure, horizon thickness, and translocation of sesquioxides. Soil profiles are strongly developed in stable parts of the landscape where rates of pedogenesis are not exceeded by those of denudation. Soil profiles indicative of Pleistocene age commonly occur only on gentle slopes in hilly terrain; they are patchy or absent on steeper slopes of similar exposure and lithology. Such contrasts in the distribution of Pleistocene and Holocene soil profiles may be interpreted as evidence for (1) long-term erosion rates on the steeper slopes that exceed rates of soil-profile development or (2) recent acceleration of erosion on the steeper slopes that has destroyed the soil profiles that were present in the past.

The Quaternary temporal terms have been assigned the following ages throughout this book (Table 1.2): late Holocene, 0–4 ka; middle Holocene, 4–8 ka; early Holocene, 8–10 ka; latest Pleistocene, 10–20 ka; late Pleistocene, 10–125 ka; middle Pleistocene, 125–790 ka; and early Pleistocene, 790–1650 ka. The 10-ka age assignment is widely used but is arbitrary. The 20 to 10 ka time span is the interval between full-glacial and interglacial climatic conditions. Recent calibration of radiocarbon ages (Bard et al., 1990) shows that the peak of full-glacial conditions may be as old as 21 to 22 ka instead of 18 ka. Unless specifically noted, the radiocarbon ages stated in this book have been corrected for isotope fractionation but not for variations in abundance of 14C in the atmosphere through use of techniques of Stuiver and Reimer (1986) or Bard and colleagues (1990). The 125 and 790–ka ages are radiometric and paleomagnetic ages that have been fine tuned by application of the astronomical clock (Edwards et al., 1987; Johnson, 1982). The 1650–ka age is at the top of the Olduvai reversed polarity event.

Evidence in many localities clearly shows that independent variables such as climate, total relief, and erodibility of surficial materials have not remained sufficiently constant to allow even an approximation of steady-state conditions. Furthermore, the impacts of humans on their environment have been so profound that human activity should be considered an independent variable conducive to changing conditions rather than attaining steady-state conditions. Historic arroyo cutting in the western United States may have been caused partly by the grazing of domestic animals and partly by construction. However, regional Holocene entrenchment of drainage nets clearly is the result of prehistoric climatic change of sufficient magnitude to cause the entrenchment and backfilling of valley floors of ephemeral streams. Late Quaternary stream terraces in New Zealand and in the Rocky Mountains of the United States reflect climatic changes in glaciofluvial systems—streams changed from aggradation during times of full-glacial climate to degradation during times of interglacial climate. In the tectonically active drainage basins of west-central California, stream and hillslope gradients are steeper than those in similar drainage basins that are less affected by base-level fall. In areas of Pleistocene glaciation in Iowa and Michigan, fluvial processes are still changing the configuration of glacial landscapes (Ruhe, 1952; Hack, 1965b).

It appears that both humid and arid landscapes typically contain relict landforms that are changing. The rate of change is related to (1) the rates of change of independent variables, (2) how far removed from equilibrium the present subsystem is, and (3) the frequency and magnitude of processes shaping the landscape. The most important changing independent variables are climate, total relief (by either tectonic or erosional processes), and erodibility of surficial materials. Thus many landforms may tend toward, but are unable to attain, steady-state conditions. Other landforms do not even tend toward a steady state.

### 1.6.2 LANDFORMS THAT DO NOT TEND TOWARD A STEADY STATE

Although most aspects of rivers tend toward a steady-state condition indicated by the base level of erosion, many elements of hillslopes and depositional environments do not (Bull, 1976b). Many landforms would not tend toward a steady state even if independent variables such as climate, total relief, and base level remained constant. In some erosional environments, self-enhancing feedback mechanisms cause progressive changes in the relative dimensions of hillslope elements. Depositional landforms commonly represent the end point of processes operating in geomorphic systems. Because deposits increase in volume with time, these landforms do not tend toward a steady state.

For hillslopes that do tend toward steady-state configurations, differential erodibility of contrasting rock types and their weathering products may be eliminated by a change in slope steepness (because the rate of erosion of a given material increases with increasing slope). For moderately contrasting materials, erosion rates of resistant materials may equal those of less resistant materials when the slope of the resistant materials becomes sufficiently steep to balance the contrast in erodibility. G. K. Gilbert and John Hack used this concept to develop the principle of dynamic equilibrium: "When the ratio of erosive action as dependent on declivities (slope) becomes equal to the ratio of resistance as dependent on rock character, there is an equality of action" (Gilbert, 1879, p. 100). If spatial variation in hillslope steepness does not reflect differences in rock erodibility, disequilibrium is present, as indicated in Figure 1.10A. Under ideal conditions a dynamic equilibrium (steady state) is approached in which all parts of a landscape downwaste at equal rates (Fig. 1.10B).

Interrelations between these variables of slope steepness, materials, and processes commonly result in slopes that do not tend toward a steady state. There is a limit to the steepness that a given lithology can attain, and most lithologies undergo changes in weathering characteristics and erodibility with change in slope and soil cover. Where erodibility is dominant over slope steepness, resistant lithologies will become progressively higher than the adjacent softer lithologies.

Two classic localities for differential erosion rates are the Table Mountains of the Toulumne (Bateman & Wahrhaftig, 1966) and San Joaquin

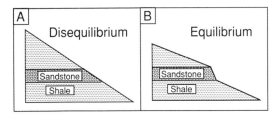

FIGURE 1.10 Diagrams illustrating Gilbert's (1879) equality of action on hillslopes underlain by shale and a more resistant sandstone bed. A. A disequilibrium slope is depicted because shale should erode more rapidly than sandstone where slope steepness of the two lithologies is similar. B. An equilibrium slope is depicted because the shale has eroded to a slope with lesser steepness than the sandstone, thus compensating for the differences of erodibility of the two rock types and causing all slope elements to downwaste at the same rate.

(Wahrhaftig, 1965) Rivers in the foothills of the Sierra Nevada of California. Andesite and basalt flows partially filled Miocene gorges; they are more resistant to weathering and erosion than the adjacent metamorphic and granitic rocks so that differential erosion has occurred. The resulting topographic inversion depicted in Figure 1.11 has left the lava flows and their underlying river grav-

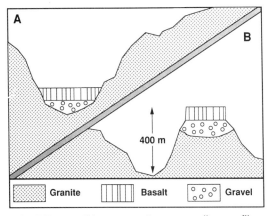

FIGURE 1.11 Diagrammatic cross-valley profiles showing the topographic inversion of the valley of the San Joaquin River, California. A. About 9.4 Ma. B. Present.

els as sinuous ridges as much as 400 m above the level of the present rivers. Ritter (1967,1972) described topographic inversion resulting from stream capture that has left fluvial gravel caps on ridgecrests.

A more common type of topographic inversion may occur in massive crystalline rocks. Wahrhaftig (1965) proposed that progressive increase in outcrop areas has occurred in the western Sierra Nevada. Because granitic rocks weather more rapidly in moist subsoils than when exposed, a random exposure of rock in a stream channel or on a hillside may result in enlargement of the outcrop. As *grus* is eroded from the edges of the outcrop, the outcrop will increase in area and height relative to the surrounding mantled hillslope. Random exposures of granitic rock coalesce with time to form steep fronts that are separated by gently sloping treads still mantled with grus.

Progressive increase in areas of steep bedrock outcrops over several million years may be considered an irreversible change because the self-enhancing feedback mechanism of rapid runoff from massive outcrops continues to erode the grus at the margins of the outcrop. Such topographic inversion requires that the drainage net, as well as the hillslopes, undergo progressive change. The steady-state model is inappropriate because a key independent variable-erodibility of surficial materials-changes in both time and space. Thus Wahrhaftig's study provides a nice example of landforms that do not tend toward a steady state.

Small moving deposits such as ripples and dunes tend toward a steady state, but most large stationary deposits do not because they are the outputs of erosional-depositional systems. Let us consider an alluvial fan. Although streamflow and depositional processes, distributary channel morphologies, and area interrelations of adjacent depositional systems (playas, floodplains, and other fans) may tend toward steady states, a fan is a depositional landform that is increasing in volume. Degradation may exceed fan aggradation if the sediment yield of the erosional part of the system is reduced or if the rate of accumulation of fan deposits is decreased in any other way. If net erosion predominates, a fan is best considered an erosional slope that is underlain by alluvium. For a fleeting instant of geologic time, the overall rates of erosion and deposition may appear to be equal, but this threshold should not be regarded as attainment of steady state. Steady state (equilibrium) implies that, once achieved, it will persist as long as the independent variables of the system remain unchanged and thresholds are not crossed. Little is gained by forcing an equilibrium conceptual model on landscapes where change is obvious.

## 1.7 Complex Response

Schumm noted that a single base-level fall at the mouth of an experimental watershed (Fig. 1.12) resulted in degradation, aggradation, and then renewed degradation of the adjacent upstream reach (Schumm, 1973; Schumm & Parker, 1973; Schumm et al., 1987). Increase in watershed relief caused by the base-level fall migrated upstream from the basin mouth as a *nickzone* that increased stream gradients. This increase in stream power caused the adjacent upstream reach to degrade to the base

level of erosion (Fig. 1.12B) before the nickzone had reached tributary parts of the watershed farther upstream. Incisement of a valley fill that was present before the perturbation resulted in formation of a fill terrace. Passage of the nickzone through upstream reaches increased bedload transport rates as valley-floor alluvium was mobilized and adjacent footslopes were trimmed. Increases in resisting power were sufficient to cause aggradation in reaches that had already attained the base level of erosion (Fig. 1.12C). Sediment yields decreased after the perturbation reached the headwaters. This decrease in resisting power caused the threshold of critical power to be crossed in downstream reaches to form a second fill terrace, whose aggradation and incisement were solely the result of the single perturbation. Each repetition of the simulated base-level fall causes a single new terrace to form (Parker, 1977). Thus the highly useful principle of complex response describes the progression of morphologic or stratigraphic changes, or both, that occur within a fluvial system after a single perturbation (that is external to the system) changes the bedload transport rate.

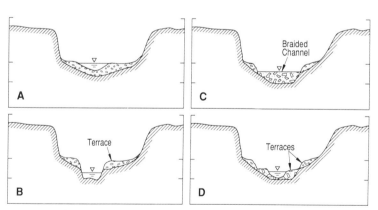

FIGURE 1.12 Cross sections of valley 1.5 m from outlet of an experimental fluvial system underlain by a cohesive mixture of 72 percent sand and 28 percent silt and clay. Modified from Figure 1 of Schumm and Parker (1973). A. Initial conditions of stream channel and broad floodplain underlain by alluvium. B. Stream-channel incision through alluvium and into underlying materials after a 10-cm base-level fall. Floodplain is left as a fill terrace. C. Aggradation creates an inset alluvial fill as bedload transport rate is increased. Unstable channel has a high width/depth ratio. D. Reduction in bedload transport rate causes stream-channel incision and lower width/depth ratio.

Complex response has tended to become an all-encompassing model for high-energy drainage basins with large potential sediment yields that have concurrent aggradation and degradation in different reaches. It has been extended to include response to perturbations such as tectonic tilting, climatic change, *coseismic landslides,* forest fires, and land use; each of these processes may overload the stream with bedload (Janda et al., 1975; Schumm, 1977; Womack & Schumm, 1977; Patton & Schumm, 1981; Pearce & Watson, 1983). However, climatic change processes that affect entire fluvial systems at the same time may have more sudden impact and greater effect than base-level falls that migrate upstream from a point source (Fig. 1.8).

Tectonic geomorphologists who consider stream terraces as time lines passing through tectonically deforming landscapes need to consider two important questions:

1. Are alluvial stratigraphies and times of terrace-tread formation sufficiently synchronous to be correlated between watersheds?
2. Are individual terrace treads *synchronous* or *diachronous?* (Sections 4.4 and 5.4.3.).

Prior to development of the complex response model, it was assumed that smooth longitudinal profiles of terrace treads indicated a single time of formation, and that scatter of radiocarbon ages of alluvium was due to analytical uncertainties. Now the pendulum has swung far to the other side to embrace opinions that alluvial fills and fill-terrace treads are diachronous. "It is not possible to correlate the terraces [of Douglas Creek] purely on the basis of elevation, nor to relate the terrace remnants throughout the valley to specific stimuli such as climatic variations or base-level changes. Moreover, there is a lack of similar terraces in nearby valleys" (Schumm et al., 1987, p. 124). One wonders if climatic and tectonic stream terraces can be recognized at all.

Resolution of dating methodologies, relative to actual time spans of terrace-tread formation, largely determines our perception of terrace treads as synchronous or diachronous. We are more likely to consider young terraces as diachronous because more field information is available and dating is more precise for these terraces than for old terraces. For these various reasons, geomorphologists should define the space and time spans they are considering (Schumm & Lichty, 1965) when discussing synchroneity of stream terraces.

A conceptual approach that occupies a philosophical middle ground is worth considering. Stream terraces should be diachronous for reasons other than complex response. Because each fluvial system consists of a different set of interacting variables, one should expect both the reaction time (time from perturbation to initiation of aggradation) and relaxation time (time from start to end of aggradation) to vary between streams derived from different watersheds. Thus it would be quite surprising if Haynes' (1968) radiocarbon ages of alluvial fills in the southwestern United States did not have significant temporal overlap. An important contribution of the complex-response model is that it encourages us to consider the time spans required for adjustment of fluvial systems relative to the time spans between external perturbations.

One may consider the complex-response model as being appropriate for describing second order responses to primary perturbations such as those caused by tectonic and climatic changes. Complex responses that occur during short time spans of 0.01 to 1 ky generally are caused by internal adjustments of fluvial systems that affect bedload transport rates; they are responsible for small *unpaired fill terraces* (Fig. 1.9) and flights of paired *degradation terraces* (Figs. 5.18–5.21) that are superimposed on larger terraces that have resulted from tectonic and climatic perturbations during time spans of 1 to 100 ky.

## 1.8 Genetic Types of Stream Terraces

Genesis of three distinct types of stream terraces can be understood through application of the concepts of tectonically induced downcutting, base level of erosion, complex response, threshold of critical power, diachronous and synchronous response times, and static and dynamic equilibrium. Climatic and tectonic stream terraces are *major* terraces below which flights of *minor* complex-response degradation terraces can form.

These three types of terraces can be summarized by describing a downcutting-aggradation-renewed downcutting sequence for high energy streams with gravelly bedload. By tectonically induced downcutting, streams degrade to achieve and maintain a dynamic equilibrium longitudinal profile at the base level of erosion. Lateral erosion bevels bedrock beneath active channels to create major straths that are the fundamental tectonic stream-terrace landform. Aggradation events record brief reversals of long term tectonically induced downcutting because they raise active channels. They may be considered as major (the result of climatic perturbations) or minor (the result of complex-response model types of perturbations). Climatically controlled aggradation followed by degradation leaves an aggradation surface; this type of fill-terrace tread is the fundamental climatic stream-terrace landform. Aggradation surfaces may be buried by subsequent episodes of deposition unless intervening tectonically induced downcutting is sufficient for younger aggradation surfaces to form below older surfaces. Raising of the active channel by either tectonic uplift or by climatically induced aggradation provides the vertical space for degradation terraces to form; first in alluvial fill and then in underlying bedrock along tectonically active streams. These are complex-response terraces because they result from interactions of dependent variables within a given fluvial system. Pauses in degradation to a new base level of erosion or minor episodes of backfilling, or both, lead to formation of complex-response fill-cut and strath, or of fill terraces. Fill-cut terraces are formed in alluvium; they are complex-response terraces because they are higher than the base level of erosion. Good exposures and dating are needed to distinguish static equilibrium complex-response minor strath terraces from dynamic equilibrium tectonic (major) straths. Strath terraces may be regarded as complex-response terraces where deg-

radation rates between times of terrace-tread formation exceed the long term uplift rate for the reach based on ages and positions of tectonic terraces.

Climatic and tectonic stream terraces are distinct because of continuity along valleys, and because aggradation-event deposits resulting from climatic changes tend to be similar in nearby valleys. Such primary-response terraces may be diachronous to varying extents that should be evaluated in terms of spatial variations of ages of stream-terrace treads and straths (Sections 4.4 and 5.2.2). Climatic and tectonic controls on the formation of major fill and major strath terraces are contrasted with complex-response fill-cut and minor strath terraces in Sections 5.3.3 and 5.4.3.

## 1.9 Allometric Change

The preceding sections have described a general lack of continuous steady-state conditions in many geomorphic systems, either because of fluctuations of independent variables in systems that tend toward a steady state or because the system does not tend toward a steady state. Perhaps-together with the critical-power threshold model-a broader and more flexible approach than the steady-state model could be used profitably in some geomorphic investigations. One possibility is the concept of allometry used in biologic and palynologic work. Although biologists and palynologists use static and dynamic types of allometric analyses, they are primarily concerned with growth of part of an organism compared with growth of the whole organism or some other part of it. To borrow the term "allometry" from biology places an undue emphasis on growth, because changes in geomorphic systems may pertain to decrease as well as increase of variables. Rather than risk confusion, it is preferable to adopt a new term when borrowing a conceptual framework of analysis from the biological sciences. Therefore, we use the broader concept of "allometric change." Allometric change is the tendency for orderly adjustment between interdependent materials, processes, and landforms in a geomorphic open system (Bull, 1975). It includes steady-state models such as dynamic equilibrium.

Let us examine the potential of the concept of allometric change for evaluating change in part of a fluvial system. A typical example is shown in Figure 1.13, which illustrates concurrent increases of stream width and discharge. Consider the case of a rising stream after an intense rain. Steady state does not exist because the stream tends to fill its channel. The concept of allometric change states that the relative rate of change of part of the system is a constant fraction of the relative rate of change of the entire system or of another measure of size of the system. For the case of the rising stream, the relative rate of change of stream width (w) is a constant fraction (b) of the relative rate of change of stream discharge (Q). If we let $dw/dt$ be the rate of change of width and $dQ/dt$ be the rate of change of discharge, then the relative rate of change of these two aspects of the system can be expressed as

$$\frac{dw/dt}{w} \text{ and } \frac{dQ/dt}{Q}$$

Thus

$$\frac{dw/dt}{w} = b\frac{dQ/dt}{Q} \qquad (1.15)$$

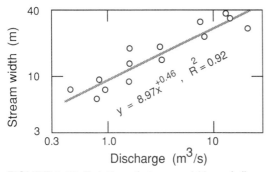

FIGURE 1.13 Relation of stream width and discharge of the Rio Galisteo at Domingo, New Mexico. From Figure 6 of Leopold and Miller, 1956.

Multiplying both sides by dt yields

$$\frac{dw}{w} = b\frac{dQ}{Q}$$

Integrating both sides yields

$$\int\frac{dw}{w} = b\int\frac{dQ}{Q}$$

$$\ln w + c_1 = b\,(\ln Q) + c_2$$

$$\ln w = \ln a + b\,\ln Q \qquad (1.16)$$

where $(c_2 - c_1)$ is combined in a constant, ln a. Taking antilogs,

$$w = aQ^b \qquad (1.17)$$

where a and b are constants and a is always positive.

Thus the rate of change of stream width, compared with the rate of change of stream discharge, may be expressed as a power-function equation. The power function is widely used in geomorphic studies because of its statistically significant fit to most data, its simplicity, and its relative ease of interpretation. Power functions are not the only means of describing allometric change; simpler or more complex equations also describe allometric relations in all empirical sciences (Gould, 1966, p. 595). Power functions, or other mathematical relationships, describe orderly changes in geomorphic systems. Interpretation of power-function equations involves comparison of both exponents for regressions of different slope and coefficients for regressions of similar slope. The explanations of the equations are allometric when analyses are made from a dynamic or static viewpoint.

Orderly adjustment between the variables of a fluvial system may be demonstrated empirically by dynamic or static allometric change, when statistically significant relations can be shown to exist between the variables. Dynamic allometric change is time dependent and refers to the interrelations of measurements made of a landform or process at different times. An example of dynamic allometric change of geomorphic processes is shown in Figure 1.13. Although twofold variations of stream width may occur for a given discharge, the power-function relation is statistically significant. Much of the scatter is caused by variables such as sediment load that vary with discharge in a manner different from that of streamflow width. An example of dynamic allometric change of a landform that does not tend toward a steady state would be the relation between the cumulative volume of lateral moraines and the distance that a glacier had moved downvalley. In this case the pairs of values of the two variables represent different times, even if the data were collected after recession of the glacier.

*Static allometric change* refers to the interrelations of measurements made at one time. Ridgecrest topographic data can illustrate static allometric change if the increase in fall with increasing slope length is measured at a single time. An example of static allometric change for a situation that does not tend toward a steady state would be an equation comparing, at a point in time, variations in the distance a monolith has moved away from a cliff of massive sandstone with distance along the fissure.

Thus, the allometric change model can be used to analyze the relative rates of change of two or more aspects of a system. It measures changes of one part of a system relative to changes of either the whole system or another part of it. Allometric change may apply either to landforms or to the processes acting on them. Because geomorphic processes or the dimensions of landforms may change with time, a variety of geomorphic interrelations can be examined by using the concept of allometric change.

The hydraulic geometry (quasi-equilibrium) model of Leopold and Maddock provides one of the more easily visualized examples of allometric change in which the interaction of variables tends toward a steady state. The hydraulic geometry described for the derivation of Equation 1.17 and illustrated in Figure 1.13 is an example of dynamic allometric change. As water and sediment discharge increases, variables such as flow width, depth, velocity, and hydraulic roughness show

concurrent changes. The orderly (allometric) change that occurs between these interdependent processes "demonstrates the strong tendency toward establishment of a quasi-equilibrium operating in channels carrying water and sediment" (Leopold & Maddock, 1953, p. 51). Dynamic allometric analysis provided Leopold and Maddock insight into how streamflow variables interact at a station. Hydraulic geometry in the downstream direction is an example of static allometric change because the analysis uses the same streamflow frequency at all gaging stations. Static allometric analyses showed that stream velocity generally increases in the downstream direction. Thus, hydraulic geometry is an example of the use of the allometric change conceptual model for prediction in time (dynamic) or space (static).

The allometric change approach accommodates both the tendency toward and the attainment of steady state, and also those processes or landforms that do not tend toward a steady state. Steady state would be a condition of zero rate of change of both variables and would be represented by one point on a graph portraying dynamic allometric change. Steady state does not apply to static allometric change because only a single time is considered.

## 1.10 Comparison of Conceptual Frameworks

Steady-state and allometric change concepts of adjustment in geomorphic open systems are different. Several types of adjustment in geomorphic systems are summarized in Figure 1.14. The change in the hypothetical variable plotted on the ordinate may be for a process or a morphologic feature. Examples would be streamflow width, velocity, and slope, or the rate of erosion at a point on a hillslope. The horizontal lines represent steady-state conditions. In Figure 1.14A, the introduction of a perturbation causes a departure from steady state. The variable fluctuates above and below the steady-state line in response to self-arresting feedback mechanisms. Because the system is stable,

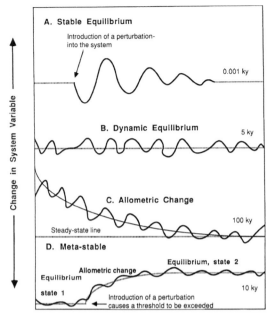

FIGURE 1.14 Types of adjustment for changing and unchanging geomorphic systems. A. Stable equilibrium. B. Dynamic equilibrium. C. Allometric change. D. Meta-stable equilibrium.

the effects of the perturbation become less with time and the interrelations between variables return to the same steady-state condition. An example might be alternating scouring and backfilling at a point downstream from a pile of alluvial material that has slipped into the channel from the undercut bank of a meandering stream. Alternating scour and backfill occur because of fluctuating turbulence. Variations of turbulence in scour depth become progressively less with time as streamflow removes the slipped material and the reach returns to the same steady-state condition that existed before the perturbation.

Figure 1.14B portrays a situation where dynamic equilibrium has been attained. The horizontal nature of the overall plot shows no net change with time. The horizontal line represents the mean of the short-term fluctuations. For short time spans the system variable oscillates due to the presence of interacting variables. Fluctuations, or changes

of variables, are largely excluded from the dynamic-equilibrium model by emphasis on means of variables. An example would be the maximum depth of streamflow along a cross-valley transect that crosses a meandering stream. Maximum streamflow depths occur at the outsides of meander bends and minimum depths occur at inflection points between the bends. Maximum depths at the transect will change as the meandering stream pattern migrates downvalley while in a condition of steady state. The mean depth remains unchanged, and alternating increases and decreases of maximum depth record the downvalley shift of the channel pattern with time.

The allometric change concept is portrayed in Figure 1.14C. Change is a dominant feature of the plot for both the long-term and short-term trends. Short-term changes in variables are merely interim adjustment states of short duration, thereby focusing attention on the longer term changes in the variables that caused the changes in the dependent variable. The system portrayed here tends toward a steady state because the overall trend of the variable approaches the hypothetical steady-state condition (represented by the lower horizontal line) in an asymptotic fashion. An example is the decrease in slope that would occur after a pulse of uplift has steepened a meandering reach similar to that described in Figure 1.14B but where the rate of downvalley meander migration is 10-fold slower. Maximum streambed slopes occur at inflection points between the bends, and minimum slopes occur in pools at the bends. Thus the short-term fluctuations in slope result from downvalley meander migration, and long-term decrease in slope occurs as this reach of the stream degrades to the same altitude that was present before uplift.

It should be kept in mind that when one defines adjustment for a given hillslope or stream few clues may be present about the direction and rate of change compared with the intermediate and long-term rates. This is particularly true when rates of geomorphic processes have been affected by humans.

A threshold and period of allometric change

between two periods of dynamic equilibrium is shown in the meta-stable equilibrium example of Figure 1.14D. An example for a meandering stream would be a given meander wavelength and radius of curvature that is upset by a climatic perturbation. A climatic change causes changes in discharge of water and sediment, and the stream responds by crossing a threshold, passing through a period of dynamic allometric adjustment, and attaining a different set of constant meander dimensions that reflect attainment of equilibrium conditions for the new water and sediment discharge inputs.

Viewpoints vary widely as to what constitutes equilibrium. For example, Chorley and Kennedy (1971) used definitions different from those of Figure 1.14. They regarded the situation of Figure 1.14B as "steady-state equilibrium," whereas I think it represents fluctuation about a given equilibrium state. They regarded the situation shown in Figure 1.14C as dynamic equilibrium, whereas I believe it represents a continuing short-term and long-term change. It seems that geomorphologists' selections of appropriate models are strongly influenced by their background and associates and by the nature of the problem at hand. Different models should not be considered mutually exclusive; instead one should compare models and then make an appropriate selection or modification.

Geomorphologists use one of the several conceptual frameworks during their investigations, and the time span represented by their data influences the type of model used (Schumm & Lichty, 1965). The equilibrium (dynamic equilibrium, steady state), quasi-equilibrium, and allometric-change models all emphasize interdependence of landscape elements in an open system. These models all accommodate the tendency toward steady state or attainment of it.

The Davisian model (Davis, 1889) is time dependent, because it emphasizes changes of landscape during stages of youth, maturity, and old age in a cycle of erosion. Those who are concerned with changes over geologic time spans may prefer a Davisian-type model that includes the tendency

for landmasses to erode to altitudes near base level. The landscape is understood by examination of its history, and the model has only one ultimate "equilibrium" condition (Chorley, 1962). Although base levels commonly do not remain stable for sufficiently long spans to allow the development of peneplains, the tendency for mountains to erode nearly to base level is valid and is a concept readily accommodated by the allometric change approach. The allometric change concept handles the problem of concurrent base-level changes by analysis of the adjustments within a fluvial system while the independent variable of total relief is decreasing with time.

Many geomorphologists prefer the equilibrium model, which emphasizes the interrelations of processes, lithologies, and landforms and deemphasizes the importance of time. The chief advantage of this model compared with the Davisian model is that it allows for intermediate equilibrium conditions for those elements of geomorphic open systems that tend toward a steady state. It also places much greater emphasis on the understanding of geomorphic processes than does the Davisian model. Hack (1960,1965a) has shown that the concurrent interaction of lithology, climate, and geomorphic processes explains landforms previously considered to be end products of denudation during geologic time spans. The model also explains drainage adjustment in the Appalachian Mountains (Hack, 1973).

The dynamic equilibrium and quasi-equilibrium models both use the word "equilibrium;" so it is easy to assume that the two models are roughly the same. However, considered from the allometric viewpoint, substantial differences in emphasis and method are apparent. The dynamic equilibrium model of Hack emphasizes the tendency toward time-independent (steady-state) landforms and the attainment of them. Geomorphic processes and materials are important in the dynamic equilibrium model, but the emphasis is on self-arresting feedback mechanisms that tend to create a steady-state condition. Leopold and Maddock preferred to name their model "quasi-equilibrium" because of the large amounts of scatter about their regressions of streamflow variables and because one cannot be certain whether all aspects of the system are balanced. For example, mean velocity increases (in the overall sense) in the downstream direction but varies markedly from station to station.

The quasi-equilibrium model also emphasizes the adjustment of variables with time. The filling of a stream channel during an increase in discharge is a situation of change, not equilibrium. Quasi-equilibrium generally is used in the context of interdependent adjustment of many streamflow variables in such a way as to accommodate the discharge of sediment and water in a manner that is in accord with minimum expenditure of total energy and a condition of maximum entropy. Furthermore, the quasi-equilibrium model utilizes the two modes of allometric analysis. Hydraulic geometry at a stream gauging station (Fig. 1.13) is dynamic allometric change and the variation of hydraulic characteristics in the downstream direction for a given flow frequency is static allometric change.

Allometric change is a unifying concept that not only includes the desirable aspects of the Davisian, dynamic equilibrium, and quasi-equilibrium models but also accommodates those processes and landforms in geomorphic open systems that do not tend toward a steady state. For elements that do tend toward a steady state, allometric analyses provide a way of describing interdependent adjustments among variables before steady state is attained. For most processes and landforms, a broader perspective of the interrelations of materials, processes, and landforms can be obtained by using the concept of allometric change.

The following chapters use the concepts introduced in this chapter. The systems approach used in this book includes independent and dependent variables as inputs and outputs, stream power, feedback relations, thresholds, stability and instability, response times, and complex responses. In particular, the threshold of critical power in streams will be used, because streams are the connecting link between erosional and depositional subsystems that have been affected by late Quaternary climatic change. At times the thread of words will seem to follow the ideas of William Morris Davis (1899) and Walther Penck (1953), but for most study sites, applications of concepts of equilibrium, thresholds, and response times are more useful. At other times, clear-cut distinctions between conceptual frameworks will seem rather meaningless and interpretation of past events and prediction of future events rather uncertain (Schumm, 1985). The reader also may consider the concepts of instability in changing chaotic systems such as that part of bifurcation theory referred to as catastrophe theory (Zeeman, 1976; Graf, 1979). Graf (1979,1982b) has used catastrophe theory, allometric change, and spatial interaction laws to analyze arroyo cutting.

The conceptual frameworks of Chapter 1 are applied first to the arid to semiarid deserts of the American southwest in Chapter 2. Subsequent chapters introduce new concepts in climatic settings that range from extremely arid to humid realms where the effects of changing climates are much different from those discussed in Chapter 2.

# 2

# Impact of Pleistocene–Holocene Climatic Change on Desert Streams

Climatic change during the latest Pleistocene and early Holocene (see Table 1.2 for definitions of temporal terms) caused profound changes in the hills, streams, piedmont landforms, and soils of the deserts of the American southwest and adjacent Mexico (Fig. 2.1). Much of this region is drained by the lower Colorado River. The scope of this chapter is diverse. It includes paleoclimatology, identification and dating of aggradation events, and geomorphic responses to Pleistocene–Holocene climatic change with an emphasis on process-response models, response times, and feedback mechanisms between dependent variables. The conceptual models discussed in Chapter 1 are used to analyze landscape change in response to varying independent and dependent variables. Botanical and synoptic paleoclimatic reconstructions are compared with the sequence of late Quaternary fluvial geomorphic events. What types of evidence for climatic change are obvious? How sensitive are desert hills and streams to Pleistocene–Holocene climatic change? What were the response times of plants, streams, and soils to climatic perturbations? We need answers to such questions to better understand the behavior of fluvial systems when climate changes.

## 2.1 Independent Variables of the Fluvial Systems

Many independent variables (Table 1.1) are fairly constant inputs to fluvial systems in the study region, which makes it easier to discern the effects of major changes in the independent variable of late Quaternary climate. Structural basins in the Death Valley and Gulf of California areas are tectonically active. Elsewhere, few streams have been affected by local base-level falls caused by faulting. Faulted Holocene alluvium has not been found, and only four Pleistocene fault scarps, less than 4 km long and 10 m high, have been described. The mountains in the study area are less than 1200 m high and most are lower than 800 m, so altitudinal effects on microclimate and plant communities are minimal.

A variety of metamorphic, plutonic, volcanic, and sedimentary rocks form the mountains. This

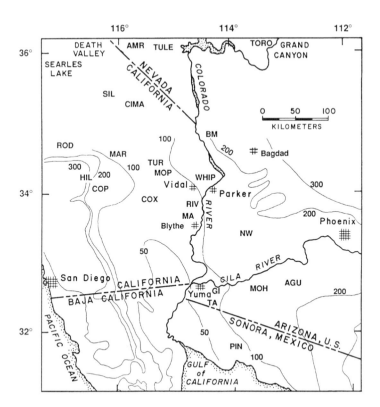

FIGURE 2.1 Southwestern deserts location map. Isopleths of equal annual precipitation, in millimeters, are from Turner and Brown (1982). AGU, Aguila Mountains; AMR, Amargosa Desert; BM, Black Mountains; CIMA, Cima volcanic field; COP, Copper Mountains; COX, Coxcomb Mountains; GI, Gila Mountains; HIL, Hildago Mountain; MA, Big Maria Mountains; MAR, Marble Mountains; MOH, Mohawk Mountains; MOP, Mopah Range; NW, New Water Mountains; PIN, Sierra Pinacate; RIV, Riverside Mountains; ROD, Rodman Mountains; SIL, Silver Lake; TA, Tinajas Altas Mountains; TORO, Toroweap; TULE, Tule Springs; TUR, Turtle Mountains; WHIP, Whipple Mountains.

diversity is useful because lithologies differ in their sensitivity (rapidity and magnitude of response) to climatic change (Chapter 3). For example, weathering, vegetation density, and fluvial processes change more on hillslopes underlain by quartz monzonite than on slopes of andesite.

Parent materials for piedmont soil profiles also are highly variable. Alluvium containing limestone or basalt clasts may be conducive to rapid accumulation of calcium carbonate in soil profiles (Lattman, 1973). Parent materials and soils that contain more than 15 percent calcium carbonate tend to inhibit the development of argillic B horizons (see Table 2.3 for abbreviations of soil terms) by flocculation of clay (Birkeland, 1984a; Gile, 1975, Gile et al., 1979). Lithologic controls on the rates of accumulation of pedogenic carbonate and clay were avoided in this study by concentrating on soils in gravels derived initially from noncalcareous metamorphic, granitic, and felsic volcanic rocks. Parent-material texture was kept fairly constant by comparing only those soils developed in gravelly alluvium. Distant and local sources of atmospheric dust and salts are critical to soil-profile development in arid (Section 2.5.1.2.1) and humid (Section 5.2.3.2) regions (Table 2.1).

Although humans have built cities in and farmed the valley of the Colorado River, they have had little impact on the desert mountains and piedmonts. Their most noticeable activities in these areas are the building of numerous dirt and paved roads and intense use of off-road vehicles. The area is too arid for grazing of domestic animals, and most mines are small.

For all of these reasons, the hot southwestern deserts of the United States are a good place to examine geomorphic responses to climatic change.

TABLE 2.1  Classification of climates.

| Precipitation | | Temperature | |
|---|---|---|---|
| *Class* | *Mean Annual (mm)* | *Class* | *Mean Annual (°C)* |
| Extremely arid | <50 | Pergelic | >0 |
| Arid | 50–250 | Frigid | 0–8 |
| Semiarid | 250–500 | Mesic | 8–15 |
| Subhumid | 500–1000 | Thermic | 15–22 |
| Humid | 1000–2000 | Hyperthermic | >22 |
| Extremely humid | >2000 | | |
| *Class* | *Seasonality Index (Sp)*[a] | *Class* | *Seasonality Index (St)*[b] *(°C)* |
| Nonseasonal | 1–1.6 | Nonseasonal | <2 |
| Weakly seasonal | 1.6–2.5 | Weakly seasonal | 2–5 |
| Moderately seasonal | 2.5–10 | Moderately seasonal | 5–15 |
| Strongly seasonal | >10 | Strongly seasonal | >15 |

[a] Precipitation seasonality index (Sp) is the ratio of average total precipitation for the three wettest consecutive months (Pw) divided by average total precipitation for the three driest consecutive months (Pd).

$$Sp = \frac{Pw}{Pd}$$

[b] Temperature seasonality index (St) is mean temperature of the hottest month (Th) minus mean temperature of the coldest month (Tc), in °C.

$$St = Th - Tc$$

A review of the nature of present and past climates will underscore the potential of the region for these studies.

### 2.1.1 REGIONAL CLIMATOLOGY

Before examining the nature and effects of past climates, we need to understand the present climate and the spatial boundaries between regions of different dominant atmospheric circulation. Weather in the southwest deserts can vary abruptly. Monotonous warm, dry weather associated with mid-latitude subsiding airmasses is interrupted by winter storm fronts from the Aleutian low-pressure center in the northern Pacific Ocean, by occasional outbursts of polar continental air, and by the *summer monsoon* (Bryson & Lowry, 1955). The monsoon derives tropical moisture from both the Gulf of Mexico and the Pacific Ocean through the Gulf of California (Hales, 1974). Casual desert visitors decry an apparent lack of seasons, but the relative influence of these types of circulation define different seasonalities for local climates that greatly influence plant communities and geomorphic processes in the Sonoran, Mojave, and Great Basin deserts.

D. L. Mitchell (1976) divided the western United States into climatic regions on the basis of dominant summer and winter airmasses. He defined climatic regions using *equivalent potential temperature*, which is calculated from monthly values of maximum temperature, relative humidity, and barometric pressure. This multivariable parameter reduces the influence of altitude on airmass characteristics at the land interface.

Zones of maximum rate of change—closely spaced *isopleths*—on equivalent-potential-temperature maps reveal winter and summer patterns for six climatic regions (Fig. 2.2). During the summer, cool maritime air affects the west coast; warm, moist monsoonal air lies over most of Arizona and Colorado and all of New Mexico; and warm dry air lies over the rest of the region. Winter climatic regions are much different. Intrusion of cold arctic air into the north central United States occurs frequently (northeastern corner of map in Fig. 2.2). The major east-west boundary separates westerly Pacific Ocean airflow on the north from a high-pressure dome over southern Nevada. The orographic influence of the Sierra Nevada in California and the Colorado Rocky Mountains is revealed by the way in which this boundary skirts the northern edge of these lofty mountains. Areal distributions of plant species support Mitchell's contention that the equivalent-potential-temperature boundaries are significant climatic boundaries.

The regional climate results from the interplay of three different types of airmass circulation. Regions of subsiding airmasses occur globally at about latitude 30 degrees and result in subtropical dry belts along western sides of continents. These deserts are characterized by atmospheric stability and high temperatures. In southwestern North America, the belt of prevailing westerly winds occurs to the north of the subtropical dry belt and the strongly seasonal Mediterranean climatic pattern occurs in the transition area between these two shifting belts. Lengthy winter rainy seasons begin in the Pacific coast region when the westerlies belt moves southward; spring to fall droughts set in when high atmospheric pressures associated with the dry belt expand northward. This Mediterranean type of climate pattern is restricted to California by the Coast Ranges, Transverse Ranges, and Sierra Nevada; these mountains cause pronounced rainshadows.

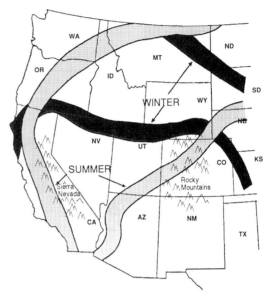

FIGURE 2.2 Major equivalent potential boundaries and climatic regions of the western United States as defined by Mitchell (1976).

During the summer the dry belt weakens sufficiently to allow two other types of airmass circulation to bring moisture into the study region. Monsoonal-type conditions occur in July and August, when moist tropical air moves northwest from the Caribbean Sea and north from the Pacific Ocean and causes thunderstorms in the Sonoran, Mojave, and Great Basin deserts. From August to October, tropical storms may move northward from off the west coast of Mexico, especially during summers of above-normal sea-surface temperatures (Douglas, 1976,1981).

The three meteorologic regimes produce different types of rainfall. Winter frontal storms that move southward along the west coast of North America and then cross the continent bring widespread cold, gentle rains and occasionally moderately intense rain to the study region. Rainfall is much warmer and is particularly intense during local summer thunderstorms. Generation and northward movement of subtropical storms may produce intense widespread rains that cause major floods during the late summer, but this rainfall source is erratic and storms do not materialize in many years.

Another important climatic factor is the position and configuration of the jet stream. Zonal airflow prevails when the jet stream heads straight east across geographic meridians, a situation that favors the east-west winter regional climatic boundaries of Figure 2.2. Meridional airflow prevails when the jet stream loops far to the south and then back to the north, a situation that favors transport of maritime tropical air into the deserts. Variations in the proportion of zonal and meridional airflow affect temperature, precipitation, and seasonality.

A simple, quantitative classification of climate is introduced in Table 2.1. Previous classifications have sometimes generalized climatic diversity by including extremely humid with humid and extremely arid with arid, but each category in Table 2.1 is characterized by major differences in geomorphic and pedogenic processes. The temperature terms are from Soil Taxonomy (Soil Survey Staff, 1975); mean annual air temperature at a site approximates soil temperature at a soil depth of 50 cm. The seasonality indices for precipitation and temperature describe the important factor of monthly climatic variability.

Along the lower Colorado River the present climate is moderately seasonal arid and strongly seasonal thermic to hyperthermic (Table 2.1). Summer temperatures can be as high as 51°C, but mild temperatures prevail during the winters. Mean annual values of precipitation and temperature for Yuma (Fig. 2.3A) are 65 mm and 21°C (Sellers & Hill, 1974). June is usually rainless but the wetter months receive about two and one-half times the mean monthly rainfall. Rainfalls that equal or exceed 25 mm in one day occur about every 5 years, but summer thunderstorms have caused local rainfalls in excess of 80 mm at several stations. Floods generated by these infrequent events do much of the erosion and sediment transport work.

Mean annual precipitation increases to the east

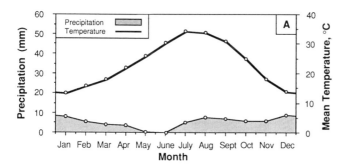

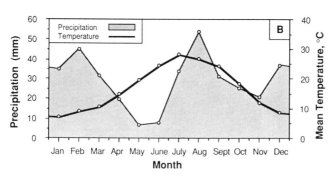

FIGURE 2.3 Monthly variations of precipitation and temperature at representative weather stations in western Arizona. Data from Sellers and Hill, 1974. A. Yuma, Arizona; altitude 59 m; moderately seasonal arid and strongly seasonal thermic. B. Bagdad, Arizona; altitude 1132 m; moderately seasonal semiarid and strongly seasonal thermic.

and west of the Colorado River to about 120 mm at the Aguila Mountains in southwestern Arizona and the Coxcomb Mountains in southeastern California. Localities such as Bagdad, Arizona (Fig. 2.3B), and the Rodman Mountains of the central Mojave Desert, and Kyle Canyon in the Spring Mountains of southern Nevada, have a moderately seasonal semiarid and strongly seasonal thermic climate. Seasonal and longer droughts influence the fluvial systems by controlling plant types and density and the depths and rates of hydrolytic weathering in soil profiles. Soil moisture is sufficiently high during the cool winter rainy season to promote plant growth and soil leaching more effectively than during the hot summer months.

The *soil-leaching index* is a useful way of determining the effectiveness of available moisture. McFadden (1982) used Arkley's (1963) method of calculating effective soil moisture–the annual sum of monthly precipitation minus monthly evapotranspiration for those months with positive values. The result is expressed as a leaching index. The leaching index at Parker, Arizona (mean annual precipitation of 97 mm and mean annual temperature of 22°C) is only 10 mm, compared with 260 to 600 mm for the semiarid to subhumid, thermic to mesic strongly seasonal climate of the San Gabriel Mountains (Chapter 4), and with more than 10,000 mm for extremely humid, frigid parts of the Southern Alps of New Zealand (Chapter 5).

The meager amounts of available soil moisture are reflected in the plants of the arid parts of the region. Vegetation is sparse and trees are restricted mainly to stream courses. Infiltration of ephemeral streamflows supports widely spaced small riparian trees [ironwood (*Olneya tesota*), blue palo verde (*Cercidium floridum*), smoke tree (*Dalea spinosa*), and catclaw accacia (*Accacia greggi*)]. Widely scattered bushes, including creosote (*Larrea tridentata*) and salt bush (*Atriplex canescens*), and annual grasses grow on the piedmonts and colluvial slopes. Plant cover on the hills is roughly 30 to 40 percent. Well-developed desert pavements are practically devoid of vegetation; densely packed

stones and high salt contents (Musick, 1975) are not conducive to germination and growth of plants.

The plants of these desert hills and piedmonts were not always widely scattered thermophilic desert shrubs. Fossils of plants that grew during the past 50 ky are the basis for determining general changes in late Quaternary climates that are scarcely believable to those who now travel across these baking plains.

### 2.1.2 PALEOCLIMATOLOGY

#### 2.1.2.1 Introduction

Late Quaternary-inferred positions of the boundaries of equivalent potential temperature (Fig. 2.2) contrast with the present positions. Four types of climate have characterized the region during the past 25 ky—full glacial, transitional, monsoonal, and interglacial. Winter and summer airmass boundaries probably were displaced toward the south during times of full-glacial climate. This shift would tend to increase winter precipitation and decrease monsoonal precipitation in the southwestern deserts. Studies of dated macrofossil plant assemblages are a powerful tool for inferring such climatic changes.

In recent years paleobotanical work has played an increasingly important role in the evaluation of the time, magnitude, and type of climatic change in the southwest (Mehringer & Ferguson, 1969; Wells, 1976; Wells & Berger, 1967; King, 1976; Van Devender, 1973,1977,1987; Van Devender & Spaulding, 1979; Cole, 1982,1985; Spaulding et al., 1983; Spaulding, 1985,1990). A variety of interpretations have been made regarding the time, type, and magnitude of latest Pleistocene and Holocene climatic change. This section relies mainly on the work of Van Devender and the model of Spaulding and Graumlich (1986).

Paleobotanists and climatic modelers have combined talents to better understand the history and causes of late Quaternary climatic change. Their work reveals both regional changes and local diversity. Not all parts of the American southwest became "hot" or "dry" at the same time. Such simple terms should be used to summarize climatic trends only with the realization that other factors such as drought frequency, frost, length of growing season, and cloudiness also may play important roles in deciding which plants were dominant at a particular time or fossil locality.

### 2.1.2.2 Applications of the Astronomical Theory of Climatic Change

Climatic modelers and paleoclimatologists have been especially interested in the astronomical theory of climatic change (Milankovitch, 1941; Imbrie & Imbrie, 1979). Temporal variations in the distribution of solar radiation input to different parts of the earth may be regarded as the ultimate independent variable that controls temperature. Such variations were the underlying cause of the Pleistocene–Holocene (Berger, 1979,1980) and much older nonglacial (Olsen, 1986) climatic changes. The marine oxygen-isotope record continues to provide convincing and detailed evidence to support the theory that the earth's global climate changes result from seemingly minor astronomical perturbations (Pisias et al., 1984; Martinson et al., 1987). Variations in the rhythmic nodding of Earth's rotation axis and precessional wandering of the direction in which the axis is pointed have caused metronome-like variations of climate that are especially well delineated in marine records. Variations in abundance of oxygen isotopes in the continental record of cave deposits (Winograd et al., 1988) suggested that the pre-Holocene major interglacial ended at 140 ka instead of at 120 ka suggested by the marine record. Renewed efforts by marine scientists (Ku et al., 1990; Bard et al., 1990) provided new ages in the 118 to 133 ka range that once again support the idea that the oceanic record is synchronized to the climatic variations caused by variations in Earth's orbital parameters. The geomorphic responses caused by changes from ice-age to present climates–the subject of this book–provide new and interesting ways of testing the astronomical theory of climatic change

in terrestrial environments. Thus it is appropriate that we briefly summarize aspects of the orbital theory of climatic change.

The earth's orbital parameters change slightly with time because of variations in the combined gravitational effects of other planets, the sun and the moon. These variations may cause contrasts of seasonal solar radiation input outside the atmosphere of as much as 30 percent at parts of the outer edge of the earth's atmosphere (Bradley, 1985). The three main orbital parameters of the earth that cause climatic change are (1) variations of eccentricity (the slightly elliptical path) of the earth's orbit around the sun, (2) obliquity of axial tilt of the earth to the plane of the orbit, and (3) precession of the seasonal equinoxes (progressive changes in seasonal timing of Earth's orbital perihelion and aphelion that result from a wobble in Earth's axis of rotation). Each of these parameters is cyclic, with periods of about 96 ky (eccentricity cycle), 41 ky (obliquity cycle), and 23 ky (precession of the equinoxes), respectively (Fig. 2.4A). The orbital variations do not cause changes in total energy received by the earth from the sun, but they do cause variations in seasonal insolation totals. Insolation maxima in equatorial regions produce heat that is redistributed poleward by the oceans and atmosphere. Temporal variations in solar energy inputs at different latitudes change the thermal gradients that drive the complex circulation systems, and may induce self-enhancing feedback mechanisms.

Size and persistence of snow and ice cover is an important insolation climatic variable (Walsh, 1984). Snow chills airmasses that pass over it by 3 to 10°C, and fresh snow reflects twice as much sunlight as bare land or ocean. Snow and ice cover reinforces troughs of cold air in the upper atmosphere (Lamb, 1972) that tend to enhance cyclogenesis (generation of anticyclonic disturbances in the northern hemisphere) along their eastward flanks, thereby providing a self-sustaining source of additional moisture for a growing ice sheet. The chilling effect of an ice mass 4000 km wide and

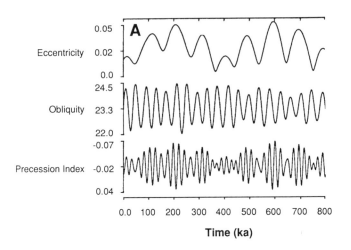

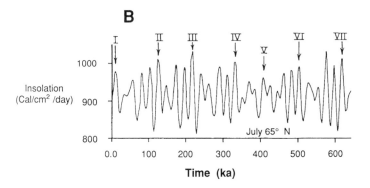

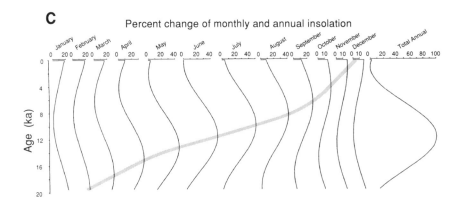

3 km thick on North America during the late Pleistocene was substantial; the mass required a long time to melt after the initiation of transitional and Holocene climates. During this time the ice continued to modify regional climates.

The effect of snow and ice on the strength of monsoonal airflow was recognized a century ago (Blanford, 1884). Summer heating of large continental interiors creates thermal lows; the resulting atmospheric pressure gradients promote incursion of tropical moisture from nearby oceans. Winters with unusually heavy snows require additional spring melting time and thereby delay and weaken summer heating of landmasses. Thus, it is not surprising that monsoonal airflow was weak when ice capped much of North America.

A variety of marine and continental paleoclimatic data support the growing realization that Earth's orbital variations played an important role in Quaternary climates and dependent geomorphic subsystems (Berger et al., 1984). The net effect of changes in the earth's orbital parameters is shown in Figure 2.4B. Major interglacials seem to occur about every 100 ky at times of marked increase in solar radiation to the northern hemisphere. Net solar input has opposite variations in the southern hemsphere. Change in the northern hemisphere appears to be sufficiently strong to cause roughly synchronous climatic change in the southern hemisphere, where landmasses (north of Antarctica) with perennial snow cover are virtually absent. The mechanisms that caused synchronous climatic change between the two hemispheres are not understood (Broecker, 1984).

Increase in annual solar radiation received by the northern hemisphere at about 15 to 11 ka was sufficient to trigger changes that led to a sequence of distinctive climates that coincided with the temporal transition from the Pleistocene to the Holocene. Figure 2.4C shows the variations in monthly solar radiation intensity during the past 20 ky. Summer increases in landmass heating that would enhance monsoonal airflow are apparent, starting at approximately 15 ka and peaking at about 12 ka. The return of monsoon rains probably encouraged growth of succulents and grasses in the northern Mojave Desert. Most of the orbital climate forcing was the result of precessional changes that continued into the Holocene; their trend is shown by the shaded line in Figure 2.4C. For example, increased heating during September at about 6 ka would favor growth of certain high altitude plants by extending their growing seasons (Davis, 1984; Davis et al., 1986). The response of a given plant community is complex because climatic reaction and relaxation times (Fig. 1.3) to orbital forcing parameters on oceans and ice masses may continue with concurrent changes in solar radiation.

Oxygen-isotope compositions of surface-dwelling planktonic foraminifera in cores of sediment from the Gulf of California seem to reflect brief events of lowered salinity beetween 15 and 11 ka (Keigwin & Jones, 1988). These radiocarbon-dated episodes of lowered salinity (calendric

FIGURE 2.4 Astronomical influences on solar radiation received by the earth. A. Variation in orbital eccentricity of the earth in degrees, obliquity of axial tilt in degrees, and precession of equinoxes during the past 800 ky. From Imbrie et al., 1984. B. Summed orbital parameters showing variations in summer insolation at 65 degrees N (from Broecker, 1984). Arrows point to the first peaks of triads of summer insolation peaks that coincide with maxima in the 100-ky eccentricity cycle. The marked increase in solar radiation intensity that peaked at about 11 ky was mainly the result of precessional effects. C. Insolation changes during the last 20 ka at 45 degrees N as percentages of total annual change (1973 langleys/day) during late Quaternary. Calculations are based on formulae of Berger (1979). The peak in total annual insolation shown at the right is the 11-ka peak of Figure 2.4B. Monthly maximum insolation changes as a result of precessional effects; the trend line through the peak for each month is shown by the patterned line. Modified from Davis et al., 1986.

ages of approximately 18.2 to 12.6 ka using the corrections of radiocarbon ages described by Bard et al. [1990]) may correlate with times of increased freshwater discharge by the Colorado River and other streams in response to increases of monsoon rainfalls predicted by the astronomical theory of climatic change.

### 2.1.2.3 Climatic Secrets Revealed by Fossil Plants Preserved in Packrat Middens

Plants are a key variable in geomorphic systems, and much of this book is devoted to exploring the many ways in which changes in density of vegetative cover and in species diversity affect all aspects of fluvial-system operation. Plants are sensitive to many types of climatic variations, but because they are a dependent variable they are also involved in feedback mechanisms and they interact with many other variables besides climate. This complicates our analyses and makes our conclusions less certain. Two major advantages encourage us to utilize plant macrofossils and pollen: (1) they can be compared with modern environments of similar plant associations, assuming the modern analogues are representative of past conditions, and (2) many plant fossils can be radiocarbon dated.

Important paleoclimatic data have been obtained from middens (Fossilized homes of this woodrat) of the rodent *Neotoma*. *Neotoma* is commonly known as the *packrat* because of its habit of collecting bits of vegetation and other objects and storing them in rocky crevices and small caves. Packrats live in many climatic and vegetative settings. By changing their eating habits they were able to adapt to the major climatic change in temperate semiarid mountains that became hot deserts. Packrats are not dependent on liquid water because they obtain water from their food.

Fossil plants preserved in packrat middens (Fig. 2.5) provide valuable information about ancient plant communities and climates. Urine excreted by packrats cemented the plant fragments into hard layers and ledges that were easily preserved in dry desert caves. Analyses of vegetation stored in

packrat middens have two advantages compared with pollen studies, which are a primary paleoclimatic tool in humid regions. First, the plants grew in the immediate vicinity of the sample locality. Leaves, seeds, and stems from bushes, succulents, and trees probably were collected within 100 m of the midden. Second, fragments of a single species, such as a stem or cone of juniper, may be dated.

Analyses of packrat middens from many mountain ranges in the study region (Fig. 2.1) allowed Van Devender (1973,1977,1987; Van Devender & Spaulding, 1979, Van Devender, Thompson, & Betancourt, 1987) to document changes in plant communities from conifers to cacti (Fig. 2.6). They estimated the time of disappearance of the latest Pleistocene plant community by determining the youngest ages of fossil juniper and pinyon. Singleleaf pinyon pine and Joshua tree decreased in abundance at about 13 ka and disappeared from the Whipple Mountains at about 11 ka (calendric ages of approximately 15.2 and 12.6 ka using the corrections of radiocarbon ages described by Bard et al. [1990]). The California juniper woodland persisted until about 8.90 to 8.5 ka, when resident species were replaced by desert scrub. The abundance of whipple yucca, ragged rock flower, bigelow beare grass, mormon tea, and mojave sage declined abruptly as desert thermophiles became dominant. The Whipple Mountains midden data show that juniper *(Juniperus californica)* and Joshua tree *(Yucca brevifolia)* grew as low as 320 m during the late Pleistocene and pinyon pine *(Pinus monophylla)* grew as low as 510 m. None of these plants now are present in the 1200-m-high mountains.

Temperature and precipitation estimates can be made by analysis of packrat middens. Van Devender (1973) suggested that the overall climatic change indicated by the Whipple Mountains data consisted of about 50 percent less precipitation and 3°C warmer mean annual temperature during the Holocene. Fossil annual plants from middens indicate that most of the precipitation decrease occurred during the winters and most of the temperature increase occurred during the summers. Variations

FIGURE 2.5 Plant fossils from a packrat midden at an altitude of 360 m in the Whipple Mountains, southeastern California. These juniper leaves have a radiocarbon age of 10.9 ± 0.18 ka.

of stable hydrogen isotopes in cellulose from modern plants and fossil packrat middens from the Great Basin and Colorado Plateau attest to large changes in temperature (Long et al., 1990). Deuterium in cellulose was at a maximum at about 7 ± 2 ka. Growing season temperatures, were 3 to 4°C cooler than present from 30 to 18 ka (calendric ages of approximately 32.2 to 21.5 ka using the corrections of radiocarbon ages described by Bard et al. [1990]) and 3 to 4°C warmer than present by the early Holocene. This 6 to 8°C total increase in temperature, when combined with changes in precipitation, resulted in the recent major aggradation event.

In 1987 Van Devender used present climate ranges of species found in late Pleistocene and in Holocene packrat middens of the Puerto Blanco Mountains 150 km southeast of Yuma, Arizona, to estimate paleoclimatic condition. The present climate is arid (230 mm) and thermic; most rain falls during the summers. Summer droughts and winter freezes are important factors controlling the northern limit of subtropical desert scrub and other species represented by the 103 taxa of the Puerto Blanco site. The late Pleistocene California Juniper-Joshua tree woodland indicated about 280 mm of rainfall per year, mainly during the winters; July temperatures may have been 8 to 11°C cooler. Warmer summers and increased monsoonal rainfall (annual rainfall of about 330 mm) by the early Holocene allowed acacia and mesquite to grow on south-facing slopes, with a few juniper. The presence of saguaro cactus indicated a decrease in hard freezes. Mid-Holocene climates were marked by a continuation of above present levels of summer rains. Rainfall decreased in the late Holocene, and summer droughts restricted most trees to the wetter microenvironment of streamcourses.

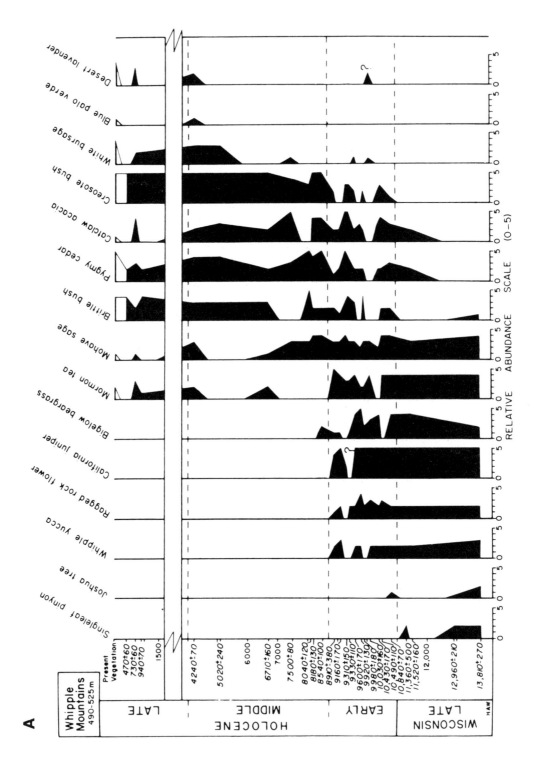

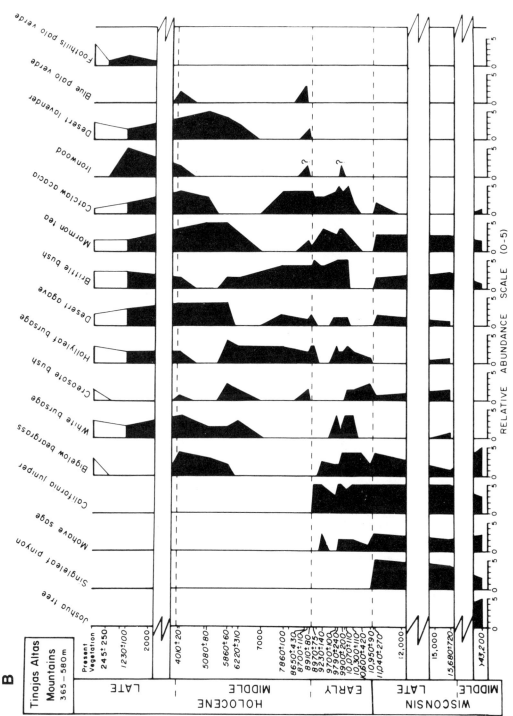

FIGURE 2.6 Chronology of latest Pleistocene and Holocene changes in the plant communities, based on analyses of dated packrat middens by Van Devender et al. (1987). A. Whipple Mountains, 21 radiocarbon dates. B. Tinajas Altas Mountains, 25 radiocarbon dates.

Several types of errors decrease the precision with which times of the beginning or end of a period of climatic change may be inferred from plant fossils collected from packrat middens. The laboratory analytical error for the radiocarbon dates typically is 100 to 300 years. Modern middens commonly represent 80 to 90 percent of the flora growing within 30 m of the nest (Cole, 1985,1986). Middens that have been used for thousands of years are bioturbated, which makes representative sampling of plant fossils of a given age more difficult. A more ambiguous uncertainty is the length of time needed for a climatic change to cause replacement of plant species in the vicinity of the middens sampled for paleobotanical studies.

Competition between resident (R) and invading (I) plant species can be described by a *plant-species threshold* where RS and IS are factors which, if increased, favor growth and reproduction of resident and invading species, respectively. The rate of net gain or loss is a function of the contrast between RS and IS. If

$$\frac{RS}{IS} = 1.0 \qquad (2.1)$$

no change in the relative abundance of plant species will occur because threshold or equilibrium conditions prevail. Resident species are quickly replaced by invading species at sites that have become marginal for the resident species as a result of climatic change.

$$\frac{RS}{IS} << 1.0$$

Conditions of other microenvironments on the same rocky desert hillside may continue to favor resident species. They may in part result from self-enhancing feedback mechanisms, such as plant-induced controls of soil acidity and moisture retention.

$$\frac{RS}{IS} > 1.0$$

Such microenvironments are regarded as *refugia* where plants can survive a climatic change. Even on a single hillslope of a given desert mountain range, plant-species change should be regarded as a *gradational threshold,* much like those proposed by Begin and Schumm (1984) for some aspects of fluvial systems. Reduction of vegetation density has the important concurrent effect of tending to increase soil-erosion rates.

In summary, the reaction time (Fig. 1.3) for a desert hillside plant association to the Pleistocene-Holocene climatic change would be rapid, but the response time to the climatic perturbation could be long. It appears that paleobotanical analyses of packrat middens may be useful in estimating times of climatic change to within 0.5 to 1 ky.

Late Pleistocene and early Holocene ranges of several plant species were dramatically different from their modern ranges. Regional paleobotanical work indicates that plant communities grew at least 410 m lower during times of full-glacial climate. Fossil plants from the middens show that latest Pleistocene plant communities persisted in the southwest deserts until about 11 ka. Although the climate has not been sufficiently wet or cool for pinyon to grow since 11 ka, juniper woodlands continued to grow in the southwestern deserts until about 8.9 ka in the Whipple, Tinajas Altas and Puerto Blanco mountains and until 7.9 ka in the New Water Mountains. Since then only desert scrub plant communities have grown in the lower Colorado River region. King's (1976) work on packrat middens in Lucerne Valley, southwest of the Rodman Mountains, shows similar times of change in plant communities. The youngest pinyon-juniper woodland was dated 11.8 ka, the juniper woodlands were replaced by desert scrub about 7.8 ka; the climate-controlled change in altitude of plant comomunities was at least 365 m. Both studies suggest that two types of climatic change occurred between 12 and 8 ka.

Relatively dry conditions north of latitude 36 degrees N were due partly to topographic modulation of the climate by the Sierra Nevada rainshadow (Spaulding & Graumlich, 1986). Changes in plant macrofossil assemblages from packrat middens (Fig. 2.7) in the Amargosa Valley of the northern Mojave Desert clearly show the transi-

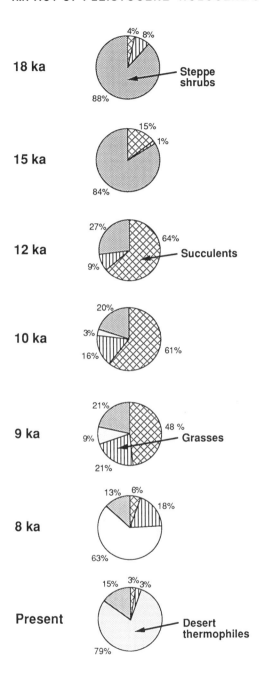

tional nature of late Quaternary climatic changes. Steppe shrubs such as *Artemisia* were dominant during cold, dry full-glacial climates at 18 ka (calendric age of approximately 21.5 ka using the corrections of radiocarbon ages described by Bard et al. [10mm rains were minimal during this time of dominantly zonal airflow, but cooler temperatures increased effective precipitation. *Effective precipitation* is the amount of water avilable to support plants, weather rocks and form soil profiles, and supply runoff from hillslopes—it is a function of temperature, precipitation, and windiness. The latest Pleistocene climate between 15 and 10 ka was transitional between glacial and interglacial conditions. The temperature increased and the return of summer monsoonal rainfall from the southeast was not affected by the Sierra Nevada rainshadow: summer rains favored growth of succulents and decrease of steppe shrubs. The early Holocene was a time of increase of grasses and the appearance of desert thermophiles; between 9 and 8 ka thermophiles became dominant. Between 10 and 8 ka woodland had retreated from low-altitude sites as effective precipitation decreased. By 8 ka succulents had decreased from 48 to 6 percent of the vegetation in response to decrease in effective precipitation.

In marked contrast, conifer woodlands continued to flourish at 400-m-lower altitudes in the Sonoran Desert, well to the south of the Sierra Nevada barrier to westerly airflow. Lesser rainshadow effects of the Coast and Transverse Ranges promoted the growth of pinyon pine during full glacial and late glacial climates, perhaps combined with a shift of winter storm tracks to the south (Craig et al., 1984; Kutzbach & Guetter, 1986). Enhanced monsoonal rainfall between 12 and 9 ka delayed the dominance of desert thermophiles in the Sonoran Desert until 8 ka. Thus "where

FIGURE 2.7 Relative abundance of four principal types of plant macrofossils from packrat middens below 1200 m altitude in the Amargosa part of the Mojave Desert. The times of maximum abundance of each plant type during the 18 ka since the time of full-glacial climates are shown by arrows. From Spaulding and Graumlich, 1986.

desertification might be expected earliest, at low elevations and latitudes in the Sonoran Desert, one finds instead the persistence of woodland long after its demise to the north'' (Spaulding & Graumlich, 1986).

Each packrat midden study area seems to reveal different combinations of controlling variables. Conifers did not grow in the presently extremely arid area west of Yuma, even during times of full-glacial climate (Cole, 1986). Other species reveal times of climatic change that are similar to those indicated by the Whipple Mountains and Tinajas Altas sites (Fig. 2.6). Areas that presently are extremely arid and hyperthermic may have served as full-glacial refugia for thermophiles such as creosote bush.

Monsoonal rains may have peaked in the early Holocene in the Mojave Desert, and continued at a high level in Sonoran Desert sites closer to the source of summer rains, such as the Puerto Blanco Mountains. Such a temporal rainfall pattern would mimic the west-to-east increase in summer rain from southern California to New Mexico, as shown by comparison of Figures 4.3, 2.3A,B, and 3.23A.

Other paleobotanical studies in the western United States show the regional nature of late Quaternary climatic change. A major woodland change occurred in southwestern New Mexico at 10 ka, and the onset of drier climate occurred at 8 ka (Markgraf et al., 1983). In central Colorado, pollen analyses indicate a change from cool-moist to warm-moist conditions at 10 ka (Markgraf & Scott, 1981). However, a pollen record from the southern Sierra Nevada indicates drier than present conditions between 11 and 7 ka (Davis et al., 1985).

### 2.1.2.4 *Modeling of Paleoclimates*
Late Quaternary climatic change has been evaluated using weather (Sellers, 1983,1984; Craig et al., 1984), vegetation (Spaulding & Graumlich, 1986) and soils (Mayer et al., 1988) data or a combination of marine and continental data (COHMAP, 1988).

Paleobotanical data are useful for testing predictions made by modeling of paleoclimates. Three-dimensional computer models trace climatic evolution. They simulate atmospheric circulation and the hydrologic cycle by taking into account factors such as orographic influences, bathymetry, sea-surface temperature, distance from ocean, and wind direction. Craig modeled changes from full-glacial to present climatic conditions for the southwest deserts and adjacent southwestern California. His equations predict climatic changes from 41 independent variables on the basis of stepwise regression analyses. Test predictions were made of historical conditions at weather stations and of pluvial-lake formation. Monthly maps of changes from full-glacial to present conditions of temperature, precipitation, evapotranspiration, and runoff indicate that changes in climate were more variable, both temporally and spatially, than previously recognized.

Other simulations of climatic change in the southwestern deserts indicate a southward shift of the winter storm track during full-glacial times (Spaulding & Graumlich, 1986; Kutzbach & Guetter, 1986). This shift in regional climatic boundaries was accompanied by significant increases in January precipitation and decreases in July precipitation. Simulations of the India–Africa monsoons indicate marked strengthening of summer precipitation at 12 to 6 ka (Kutzbach & Otto-Bleisner, 1982; Kutzbach, 1983; Kutzbach & Guetter, 1984). Compared with the present, the earth's orbital parameters at 9 ka caused 7 percent more summer radiation and 7 percent less winter radiation at all northern hemisphere latitudes. This increase in seasonality of solar radiation input caused increased seasonality of large mid-latitude landmasses such as Asia and southwestern North America, which intensified the monsoonal airflow. The resulting seasonal changes in effective precipitation have been recognized in the fossil plant communities by Cole, Spaulding, and Van Devender.

The types of climatic change that occurred during the Pleistocene–Holocene transition are open to debate. Furthermore, the relative changes in precipitation and temperature needed to cause certain changes in vegetational, fluvial, pedogenic, and lacustrine subsystems are indeterminate. Van Devender provided an example of the type of climatic change by comparing the present Whipple Mountains' climate with the climate of the nearest pinyon-juniper and juniper woodland climates. Brakenridge (1978) used a different model; he pointed out that many combinations of change in precipitation and temperature may produce the same end effect. For example, juniper woodlands grow along the Snake River plain in southern Idaho where present annual precipitation is similar to that of the Whipple Mountains but temperatures are about 8°C cooler thereby reducing evapotranspiration by about 40 percent.

This situation regarding the *equifinality* (similar effects caused by several different combinations of variables) of cause and effect of Pleistocene–Holocene climatic changes has resulted in two schools of thought. The model used by Antevs (1954,1955), Snyder and Langbein (1962), Flint (1971), Van Devender (1977), and Wells (1979) emphasizes decreases in precipitation at the beginning of the Holocene, whereas Galloway (1970), Brakenridge (1978), and McFadden (1982) concluded that precipitation change was not necessary because change in temperature can account for the changes in the systems that they evaluated. For example, McFadden's modeling of soil-profile formation indicates that the differences between late Pleistocene and mid-Holocene soils of the Whipple Mountains piedmont can be accounted for by an 8°C warming, which would decrease the leaching index from 23 to 10 mm. Information regarding the effects of change of temperature on runoff in fluvial systems is provided by the graphs of Langbein and others (1959), which show that increased temperature leads to decreasing runoff from a given mean annual precipitation.

The Pleistocene–Holocene transition was a time of major change in climate no matter which of these two models one favors. Effective precipitation decreased dramatically in two stages at about 11 ka and 8 ka. Annual runoff from the hills of the lower Colorado River region may have de-creased by roughly two-thirds by 8 ka as the climate became warmer, drier, or both. Climatic variations since about 8 ka were minor in comparison. H.T.U. Smith's (1967) study of eolian systems in the Mojave Desert suggests a major interval of wind action between 7.5 and 4 ka (during the altithermal interval of Antevs, 1948, 1955) that was followed by a relatively less arid climate.

Late Quaternary changes in plant communities were part of overall adjustments of fluvial systems to the impacts of changing climate. Quaternary climatic variations consecutively affected the density and type of plant cover and weathering rates on hillslopes, hillslope sediment yield, and erosional or depositional modes of operation of streams that deposited or entrenched alluvium in response to changes in their watersheds. Although vegetation, hillslope, stream, and pedogenic subsystems have vastly different response times and types of thresholds when they undergo climate-controlled modes of operation, there is a consistent sequence in the times of change of system behavior. For example, hillslope sediment yield increases after a climatically induced decrease in density of protective plant cover. Thus dated plant fossils from packrat middens provide useful information for constraining maximum ages of late Pleistocene and early Holocene alluvial fans and stream terraces. The 13 to 11-ka plant-community transition occurred just before a major aggradation event. Although the preceding discussions provide considerable insight about past and present climates, one more regional control of the piedmont streams needs to be considered.

### 2.1.3 THE COLORADO RIVER AS A REGIONAL BASE-LEVEL CONTROL

The Colorado River heads in the Rocky Mountains, flows through the Grand Canyon, and crosses the deserts of the southwestern Basin and Range Province to the Gulf of California (Fig. 2.1). Base-level rise in nearby basins of internal drainage generally has favored burial of Quaternary piedmont deposits, but a different situation prevails along the lower Colorado River. Climatic change caused alternating aggradation and degradation by the river. Although the altitude of the river has fluctuated more than 50 m, net downcutting of roughly 100 m occurred during the Quaternary. Changes in the altitude of the river have constituted rises and falls in the base levels for the piedmont streams tributary to the river. Recent river downcutting promotes downcutting of tributary streams through the process of upstream migration of steepened reaches and headcuts. Deepening of tributary valleys promotes erosion of piedmont alluvium. Backfilling of the river valley promotes piedmont aggradation but, as is pointed out by Leopold and Bull (1979), a local base-level rise causes deposition for only a short distance upstream from the perturbation. Climate change-induced variations of sediment yield on the hillslopes are much more important than base-level rise in causing valley-floor aggradation (Section 2.5.2.2). Each subsequent base-level fall of the river contributed to a new episode of entrenchment of piedmont ephemeral streams, which created new space in which low inset terraces could form.

Glacial and periglacial processes in the Rocky Mountains introduced sediment into the Colorado River much faster than it could be transported to the ocean (resisting power greatly exceeded stream power). Despite a much lower sea level than at present, the valley of the Colorado River was backfilled during times of large input of detritus, because the river was far to the aggradational side of the critical-power threshold. With the advent of Holocene climates, decreases in glacial and periglacial processes in headwaters reaches caused reduction of bedload. Resisting power became less than stream power, and downcutting removed 20 to 80 m of valley fill. Locally a new major strath surface is being beveled into the Bouse Formation of Miocene and Pliocene age.

Longitudinal profiles of stream terraces may be regarded as geomorphic time lines along the valley of the Colorado River (Fig. 2.8); they provide insight regarding climatic and tectonic influences.

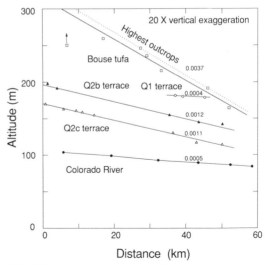

FIGURE 2.8 Longitudinal profiles of the Colorado River, Quaternary fill and strath terraces of the river, and the highest altitudes of outcrops of tufa in the basal Bouse Formation in the reach 60 km downstream from Parker, Arizona. See Table 2.13 for estimates of terrace ages.

The present river and the older than 1.2-Ma Q1 terrace are strath (type 1 dynamic equilibrium) gradients; both have about the same slope and represent times of attainment of the base level of erosion. Two late Pleistocene climatically induced fill terraces, Q2b and Q2c, are twice as steep as the present valley; each of these aggradation surfaces represents a threshold at the end of a bedload aggradation event. The present river is meandering, but the type of channel pattern at the time of deposition of fill-terrace tread deposits most likely was braided. Steep braided streams are more efficient in transporting large bedloads (Schumm & Khan, 1972). The 90 m of post-Q1 tectonically induced downcutting suggests a regional uplift rate of only about 0.05 m/ky. Stromatolitic tufa in the estuarine basal Bouse Formation (Metzger, 1968; Olmsted et al., 1973; Metzger et al., 1973) is 5.5 Ma (Shafiqullah et al., 1980). I assume that the tufa member of the Bouse was deposited at a single sea-level position in the ancestral Colorado River estuary. Subsequent downcutting has exhumed tufa outcrops on rocky hillsides near the river. Altitudes of the highest tufa outcrops progressively decrease downvalley but at a much greater rate than the stream terraces. Apparently, the tufa has been raised 170 to 260 m by regional uplift that increased toward the north. The maximum mean uplift rate has been only 0.03 m/ky, or 0.05 m/ky if tilting occurred before Q1 terrace time. Thus, Figure 2.8 suggests very slow and nonuniform regional uplift during the Quaternary, but no local faulting.

The Colorado River has two climate-controlled modes of operation—aggradation and degradation—associated with glacial and interglacial climates in the Rocky Mountains. The 400 km of the river upstream from its delta was alternately backfilled and downcut, which raised and lowered the local base levels for the streams tributary to the river. It is unlikely, however, that the ages of aggradation surfaces of the two types of fluvial systems were synchronous. Both the local streams and the river had different reaction and response times (Fig. 1.3) determined by times of maximum hillslope sediment yield and transport of bedload to downstream reaches. Differences in geomorphic responses to climatic changes were functions of drainage-basin size, types and intensities of geomorphic processes, and sensitivity of lithologies to climatic change. All these factors determined how far removed specific reaches of stream subsystems were from threshold or equilibrium conditions at given times after a climatic perturbation.

The following discussion focuses on alluvial geomorphic surfaces that resulted from aggradation events. Depositional subsystems on desert piedmonts contain diverse and abundant evidence of climatic change that is as interesting as the fossilized plant fragments of vegetation from hillslope subsystems found in packrat middens.

## 2.2  Criteria for Mapping Desert Piedmonts

Mapping of basin fill is much like traditional geologic mapping, which defines formations on the basis of distinctive characteristics that clearly separate mappable units from one another. Guidelines such as fossils and sedimentary petrology are rarely used. Instead, those studying basin fill utilize process-oriented geomorphology, sedimentology, and pedology to define synchronous and diachronous alluvial geomorphic surfaces. Multi-, parameter-based correlations are a useful tool when working with alluvial geomorphic surfaces of desert piedmonts (McFadden et al., 1989) or postglacial terraces in a humid terrane (Robertson-Rintoul, 1986).

Section 2.2 describes characteristics for mapping desert piedmonts in the context of surficial processes. Selected maps of the lower Colorado River region are presented in Section 2.3, and Section 2.4 discusses recent advances in geochronology. These sections form the basis for the remainder of Chapter 2, which explores landscape changes that have resulted from late Quaternary climatic change.

### 2.2.1  QUATERNARY ALLUVIAL GEOMORPHIC SURFACES

First we consider the basic mappable unit, which by definition has regional extent. A *geomorphic surface* (Ruhe, 1956) is a mappable landscape element formed during a discrete time period; it has distinctive materials, topographic features, soil profile, and weathering characteristics. A volcanic geomorphic surface might consist of a lava flow, or series of flows, formed during one eruption. This book discusses *alluvial geomorphic surfaces* such as alluvial fans and stream terraces formed at climatically induced times of deposition during the Quaternary. Each age of alluvial geomorphic surface in the valleys tributary to the Colorado River has a distinctly different topography, soil profile, and sedimentology. Figure 2.9 shows the characteristic heights of terraces above active stream channels and soil-profile data; Figure 2.10 is an oblique aerial view of diverse alluvial geomorphic surfaces east of Yuma, Arizona.

Nine alluvial geomorphic surfaces of the study

region record outputs of fluvial systems; topographic and pedogenic characteristics are summarized in Table 2.2. Q4a and Q4b surfaces consist of the active stream channels and those channels that have been abandoned for such a short time that visible rock varnish has yet to form on surficial cobbles. Q1 is a general category that includes surfaces so old and dissected that planar surfaces, desert pavements, and original soil profiles typically have been removed by erosion. The remaining six geomorphic surfaces of intermediate age (Q2 and Q3) are the main subject of this chapter; three were deposited in the latest Pleistocene and Holocene and three in the mid to late Pleistocene.

Part of the evidence that the six main geomorphic surfaces are the result of climatic change consists of their regional extent. All six occur at widely scattered localities in an area of more than 100,000 km², and the late Pleistocene (Q2c) and mid-Holocene (Q3b) geomorphic surfaces occur on most piedmonts in a region that extends from southeastern Arizona to Death Valley. Before discussing the ages, origin, and climatic significance of these alluvial geomorphic surfaces it is appropriate to describe their mappable characteristics and ages.

### 2.2.2 TERRACE HEIGHT AND FAN DISSECTION

Differences in the topographic characteristics (Fig. 2.9, Table 2.2) of the various alluvial geomorphic surfaces are useful for mapping Quaternary alluvium. However, height above stream channels should not be used as the sole criterion for correlation of surfaces presumed to be the same age.

Heights of terraces above adjacent stream channels that head in mountainous terrain commonly are used for assigning relative ages to stream terraces because older terrace treads generally are higher than younger treads. However, burial of Q2c treads by Q3 deposits is common, which serves as an important reminder that stratigraphic superposition of alluvium is indicative of relative ages that are the inverse of the topographic superposition of the surfaces. Heights of fill-terrace

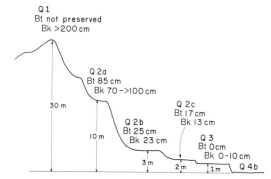

FIGURE 2.9 Diagram comparing heights above main stream channels and thicknesses of Bt and Bk (see Table 2.3) soil-profile horizons for piedmont geomorphic surfaces south of the Whipple Mountains, southeastern California. See Table 2.2 for summary of the Quaternary alluvial geomorphic surfaces.

treads vary greatly between mountain ranges, or even between adjacent stream systems issuing from a range. The best way to use terrace heights is to find a reach where a complete sequence of terraces can be identified and then to correlate these to terraces in upstream and downstream reaches where terrace preservation is not as good.

Depth of incision of channels originating on piedmont surfaces is a characteristic measured readily in the field and assessable through studies of topographic maps and aerial photographs. The proportion of a geomorphic surface that remains as a desert pavement is largely a function of the erodibility of the alluvium, depth of incision of main channels, and length of time available for pavements to be destroyed on the intervening flat-topped ridgecrests.

Topographic patterns vary with age. All the piedmont deposits were laid down by networks of braided streams but the character of the streams varied with time, and post-depositional modifications have resulted in four basic types of topography. These types are shown in Figure 2.11.

Active and recently abandoned stream channels have a bar-and-channel pattern typical of braided

FIGURE 2.10 Oblique aerial view of the piedmont east of the Gila Mountains, Arizona. The ridge-and-ravine topography of the highly dissected Q1 alluvial geomorphic surface is mainly in the foreground and right background. The smooth black desert pavements of the Q2 surfaces are mainly in the right middle ground; the moderately dissected Q2b surface is the most widespread, but Q2c and Q2a surfaces also are present. The Q3a, b, and c surfaces to the left of center are undissected by streams originating on the surfaces and have a bar-and-swale topography that gives a plumose texture to the aerial view. The Q4 surfaces are associated with 3- to 7-m-high trees whose growth is restricted to the active washes. Photo by William C. Tucker. (Figure 2.20 is a vertical aerial view of this piedmont; Figure 2.24 is a map of the alluvial geomorphic surfaces.)

TABLE 2.2 Characteristics of Quaternary alluvial geomorphic surfaces on piedmonts of the lower Colorado River region.

| Geomorphic Surface[a] | Postulated Climate | Topography and Rock Varnish Color | Soil-Profile Horizons[b] | | |
|---|---|---|---|---|---|
| | | | A | Bw,Bt | Bk,K |
| Q4b (torriorthents) | Arid | Bouldery bars and channels of active streams | None | None | Unweathered sandy gravel |
| Q4a (torriorthents) | Arid | Abandoned bouldery bars and stream channels | None | None | Unweathered sandy gravel |
| Q3c (torriorthents) | Arid | Undissected, bouldery bar-and-swale pavements 0–2 m above stream channels. Rock varnish 7.5YR 3/4 to 3/3 | 10YR 7/4 silt, vesicular, 0.5–2 cm thick | None | Bk at 1–15 cm, I, coatings <0.1 mm |
| Q3b (torriorthents) | Arid | Undissected, bouldery bar-and-swale pavements 0–4 m above stream channels. Rock varnish 7.5YR 3/4 to 3/3 | 10YR 7/3 silt, vesicular, 5–12 cm thick | None | Bk at 5–20 cm, I, coatings 0.1–0.5 mm |
| Q3a (torriorthents, camborthids, calciorthids) | Arid to semiarid | Undissected, bouldery bar-and-swale pavements 1–>10 m above channels. Rock varnish 7.5YR 3/3 to 3/2 | 10YR 7/5 silt, vesicular, 8–16 cm thick | Bw or none | Bk at 8–30 cm, I–II, coatings 0.5–1.0 mm |
| Q2c (haplargids, camborthids) | Semiarid | Slightly dissected, planar pavements 1–6 m above channels. Rock varnish 7.5Yr 3/2 to 2/3 | 7.5YR 7/5 silt, vesicular, 8–18 cm thick | 5YR 5/6 to 5/8 clayey gravel, Bt 8–30 cm thick | Bk at 20–55 cm, III, coatings 1–6 mm |
| Q2b (haplargids) | Arid and semiarid | Moderately dissected, smooth pavements 2–8 m above channels. Rock varnish 7.5YR 2/3 to 2/2 | 7.5YR 7/4 silt, vesicular, 8–20 cm thick | 5YR 5/6 to 5/7 mottles, Bt 20–60 cm thick | Bk or K at 40–90 cm, III, coatings 2–17 mm |
| Q2a (paleargids) | Arid and semiarid | Dissected remnants of pavements 4–10 m above stream channels. Rock varnish 7.5YR 2/2 to 2/1 | 7.5YR 7/3 silt, vesicular, 10–20 cm thick | 2.5YR 4/4 to 5YR 4/6 mottles, Bt 95 cm thick | Bk or K at 70–200 cm, IV, coatings 12–34 mm |
| Q1 (petrocalcic, paleorthids) | Arid and semiarid | Ridge and ravine landscape. Ridges 3–40 m above channel. Rock varnish 7.5YR 3/2 to 2/2. Rare dissected smooth surfaces | 7/5YR 8/3 silt, vesicular, 4–12 cm thick | Removed by erosion | Largely removed by erosion; K locally V and >2 m thick, coatings 20–50 cm |

[a] Names from Soil Survey Staff, 1975, are defined in Glossary.

[b] See Table 2.3 for definitions and Table 2.4 for stages of carbonate morphology, I–VI.

streamflow (Fig. 2.11A). The size of the bars is a function of stream discharge and bedload size.

The Holocene surfaces consist of braided-channel networks formed by streams similar to the active channels. The bouldery Holocene alluvium shown in Figures 2.12A, B, and D has several surficial features that clearly indicate lack of streamflow modification for thousands of years. Gravel bars are common and prominent, but the intervening fine-grained materials in former sandy channels have been smoothed by subtle processes such as raindrop impact, removal of sand and silt by wind and sheetwash, and heaving caused by the wetting and drying of vesicular clayey silt of the Av soil horizon (see Table 2.3) that immediately underlies desert pavements. In Figure 2.12B, the former channel is littered with small angular fragments of schist and gneiss and the bars contain varnished and weathered cobbles and boulders of schist and gneiss. The undulating topography of Q3 surfaces is best described as a bar-and-swale pattern; it is responsible for the plumose texture shown in Figures 2.10, 2.12D, and 2.20.

The late Pleistocene geomorphic surfaces have a different topographic pattern (Figs. 2.12A,C,D). Exceptionally smooth desert pavements are typical of the Q2 surfaces and especially of the Q2c surface; all are incised by dendritic drainage nets. The genesis of these pavements is much different from that of the Q3 surfaces (Section 2.2.4.2).

The fourth type of topographic pattern occurs where erosion has destroyed planar and undulating surfaces that have been little modified since their deposition. Dissection of the Q1 alluvium has made a ridge-and-ravine (ballena) topographic pattern. Even the surfaces of dissected alluvium are underlain by deposits whose sedimentology and stratigraphy reflect climate-induced changes in erosion, transport, and deposition.

## 2.2.3 STRATIGRAPHY AND SEDIMENTOLOGY

Most studies of landforms tend to emphasize events associated with culmination of glacial and interglacial conditions (Porter, 1989). Most aggradation events in the lower Colorado River region

occurred during interglacial climatic environments (Table 2.3) which represent only a small portion of the late Quaternary. Q1 deposits may represent tectonically active conditions modulated by the effects of concurrent climatic change. Each post-Q1 aggradational event tended to deposit climatic alluvial fans that are quite thin (Fig. 2.17B) compared with tectonic alluvial fans.

Holocene as well as active-channel deposits consist of poorly sorted water-laid gravel with a sandy matrix. Debris-flow beds occur in some Q3a and Q3b deposits. Ephemeral streams generally flow across piedmonts once or twice a year, and streams that drain watersheds larger than 10 km² may not flow at all in some years. The braided streams lack pool-riffle sequences and have width/depth ratios of more than 40. High-water lines are about 0.5 m above the dry streambeds even along large washes. Granitic cobbles in streambeds are rounded but generally are not smooth; volcanic boulders have smooth abrasion surfaces and rocks such as schist and quartzite tend to produce subangular gravel that reveals the influence of foliation and joints.

The contrast between the topographies, pavements, and deposits of Q2c and Q3b ages is remarkable. The late Pleistocene deposits are well-sorted, tightly packed gravel, particularly where the source rocks are schist. Figures 2.12A, 2.13, and 2.14 show pairs of deposits. Each pair was

TABLE 2.3 Climatic environments associated with alluvial geomorphic surfaces.

| Geomorphic Surfaces | Climatic Environment of Deposition |
| --- | --- |
| Q4, Q3c | Late interglacial |
| Q3b | Culmination of interglacial |
| Q3a | Transition from full glacial to interglacial |
| Q2c | Pulse of deposition in period of warming during average glacial conditions |
| Q2b | Culmination of interglacial |
| Q2a | Culmination of interglacial |
| Q1 | Both glacial and interglacial |

FIGURE 2.11 Topographies of different ages of alluvial geomorphic surfaces on piedmont south of the Whipple Mountains. A. Braided-stream deposits in an active wash. B. Bar-and-swale topography of Q3b surface. The book and hat are on gravel bars; most cobbles are weathered but are still in depositional positions. The rock hammer is in swale that has a pavement of well-sorted fine-grained gravel. C. Carbonate-coated gravel from the white Bk horizon contrasts with the smooth black pavement of a Q2b surface, whose deposits are distinctly finer grained and better sorted than the Q3b deposits. D. All of this view is of ridge-and-ravine erosional topography of the Q1 surface. Large boulders are common in the surficial lag gravel.

FIGURE 2.12 Contrasts in desert pavements of latest Pleistocene (Q2c) and mid-Holocene (Q3b) ages. A. East side of the Riverside Mountains, with smooth Q2c pavement in the foreground and rough Q3b bar-and-swale topography in the middle ground. The high fan remnant in the background is of Q2a age and has case-hardened exposures of calcic soil horizons. For an aerial view see Figure 2.12D; map and cross sections are shown in Figure 2.22. B. East side of the Tinajas Altas Mountains. White desert pavements of Q2c and Q2b age have formed as a lag-gravel concentration of white phenocrysts derived from weathering of porphyritic quartz monzonite. Dendritic drainage nets have been cut into the clayey grus. Areas of Holocene and active wash deposition occur on the lower half of the piedmont. The dirt road near the left side provides a scale. C. Tightly packed Q2c pavement east of the Riverside Mountains. The block of unvarnished gneiss shows the relative darkening of pavement stones by rock varnish. 15 cm scale. D. Oblique aerial view of piedmont east of the Riverside Mountains. Dissected smooth black pavement of island-like remnant of Q2c surface is surrounded by flood of Q3b alluvium. Plumose texture of Q3b surface is formed by the bar-and-swale topography. Photo 2.12A was taken to the left of the dirt road on the Q2c remnant.

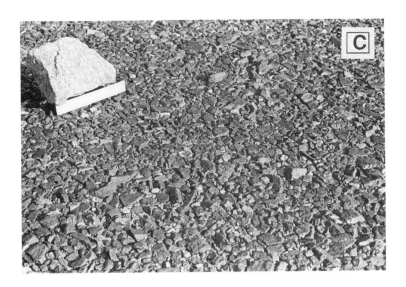

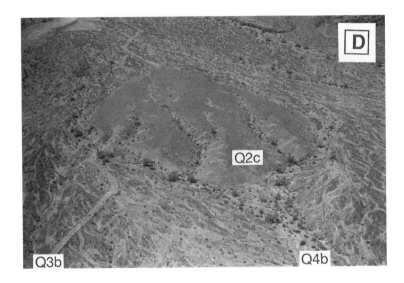

derived from the same drainage basin and reflects profound changes in the outputs of the fluvial systems. Deposits derived from the granitic rocks of the Gila Mountains illustrate the sedimentologic effects of climatic change. Late Pleistocene de- posits consist of well-sorted, cross-bedded sands of grus and clayey grus with thin gravel layers, and the deposits typically are orange (7.5YR 7/6) to dull brown (7.5YR 6/3). (Abbreviations used for colors are described in Table 2.5.) Mid-

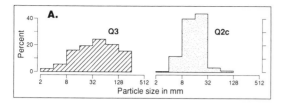

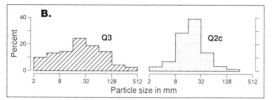

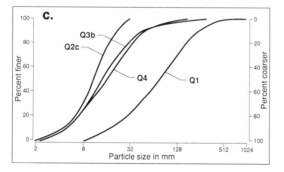

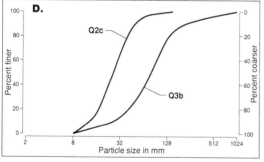

Holocene deposits are poorly sorted, massive, sandy gravels that are light gray (10YR 7/1) to dull yellow-orange (10YR 7/4). Section 3.2.2.3.2 shows that not only the particle-size distributions and topographies change with time but that lithologies that are resistant to hydrolytic weathering comprise more of the stream gravels during times of wetter climates.

Particle-size distributions of gravels in the alluvium and pavements of Holocene and late Pleistocene geomorphic surfaces are compared in Figure 2.13 for alluvium that consists mainly of schist and gneiss. Surfaces of desert-pavement rocks with a granitic fabric are weathered but appear to have undergone minimal reduction in size. The A, B, and D parts of the figure are for pairs of samples from the same drainage basins. The contrast in size and sorting of gravel in the alluvium histogram of Figure 2.13A is to be noted. The pavement of the late Pleistocene (Q2c) surface consists of fragments in five phi size classes, but the Holocene pavement consists of fragments that are spread over eight *phi size classes*. A similar situation is shown in Figure 2.13B for the two ages of pavements on alluvium derived from the Gila Mountains. In Figure 2.13C, the cumulative curve for the gravel of the late Pleistocene pavement stands apart from the rest of the curves because it is finer grained and lacks the coarse-grained tail characteristic of the Holocene and other deposits— the two to three phi size classes between 32 and 256 mm. Particles as large as 300 mm in maximum

FIGURE 2.13 Particle-size distributions of late Quaternary alluvium and desert pavements. A. Histograms for adjacent Holocene (Q3b) and latest Pleistocene (Q2c) desert pavements, Riverside Mountains, southeastern California. B. Histograms for adjacent Holocene (Q3b) and latest Pleistocene (Q2c) desert pavements, Gila Mountains, southwestern Arizona. C. Cumulative curves of surficial materials from an active wash (Q4), Holocene pavement (Q3b), latest Pleistocene pavement (Q2c), and lag gravel on ridge of dissected early Pleistocene alluvium (Q1); all from the piedmont south of the Whipple Mountains. Q2 gravels typically are fine grained and well sorted; Q1 gravels have the coarsest desert pavement despite prolonged in-situ weathering. D. Cumulative curves of fill terrace alluvium of latest Pleistocene (Q2c) and mid-Holocene (Q3b) age, Copper Mountains, southeastern California. From Figure 5.2, Gerson et al., 1978.

length occur on most late Pleistocene surfaces but are not abundant enough to be included in a 100–particle count. During Q2c time, stream power was sufficient to transport boulders, but apparently few were supplied from the source watersheds. In the Copper Mountains, Q2c alluvial fill is preserved under Q3b fill, which indicates that Q3b was a much stronger climate-change-induced aggradation event. Particle-size analyses of the two deposits were made in a streambank exposure and revealed a major change in drainage basin output (Fig. 2.13D) similar to that of the distant Riverside and Gila mountains (Fig. 2.1).

Particle-size measurements that were made at many points along stream channels, fans, and terraces describe decreases of gravel size in the downstream direction. Late Holocene alluvium (Q3c) not only is coarser grained than early Holocene (Q3a) and Pleistocene alluvium, but decreases more rapidly in mean gravel size in the downstream direction than does Pleistocene alluvium at the Aguila Mountains site in southwestern Arizona (Fig. 2.14A). Particle sizes on the nearby Mohawk Mountains piedmont decrease rapidly in active channels but very little for a late Pleistocene surface (Fig. 2.14B). At the Gila Mountains site (Fig. 2.14C), the plots for the coarser grained Holocene and relatively finer grained Pleistocene pavement gravels have parallel slopes that indicate moderate rates of particle-size decrease. The plot for surficial gravel in the active channel is markedly steeper.

Interpretations of the Figure 2.14 regressions need to consider whether the samples represent threshold or equilibrium conditions, or both. Temporal coarsening of piedmont alluvium at times of threshold crossing (Fig. 2.14A) is indicative of increase in streamflow competence, in availability of coarser gravel, or in bedload supply rate. A flume study (Dietrich et al., 1989) indicates that surface coarsening develops in gravel-bedded rivers whenever stream power exceeds resisting power. Steep regression line slopes for both active-channel plots (Fig. 2.14B,C) may represent a continuation of the temporal trend indicated by Figure

2.14A or reflect sampling from streambeds on the erosional side of the critical-power threshold. Both the coarser mean sizes and the more rapid decrease in size in the downstream direction for degrading active channels may reflect the presence of boulders derived from gravelly streambanks and adjacent hillslopes that remain in the streambed as a lag deposit. Ongoing studies of painted cobbles and boulders in the stream channel of Figure 2.14C indicate that with continued channel degradation such residual boulders are transported only short distances downstream and become more abundant relative to the finer gravel.

The other regression lines are for materials sampled from deposits and surfaces that represent times of maximum aggradation–the materials being transported at the time of the threshold of critical power. Competence of ephemeral streamflows tends to decrease rapidly where the streams spread out and infiltrate into alluvial fans (Q4 of Fig. 2.12B). The resulting decreases in flow resistance and stream power are described in Sections 1.5.2.2 and 1.5.2.3. Decrease in the competence of streamflows probably is the main reason for the particle-size decreases (Fig. 2.14). The decrease may be more rapid in the downstream direction for flashy flow events of the Holocene in streams overloaded with gravel than for more sustained streamflows that may have been typical of full-glacial climates.

The third type of deposit is associated with the Q1 terrains. Q1 deposits are cemented with calcium carbonate. Instead of being 1 to 4 m thick, Q1 deposits typically have thicknesses of 5 to more than 50 m. The Q1 deposits adjacent to nearly every mountain range have larger boulders than the younger geomorphic surfaces. The concentrations of large boulders at the surface of Q1 deposits on the Whipple Mountains piedmont (Fig. 2.11D) and elsewhere are largely the result of selective weathering and removal of nonresistant lithologies that tend to concentrate resistant lithologies such as andesitic agglomerate. Such situations are readily discerned where stream erosion exposes 10 m or more of Q1 deposits that can be

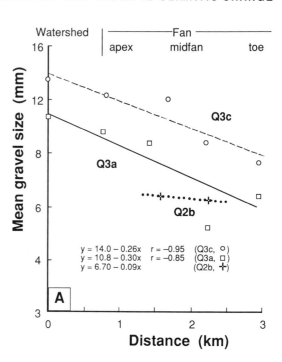

FIGURE 2.14 Decreases of mean gravel size in the downstream direction for late Quaternary alluvium and desert pavements. A. Gravel of late Pleistocene (Q2b) and Holocene (Q3a and Q3c) alluvium on the piedmont west of the Aguila Mountains. From Figure 12, McHargue, 1981. B. Surficial gravel of late Pleistocene surface (Q2) and active stream channel (Q4) on piedmont of Mohawk Mountains. From Figure 6, Mayer et al., 1984. C. Gravel fraction coarser than 8 mm for the latest Pleistocene (Q2c) and mid-Holocene (Q3b) pavements, and the active stream channel (Q4b) on the piedmont northeast of the Gila Mountains. From Figure 7, Mayer et al., 1984. Pavement data from Schenker, 1977.

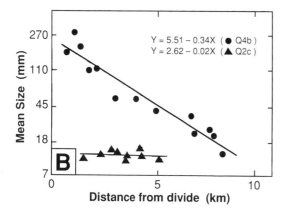

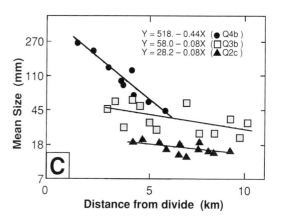

compared with the lag of resistant boulders on the overlying hillslope. The more cemented nature of the Q1 deposits compared with the younger deposits implies a much older age. The large thicknesses and coarser grained nature of the deposits suggests that the Q1 deposits were laid down millions of years ago when the mountain fronts of the region were bounded by tectonically active faults.

The variations of piedmont deposits described in this section provide paleoclimatic insight as well as distinctiveness needed to define mappable units. Soils that form on stable alluvial surfaces also are a useful criterion for mapping; in addition they can be used to assign relative ages to flights of terraces.

## 2.2.4 SOIL-PROFILE CHARACTERISTICS

Many weathering and pedogenic processes can be used in assessing the types and times of late Quaternary climatic change in the study region. Some appear to be insensitive recorders of rapid shifts because they are the result of cumulative processes over time spans that are long compared with the duration of the four types of climate that have characterized the region during the past 20 ky—full glacial, transitional, monsoonal, and interglacial. Examples are weathering rinds on surficial basalt boulders and total mass of illuviated carbonate or iron oxides within soil profiles. Accretionary processes such as rock varnishing, which provides carbon isotopes, leachable cations, and variable amounts of iron and manganese, seem much better for recording climatic change that was of sufficient magnitude to trigger a depositional episode in a fluvial system. Other parameters, such as carbonate *illuviation* (translocation of materials such as clay minerals, iron oxyhydroxides, and calcium carbonate to deeper horizons in a soil profile), provide insight into types of climate-induced changes in pedogenic processes.

Mappable units can be identified in a time-stratigraphic and time-geomorphic context because weathering and soil-forming processes in deserts are sensitive both to late Quaternary climatic change and to the passage of time (Yaalon, 1971,1975). Thus desert soils (Birkeland, 1984a; Nettleton et al., 1975; Nettleton & Peterson, 1983) provide essential data for correlation and mapping of alluvial deposits and surfaces. The surficial horizon—varnished desert pavement—is especially diagnostic, which aides in remote-sensing studies of desert piedmonts. The following subsections discuss soil layers with increasing depth, varnish, pavement, and horizons.

### 2.2.4.1 Rock Varnish

Accretion of *rock varnish* on gravel clasts of stable pavements is an important surficial pedogenic process in arid regions, and many aspects of varnished desert pavements are controlled by climate and climatic change. Brownish-black manganese-rich varnish accumulates on the tops of cobbles, and bright reddish-brown iron-rich varnish accumulates on the bottoms. Progressive accumulation of manganese oxyhydroxides is aided by manganese-oxidizing microorganisms; iron oxyhydroxides accumulate by physicochemical processes in alkaline microenvironments on the subsurface parts of the same clast (Dorn & Oberlander, 1981,1982). The presence of abundant silicate clay in the varnish layer noted by some workers (Potter & Rossman, 1977; Allen, 1978; Perry & Adams, 1978; Elvidge & Moore, 1979) may be more common in Q2a and Q2b (see Table 2.2 for soils-chronosequence abbreviations) pavements whose stones are derived partly from Bt soil horizons. Cobbles in Q3 pavements are still in depositional positions. Electron probe work by Allen clearly indicates that varnish is not derived from underlying rocks. Clay and iron oxyhydroxides on the undersides of pavement stones may be largely derived from the soil matrix, but the black manganese-rich oxyhydroxides on the tops are derived from atmospheric sources of rain and dust with some additions from the soil matrix in a concentrated blacker band just above the soil line.

Although cobbles of different lithologies vary in their degree of varnishing and their ability to retain varnish, most surficial cobbles in a given gravel bar accrete incipient amounts of varnish at about the same time after deposition. Varnish is removed rapidly by abrasion during stream transport. It is unlikely that stream-transported cobbles from the older surfaces would retain uniform varnish coatings or be deposited in their original positions.

The presence or lack of visible rock varnish can be used to distinguish between surfaces of different ages. Q4b and Q3c surfaces can be identified on the basis of lack of or incipient presence of rock varnish, respectively. Q3b and Q3a surfaces have progressively darker varnish, which is consistent with their older relative ages.

When used as a mapping criterion for distinguishing between alluvial geomorphic surfaces, varnish should be used only for a given lithology within a given piedmont. The lithology used in mapping the Whipple Mountains piedmont was a type of andesitic cobble that had a dimpled abrasion surface; this choice eliminated the need for initially breaking many cobbles to ascertain the correct lithology. Fine-grained mafic gneiss was used in mapping the Gila Mountains piedmont (Fig. 2.24), and in Israel (Section 3.2.2.3) micaceous quartzite was the most suitable rock type.

Both the value and chroma of varnished desert pavements decrease with age, so the darkest varnished cobbles occur on Q2a surfaces (Tables 2.2, 2.4). Values decrease from 3 for Q3c surfaces to 2 for Q2a surfaces, and chromas decrease from 4 to 1. These data are largely from the Whipple Mountains piedmont. If both value and chroma are used to define changes in the color of the oxides on the tops of the pavement cobbles, varnish darkness (Vd) may be defined as

$$Vd = \frac{Value + chroma}{2} \qquad (2.2)$$

The graph of change in varnish darkness with time in Figure 2.15 shows that about two-thirds of the darkening occurs in the first 100 ky and that an additional 500 ky is needed for the remaining one-third. Half of the darkening occurs in the first 50 ky. Gravel varnishes best when the surface is stable, but dissection causes erosional stability to decrease with age of geomorphic surface. Pavements occur only on the flatter ridge crests in Q1 terrains, and even these sites are undergoing sufficiently rapid rates of erosion to cause surficial gravel clasts to turn over and spread apart, which inhibits accretion of varnish.

The overall darkness of a desert pavement is a function of the degree of packing of the surficial rocks as well as of the blackness of the rocks. The percentage of bare ground was measured during the particle-size counts of the Whipple Mountains pavement by noting where the top of the vesicular A2 horizon was exposed beneath marks on an outstretched survey tape. The increase in closeness of packing of the pavement rocks between Q3b and Q2c pavements (Fig. 2.15) is primarily because the Q3b materials were deposited as loose,

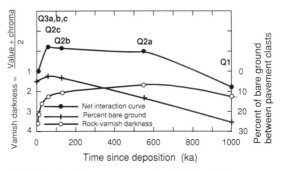

FIGURE 2.15 Interaction of variables affecting darkness of desert pavements on the piedmont south of the Whipple Mountains.

sandy gravel whereas the Q2c materials were deposited as tightly packed, well-sorted gravel. Decreases in closeness of packing of pavement gravels occur for all pavements older than Q2c.

The net effect of progressive darkening and progressive erosional instability with time is shown by the net interaction curve of Figure 2.15, which describes the overall darkness of pavements on images. Indeed, Q2a pavements of the Whipple Mountains piedmont are only slightly lighter than Q2b and Q2c pavements. The Q3b pavements are decidedly lighter and the Q1 surfaces have minimal pavement darkness. Only the general trend of the net interaction curve of Figure 2.15 applies to other piedmonts in the region. For example, on the piedmont northeast of the Gila Mountains (Figs. 2.10, 2.20) Q2c pavements are much darker than younger or older pavements. Interactions between such developmental and regressive processes are the basis for a general model of dynamic pedogenesis proposed by Johnson and colleagues (1990).

Impressive advances in the rapidly developing field of remote sensing now allow mapping of differences of relative age and composition of surficial piedmont alluvium. Both multispectral and radar images can be used efficiently to map large regions accurately because of distinctive differences in chemistry and micro-relief of desert pavements of different ages (Blom et al., 1982; Farr, 1985). Burrowing rodents commonly bring enough materials from argillic horizons to the surface of sandy soils to allow sensing of B horizon characteristics. Excellent color illustrations of the mapping of Death Valley alluvial fans by a six channel thermal infrared multispectral scanner have been prepared by Gillespie et al. (1984) and by Kahle et al. (1984). Their work is indicative of the truly impressive advances that remote sensing techniques will bring to the mapping and interpretation of desert piedmonts in the next few years.

### 2.2.4.2 Desert Pavements

The lower Colorado River region has excellently preserved desert pavements whose formation, varnishing, and destruction are functions of lithology, climate, and time. The characteristics of the barren black Q2 surfaces that sweep for 10 km down the Whipple Mountains piedmont are summarized in Table 2.4, together with the data for the Q3b desert pavements. With increasing pavement age,

rock types such as andesite that are resistant to weathering become more common relative to schist. Lithologies that tend to split and crumble rapidly, such as quartz monzonite and granitic gneiss, occur mainly on Holocene surfaces and disintegrate so rapidly that they do not retain varnish well.

Surficial cobbles and boulders of the Holocene Q3b alluvium are varnished and weathered (Table 2.8). Cobbles of metamorphic and volcanic rocks are varnished. Andesite cobbles lack obvious weathering rinds and have smooth unweathered abrasion surfaces. Schist cobbles commonly are split along foliation planes. Gneissic rocks have 1-cm-thick oxidized weathering rinds, and the shapes of the stream-transported cobbles no longer are apparent. Surficial cobbles of leucocratic plutonic rocks are so weathered and friable (stage 3 of Table 2.4) that they retain little or no varnish; some are sufficiently weathered that they can be broken by hand. Physical weathering caused by

TABLE 2.4 Characteristics of desert pavements on Quaternary alluvial geomorphic surfaces south of the Whipple Mountains.

| Surface | Percent of Surface in Pavement | Percent of Bare Ground Between Particles | Varnish on Top of Andesite Cobbles | Degree of Preservation | | Particle Size[a] | | | Max. Length[b] of Particle (mm) |
| | | | | Stream Channels | Gravel Bars | Median (mm) | Sorting | Percent >32 mm | |
|---|---|---|---|---|---|---|---|---|---|
| Q4a | 0 | — | None | Excellent to good | Excellent | 16 | Poor | 22 | 500 |
| Q3b | >90 | <5 | 7.5YR 3/4 | None | Excellent | 14,15 | Poor | 18–20 | 300–500 |
| Q2c | >80 | <2 | 7.5YR 2/3 | None | None to faint | 11–15 | Good | 0–16 | 300 |
| Q2b | 60–80 | 4 | 7.5YR 2/3 | None | None to fairly good | 13,14 | Poor to good | 6–18 | 300 |
| Q2a | ~30–40 | 4–20 | 7/5YR 2.5/2 | None | None to faint | 12–25 | Poor | 8–12 | 200–500 |
| Q1 | <1 | >20–30 | 7/5YR 2/2 | None | None | 18 | Poor | 15 | 300–1000 |

[a]The particle-size distributions of the desert pavements were measured by starting at a mark on a tape using a table of random numbers to select the starting point. The tape was then laid out normal to the first tape bearing and the intermediate particle diameters under each mark were measured for 70 to 110 particles.

[b]The largest boulder within 1.5 m of the transect described in footnote a.

crystallization of atmospherically derived salts may accelerate weathering of coarsely crystalline rock types in desert pavements.

The smooth black desert pavements of the Pleistocene Q2 alluvial geomorphic surfaces are as ubiquitous as the pavements with a Q3 bar-and-swale topography. The marked contrast of the topographies of Q2 and Q3 pavements permits ready mapping of these two general classes of surfaces on images. Field characteristics are equally useful for mapping purposes. Weathering of rocks on the late Pleistocene desert pavements (Q2c) of the Whipple Mountains provides clues about the intensity and type of weathering processes. Schist cobbles split along foliation planes and abundant platy fragments of schist are darkly varnished. Andesite cobbles typically have a brownish-black varnish and thin (<2 mm) weathering rinds. Coarse-grained gneiss is not as common as on the Holocene surfaces either because fewer gneiss cobbles were being supplied from hillslopes to streams or because weathering took place after deposition.

Many processes may concentrate stones at land surfaces to form desert pavements, and the relative importance of a given process varies from site to site. Hillslopes above an alluvial geomorphic surface provide ready sources of surficial gravel in footslope colluvium where desert pavements tend to remain at the surface. Such pavements are born at the land surface, where they remain despite episodes of loessial dust accumulation that provide silt, clay, and salts to form the underlying vesicular and argillic soil-profile horizons. An excellent detailed study of such processes was made at the Cima volcanic field (McFadden et al., 1987; McFadden & Wells, 1989), where topographic highs on basalt flows are sources of clasts that accumulate as desert pavements in adjacent hollows. Uniformly varnished pavement clasts that lack coats of pedogenic red clay show that the rocks were never deeply buried in the underlying soil.

Dispersive stresses are associated with shrink-and-swell processes in clayey soils, and the land surface may be regarded as an upper bounding horizon of zero dispersive stress. Dispersive stresses tend to keep pavement rocks at the surface. For many Pleistocene soils in hot deserts, variations in varnish development and presence of red clay coats suggest that dispersive stresses also have moved some rocks up from argillic horizons (Springer, 1958; Jessup, 1960; Denny, 1965; Cooke, 1970; Mabbutt, 1977; Dan et al., 1982).

Different processes are required to concentrate rocks at the surface of desert pavements in sandy gravel of planar alluvial geomorphic surfaces that are not subject to colluvial slope wash. Such sites are common and seem to involve, at least initially, the processes of deflation and fluvial erosion (Cooke & Warren, 1973; Ritter, 1986). Pavements are not mainly the result of deflation as envisaged by Free (1911) except in the initial stages of development on gravelly fluvial surfaces composed partly of fine sand and silt. Deflation of sand, silt, and clay-size materials is a visually obvious process along desert streamcourses after infrequent floods, but piedmont alluvium typically contains less than 5 percent silt and clay. Silt and clay in an amount greater than 20 percent in the late Quaternary soil profiles of the region attest to net introduction of eolian dust into gravel parent materials .

Atmospheric additions of fine sediment greatly reduce infiltration capacity of surficial sandy gravels, thereby increasing runoff and the relative importance of fluvial erosion. Fluvial processes of raindrop splash and sheetflow wash sand, silt, and clay into adjacent stream channels or into underlying soil horizons. This self-arresting set of processes aids in the formation of a tightly packed pavement of stones. Rillwash and eventual destruction of desert pavements become dominant when streams dissect pavements. The important long-term processes on undissected Pleistocene pavements continue to be the incorporation of atmospheric dust into the soil horizons beneath the surficial pavements of weathering rocks (Section 2.5.1.2.1).

McHargue (1981) studied the processes of pavement formation in Holocene sandy fine-grained piedmont gravels west of the Aguila Mountains in southwestern Arizona. The mean particle sizes of the pavements are .5 to 2 *phi units* coarser than the underlying parent alluvium. He calculated the amounts of selective erosion of fine sediment from the alluvium that are needed to form stable pavements, in terms of the thickness (Tk) of the original layer of alluvium.

$$Tk = \frac{\pm (Gg)Me/Df + Mr/Dg}{As} \qquad (2.3)$$

where Gg is the net change of percentage of gravel in the transition from alluvium to pavement gravel for the given phi size interval, Me is the mass eroded from the alluvium, Mr is the mass of gravel remaining in the pavement, Df is the bulk density of the fine fraction of the alluvium, Dg is the density of the gravel clasts, and As is the area

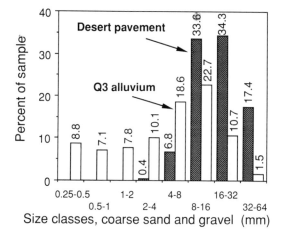

FIGURE 2.16 Histograms showing the changes in particle-size distribution that occur as alluvium is converted into desert pavement on the Q3a and Q3b alluvial geomorphic surfaces of the Aguila Mountains piedmont. The separations between the desert pavement and Q3 alluvium columns represent the changes in abundance for each particle-size class of the desert pavement and the underlying alluvium. Particles smaller than 8 mm become less common as alluvium is converted into pavement, and particles larger than 8 mm become more common. Derived from Figure 19 of McHargue, 1981.

sampled. McHargue concluded that at least 1- to 3-cm lowering of the alluvium is needed to form an incipient pavement. After 3 to 15 cm of erosion, sufficient clasts have become concentrated at the surface to form a stable pavement. Erosion rates decrease exponentially, and atmospheric dust that is trapped by the pavement is translocated to soil-profile horizons by occasional infiltrating water. This illuviation of dust initiates the long-term stage of pavement stability and soil formation. McHargue's particle-size analyses of the Q3a and Q3b alluvium and pavements indicate selective erosion and concentration of fine sediment and gravel, as depicted in Figure 2.16. Alluvium finer than 8 mm tends to be removed; larger materials tend to be concentrated in the Q3 pavements on the Aguila Mountains piedmont. McHargue's study shows that pavements form quickly in Holocene deposits that contain negligible amounts of silt and clay. Clay-rich soil horizons are not essential for the formation of desert pavements.

Lava flows are at the opposite end of the spectrum of pavement parent materials from pebbly sands. Clasts for pavement formation abound but fine materials are lacking initially. McFadden et al. (1987) described pavement-forming processes on basalt flows in the Cima volcanic field. Atmospherically derived silt, clay, and salts disaggregate surficial rubble. As in the case of alluvium, dust is washed below the surface to form stone-free vesicular and argillic horizons. Accretion of rock varnish (Section 2.4.1.2.1) and formation of cobble-weathering rinds (Section 2.4.1.2.2) proceed during long time spans on typical stable pavements on either alluvium or lava, while illuviation of new increments of dust and soil-profile formation continue beneath the pavement.

The time needed to produce smooth desert pavements from braided stream deposits depends on three factors. Size and abundance of gravel is the most important factor. Where sand is deposited by braided streams, the surficial depositional features will be gone in a decade. Fine-grained gravel, such as that characteristic of the late Pleistocene deposits and the former channels of Holocene alluvial geomorphic surfaces, also has too few large particles to resist the smoothing effects of desert-pavement processes. Such depositional surfaces are transformed into smooth desert pavements in 1 to 2 ky. Gravel bars have persisted for more than 100 ka in areas where streams deposited bouldery alluvium that become Q2b and Q2a pavements.

A second factor is the frequency and magnitude of smoothing processes such as shrink and swell caused by wetting. The third factor is the time needed for cobbles and boulders to weather into smaller fragments. Production of smaller fragments is largely a function of abundance of planes of weakness such as bedding, foliation, and jointing. Splitting of rocks by carbonate, gypsum, and halite is present locally, and near the Gulf of California this process may be more effective than chemical weathering in reducing size of surficial cobbles.

Many pavement-smoothing processes varied with late Quaternary climatic changes. Eolian processes became more important during times of maximum dryness such as from about 7 to 5 ka. An increase in frequency and intensity of monsoonal thunderstorms would increase raindrop splash. Freeze–thaw processes of surficial deposits would be enhanced by wetter and colder winters. Such processes vary spatially with changes in altitude and with dominant airmass circulation.

Incision of streams that head in pavement areas converts 1-degree planar surfaces to short 5- to 20-degree hillslopes. Local slopes steepen with increasing depth of dissection, and pavement gravels spread apart as they creep downslope, thus exposing more of the underlying Av soil horizon. The percentage of bare ground between the gravel clasts of the pavements generally is inversely related to the percentage of geomorphic surface remaining in the pavement. The most tightly packed desert pavements are of Q2c age (Fig. 2.15, Table 2.4).

### 2.2.4.3 Soil-Profile Horizons

The distinctive characteristics of soil profiles are routinely used for mapping of alluvial geomorphic surfaces, and standard soil abbreviations are used in field notes and reports. The abbreviations used to describe the mappable units were introduced in Table 2.2. Standard abbreviations for soil-profile characteristics and horizonation are summarized in Table 2.5. Table 2.6 presents stages of cobble weathering and of soil-carbonate morphology that are essential tools for the reconnaissance mapper. Abbreviations permit condensation of complete descriptions for the soils chronosequence of the Whipple Mountains piedmont in Table 2.7.

Soil-profile horizon characteristics vary in their usefulness for mapping. Desert pavements are a distinctive and varied surficial horizon for gravelly soils. The ubiquitous vesicular silt of Av horizons forms quickly and is vulnerable to erosion; thus it generally is not useful for mapping or age estimates. Information about cambic and argillic (Bw and Bt) horizons is highly useful, but these horizons also may be partly truncated by erosion. Calcic (Bk) horizons provide the most consistent information for correlation and mapping. Depth to the top of Bk or K horizons, their thickness, thickness of carbonate pebble coatings, morphology of pedogenic calcite crystals (Chadwick et al., 1988) and morphological stage of carbonate development are especially valuable parameters (Gile et al., 1965,1966; Bachman & Machette, 1977; Machette, 1985). These parameters describe distinctive features of soils and are in part time dependent; they can be used to assign relative ages (Section 2.5.1.2.3).

Pavements of Q3b geomorphic surfaces are lightly to darkly varnished, and Av and Ck horizons typically are present. The general lack of B horizons and the shallow depths of carbonate illuviation (Table 2.2) suggest that the Q3b soils have not been subjected to a substantially wetter or cooler climate than the present arid hyperthermic climate. This type of climate-controlled contrast in soil-profile characteristics is invaluable for identifying mappable soils-geomorphic units on a regional basis and for examining the influence of spatial variations of climate. Weak Bw and Bt horizons occur in some Q3b soils of semiarid thermic parts of the region.

TABLE 2.5 Abbreviations used in describing soil horizons and soil profiles.[a]

---

*Master Soil Horizons*

A           Surficial horizon that characterized accumulation of humidified organic matter intimately mixed with mineral fraction. Desert A horizons have little alteration and organic matter.

B           Horizon below A horizon. Original rock or sedimentary structures are obliterated by (1) illuvial concentration of silicate clay, iron, aluminum, humus, carbonates, gypsum, or silica; (2) removal of carbonates; (3) residual concentration of sesquioxides; (4) alteration that forms pedogenic structure.

K           Horizon below Bt horizon. Cementing carbonate coats, engulfs, and commonly separates parent material particles (Gile et al., 1979).

C           Parent material with minimal effects of pedogenic processes.

AB, BC      Transitional states—two horizons superimposed on one another.

*Subordinate Distinctions Within Master Horizons*[a]

b           Buried genetic horizons

k           Accumulation of carbonates

m           Strong cementation

q           Accumulation of silica

t           Accumulation of clay minerals

v           Vesicular pores

y           Accumulation of gypsum

z           Accumulation of soluble salts such as halite

Bk          Visible accumulations of carbonate

Bw          Incipient visible soil development relative to C horizon; has color or structure

Cox         Oxidized C horizon material; not enough to qualify as a Bw horizon

Ck          Calcium carbonate not visibly obvious but noncalcic parent material has effervescent reaction; not enough to qualify as a Bk horizon

Cu          Unweathered C horizon; no oxidation or pedogenic carbonate

*Soil Profile Characteristics*

**Colors**

YR              Intermediate hue (wavelength of reflected light) made by mixing red and yellow. Four equally spaced mixes become more yellow (2.5, 5, 7.5, 10).

Value/Chroma    Value is relative lightness of a color from black (01) to white (101). Chroma describes dilution of spectral color (YR) by gray of a given value. It ranges from gray (10) to maximum brightness (18).

10YR 6/3        An example of a complete color notation (pale brown) using a Munsell type soil color chart to describe the hue, value, and chroma for a dry or wet soil sample.

**Texture**

| egr | Extremely gravelly | S | Sand |
|-----|--------------------|---|------|
| vgr | Very gravelly | SL | Sandy loam |
| g | Gravelly | SCL | Sandy clay loam |
| co | Coarse | SC | Sandy clay |
| vco | Very coarse | LS | Loamy sand |
| f | Fine | L | Loam |

TABLE 2.5  Abbreviations used in describing soil horizons and soil profiles (*Continued*)

**Texture** (*cont.*)

| | | | | |
|---|---|---|---|---|
| vf | Very fine | SiL | Silt loam |
| | | Si | Silt |
| | | SiCL | Silty clay loam |
| | | SiC | Silty clay |
| | | ClL | Clay loam |
| | | C | Clay |

**Structure**

| *Type* | | *Grade* | | *Size* | |
|---|---|---|---|---|---|
| g | Granular | m | Massive | vf | Very fine |
| pl | Platy | sg | Single grain | f | Fine |
| sbk | Subangular blocky | 1 | Weak | m | Medium |
| abk | Angular blocky | 2 | Moderate | c | Coarse |
| p | Prismatic | 3 | Strong | vc | Very coarse |
| cr | Columnar | | | | |

**Consistence**

| *Dry* | | *Moist* | | *Wet* | |
|---|---|---|---|---|---|
| lo | Loose | lo | Loose | so,po | Nonsticky, nonplastic |
| so | Soft | vfr | Very friable | ss,sp | Slightly sticky, slighty plastic |
| sh | Slightly hard | fr | Friable | s,p | Sticky, plastic |
| h | Hard | fi | Firm | vs,vp | Very sticky, very plastic |
| vh | Very hard | vfi | Very firm | | |
| eh | Extremely hard | efi | Extremely firm | | |

**Clay Films**

| *Frequency* | | *Distinctiveness* | | *Morphology* | |
|---|---|---|---|---|---|
| vl | Very few | f | Faint; thin, low contrast | pf | Ped face |
| 1 | Few | d | Distinct; seen without magnification | po | Line pores |
| 2 | Common | p | Prominent; thick, high contrast | br | Bridges |
| 3 | Many | | | co | Colloid coats on grains |
| 4 | Continuous | | | | |

**Matrix**

*HCl effervescence Reaction*

| | |
|---|---|
| eo | None |
| e | Moderate |
| es | Strong |
| ev | Extreme |

[a]Abbreviations are mainly from the Soil Survey Manual of the U.S. Soil Conservation Service, Chapter 4, Directive 430V, issue 1, 1981. Here, v describes vesicular characteristics, instead of plinthite.

Table 2.6  Stages of cobble weathering and carbonate
morphology in soils.

| Stage | Characteristics |
|---|---|
| **Cobble Weathering Stages** | |
| 1 | Unweathered except for minor pitting of abrasion surface or incipient oxidation rinds; rings sharply to blow of hammer |
| 2 | Slight weathering characterized by moderate surficial pitting, incipient fracturing, and incipient to moderately developed oxidation rinds; emits solid sound when struck by hammer |
| 3 | Substantial weathering indicated by highly pitted surface or strongly developed fractures; ferruginous material may coat sides of fracture planes. Calcium-rich plagioclase and/or ferromagnesian minerals genrally are strongly altered. Cobble broken with dififculty by hand, emits dull sound when struck with hammer |
| 4 | Strongly weathered and fractured to saprolitic-appearing residuum. Cobble can be completely disaggregated by hand with little difficulty, emits punky sound as hammer head sinks into cobble |
| **Stages of Carbonate Morphology in Gravel** | |
| I | Thin discontinuous pebble coatings |
| II | Continuous pebble coatings |
| III | Coalesced pebble coatings, nodules between clasts |
| IV | Plugged horizon with incipient to 1-cm-thick laminar layer |
| V | Thick laminar layer and thin pisoliths, case hardened in vertical cuts |
| VI | Multiple laminar layers, brecciated, common pisoliths, recemented and case hardened |

*Source:* Cobble weathering stages are adapted from Melton, 1965, Carbonate morphology stages are from Machette, 1985, as modified from stages originally proposed by Gile et al., 1966.

One might expect the advanced state of weathering of plutonic and metamorphic rocks of Q3b pavements to be associated with well-defined subsurface weathering horizons, but such is not the case. B horizons are not present and Ck horizons consist of extremely thin and discontinuous carbonate coatings on pebbles in a depth range of only about 10 cm (Fig. 2.17A, Table 2.8). Below a depth of about 15 cm, unoxidized sandy gravel of the Cu horizon is visually the same as that in the active washes.

Q3b and Q3c surfaces are sufficiently old to have light to dark rock varnish and a bar-and-swale topographic pattern, but carbonate pebble coatings (Fig. 2.17A) are discontinuous and may be absent. Some Q3c soils east of Yuma have lightly varnished pavements but lack carbonate pebble coatings. Farther north the soil-leaching

A

FIGURE 2.17 Typical soil profiles. A. Soil profile (torriorthent) of mid-Holocene age (Q3b) near Hildago Mountain in the central Mojave Desert. Surface of quartz monzonite boulder is weathered but abrasion surface of subsurface portion is coated with stage I calcium carbonate. Atmospherically derived silt and clay engulf gravel to a depth of only 15 cm; the underlying materials are in the Cu horizon. Thirteen centimeters of the plastic scale are shown. B. *(opposite page)* Soil profile (typic haplargid) of late Pleistocene age (Q2c) on piedmont south of the Whipple Mountains. Bouldery Q2a(?) alluvium is under the rock hammer. The notebook is on cemented Q1 alluvium, which also crops out as ridge in the background. This soil profile is described in Table 2.9.

index is higher and discontinuous coatings of carbonate less than 0.1 to 0.2 mm thick are typical of the thin Ck horizons. These coatings effervesce strongly with dilute HCl but are so thin that the cobble lithology is visibly identifiable beneath the

coating; the carbonate has a rough feel when touched lightly with the fingers.

Surficial cobbles of Q3a soils are more weathered and darkly varnished than those on Q3b and Q3c surfaces. Bt horizons generally are absent in

**B**

arid areas but 1 to 2 cm of Bw material occurs as a reddish-brown matrix between the gravel clasts and under boulders beneath the Av horizon. Boulders concentrate infiltrating water and create local zones of more intense soil leaching. Bk-horizon pebble coatings are discontinuous. Northeast of the Gila Mountains and east of the Riverside Mountains continuous pebble coatings 0.2 to 0.8 mm thick are typical of the Ck horizons. Areas with higher leaching indices have continuous coatings more than 1 mm thick.

All three Q2 surfaces have argillic and calcic horizons (Tables 2.8, 2.9, 2.10). Soil profiles are well developed in the Q2c alluvium of late Pleistocene age (Fig. 2.17B, Table 2.9). Silty Av horizons differ little from the Av horizons of other geomorphic surfaces and change abruptly to argillic (Bt) horizons that are 8 to 30 cm thick. Bk horizons are typically about 13 cm thick. Carbon-

ate occurs as soft gritty nodules in a sandy matrix and as 0.5- to 6-mm coatings on the bottoms of pebbles and cobbles. The thin Bt and Bk horizons indicate soil genesis under much deeper leaching conditions than for Holocene soils.

The characteristics of the Bt soil horizons are useful for mapping. Separation of Q3 from Q2c surfaces generally is easy on the basis of topographic pattern and pavement varnish. Where doubt exists, determining the presence or absence of a Bt horizon and the representative thickness of carbonate coatings on pebbles is a simple matter. Thicknesses and maximum clay contents of Bt horizons also are useful for distinguishing among the three Q2 surfaces where moderately complete profiles have been preserved. Older Bt horizons are thicker and tend to be engulfed by carbonate coatings and nodules in the Q2a (Table 2.11) and Q2b soils (Table 2.10).

TABLE 2.7 Summary of morphologic characteristics of soils of the Whipple Mountains piedmont[a]

| Surface | Horizon | Depth (cm) | Matrix Color (dry) | Texture[b] | Structure[b] | Consistence[b] Dry | Consistence[b] Wet | Clay Films[b] | Carbonate Morphology[c] |
|---|---|---|---|---|---|---|---|---|---|
| Q4 | Av | 0–1 | 10YR 8/2 | fSL | g | so | ss,sp | None | 0 |
| | Cu | 1+ | 10YR 7/2 | gS | g | lo | ss,po | None | 0 |
| Q3b | Avk | 0–4 | 7.5YR 5/3 | SiL | 2msbk, 1cpl | so | s,p | 2fpo | I |
| | Ck | 4–14 | 10YR 7/3 | gcoS | g | lo | so,po | None | I |
| | Cu | 14+ | 10YR 7/2 | gcoS | g | lo | so,po | None | 0 |
| | Btb | 0–4 | 7.5YR 6/6 | gcoSiL | g | so | ss,sp | 2fco | 0 |
| Q2c1 | Avk | 0–3 | 7.5YR 7/3 | SiL | 2cpl | sh | s,p | 2fpo | I |
| | Btk | 3–19 | 5YR 6/4 | gSiL | 2sbk | sh | s,p | 2fpo&br | II |
| | BCk | 19–27 | 7.5YR 6/4 | gSiL | mg | sh | ss,sp | None | II |
| | Ck1 | 27–42 | 7.5YR 8/2 | gSiL | mg | sh | so,po | None | II |
| | Ck2 | 42+ | 10YR 7/2 | gS | g | lo | so,po | None | I |
| Q2c2 | Avk | 0–8 | 5YR 7/3 | SiL | 2sbk | sh | vs,vp | 3fpo | I |
| | Btk1 | 8–13 | 5YR 6/4 | gSiCL | 1g | sh | vs,vp | co | II+ |
| | Btk2 | 13–21 | 5YR 5/8 | gSiCL | g | lo | vs,vp | 4fpo&br | III |
| | Btk3 | 21–30 | 5YR 4/6 | gSiCL | g | so | vs,vp | 1fpo | III |
| | Bk1 | 30–52 | 7.5YR 7/3 | gS | g | so | ss,sp | None | II |
| | Bk2 | 52–56 | 7.5YR 5/3 | gLS | g | lo | ss,sp | None | II |
| Q2b | Avk | 0–2 | 5YR 8/2 | SiL | 2cpl | so | vs,vp | 3fpo | I |
| | Btk1 | 2–24 | 5YR 6/4 | gSL | mg | so | ss,sp | 4fpo&br | II |
| | Btk2 | 24–45 | 7.5YR 8/1 | gLS | mg | h,vh | ss,po | 3fpo&br | II & III |
| | Btk3 | 45–67 | 7.5YR 6/4 | gSL | mg | sh | ss,sp | 1fpo&br | II |
| | K | 67–120 | 7.5YR 8/1 | gLS | mg | vh | so,po | None | III |
| | Ckox | 120+ | 7.5YR 7/3 | gLS | sg | lo | ss,sp | None | I |
| Q2a | Avk | 0–3 | 5YR 8/2 | SiL | 2cpl | sh | vs,vp | 4fpo | I |
| | ABvk | 3–8 | 5YR 7/4 | SCL | 1msbk | sh | s,p | 4fpo | II & III |
| | Btk1 | 8–75 | 5YR 5/6 | gClL | 2mg | sh | vs,vp | 4dkpo&br | II & III |
| | Btk2 | 75–100 | 5YR 5/4 | gSiL | 2mg | so | s,p | 2dkpo&br | II & III |
| | Btk3 | 100–116 | 5YR 6/4 | gSiL | mg | sh,h | s,p | 2dkpo&br | II & III |
| | Btk4 | 116–130 | 5YR 7/4 | gSiL | mg | so | s,p | 3fpo&br | II & III |
| | Ckox1 | 130–185 | 7.5YR 6/3 | gLS | g | so | ss,sp | None | II |
| | Ckox2 | 185–195 | 7.5YR 6/3 | gLS | g | lo | ss,po | None | I |

[a] Analyzed and compiled by L. D. McFadden.

[b] See Table 2.3 for definitions of abbreviations.

[c] See Table 2.4 for stages of cobble weathering and carbonate morphology.

TABLE 2.8 Soil profile on mid-Holocene alluvium (Q3b) derived from schist, gneiss, and andesite of Whipple Mountains.[a]

*Location:* San Bernardino County, California, ~0.4 km north of Highway 62 in the NE 1/4 of the SE 1/4 of Sec. 2, T1N, R24E

*Physiographic position:* Pit on gravel bar, slightly varnished; slope is 1° to south; altitude 261 m

*Classification:* Typic torriorthent

| Horizon[b] | Depth (cm) | Description |
|---|---|---|
| | 4–0 | Moderately varnished, poorly sorted bar gravel; exposed tops and sides of clasts dark brown (7.5YR 3/3) and noncalcareous; base of clast gray-orange (7.5YR 6/3) and noncalcareous; weathering rind less than 0.1 mm thick; most clasts weathered to stage 1 to stage 2[c] and composed of mesocratic volcanic lithologies; schist and rare leucocratic plutonic rocks much more lightly varnished and are weathered to stage 3 |
| Avk | 0–4 | Light orange to gray-brown (7.5YR 8/3 dry, 5/4 moist) clayey silt; weak medium-to-coarse subangular blocky to weak coarse platy; soft, sticky, and plastic; many very fine to fine roots; many micro to fine, in ped, vesicular pores with common very thin clay films; strongly effervescent; stage I carbonate occurs both disseminated and as segregated small nodules at base of the horizon; abrupt, smooth boundary |
| Ck | 4–14 | Gray yellowish orange to gray yellowish brown (10YR 7/3 dry, 4/3 moist) very gravelly sand to sandy gravel; massive; loose, nonsticky, and nonplastic; few fine to medium roots; strongly effervescent; stage I carbonate occurs both disseminated and segregated as <0.1-mm-thick discontinuous coatings on bottoms and sides of large clasts or as rootlet-shaped coatings; clear, smooth boundary |
| Cu | 14– | Light yellow-brownish gray to yellow-brownish gray (10YR 7/2 dry, 4/2 moist) sandy gravel; massive; loose, nonsticky, and nonplastic; slightly effervescent, carbonate occurs chiefly as very thin, discontinuous coatings or as rootlet-shaped coatings on surfaces of large clasts |

Cobble weathering stages

Ck horizon: leucocratic plutonic rocks, 2; schist, 2

[a] Described by L. D. McFadden.

[b] See Table 2.3 for definitions.

[c] See Table 2.4 for definitions.

Pedogenic carbonate morphology shows the same changes with increasing age of surface that have been observed elsewhere in the southwestern United States (Machette, 1985). Carbonate-coated pebbles occur in both Btk and Bk horizons. Most Q2 surfaces are sufficiently dissected that scattered carbonate-coated pebbles crop out at the break in slope between a pavement and the adjacent hillslope (Fig. 2.18).

Extensive remnants of Q1 surfaces are adjacent to the Mopah Range and the south end of the Turtle Mountains. These surfaces are 30 to 60 km from the Colorado River, which accounts for their not being completely destroyed by dissection caused by erosional base-level fall. Bt soil horizons have been removed by erosion (Table 2.12) but occasional remnants of 5YR Bwk and BCk horizons are present. Figure 2.19 shows a Q1 surface ad-

TABLE 2.9 Soil profile on latest Pleistocene alluvium (Q2c) derived from schist, gneiss, and andesite of Whipple Mountains.[a]

*Location:* San Bernardino County, California; 1 km south of Highway 62 in NW 1/4 of the NW 1/4 of Sec. 15, T1N, R24E

*Physiographic position:* Streamcut exposure of fluvial deposit, surface 1.5 m above modern channel, slope is 1° south; altitude 253 m

*Classification:* Typic haplargid

| Horizon[b] | Depth (cm) | Description |
|---|---|---|
| | 2–0 | Strongly varnished gravel pavement, exposed clasts black-brown (7.5YR 2/3); bases of clasts gray-orange (5YR 6/6); noncalcareous; weathering rind as much as 0.2 mm thick; clasts dominantly mesocratic volcanics weathered to incipient stage 2[c] |
| Avk | 0–8 | Gray yellowish-orange to dark reddish-brown (10YR 7/3 dry, 5YR 4/6 moist) silty loam; moderate coarse subangular blocky; soft to slightly hard, very sticky and very plastic; many microfine to fine vesicular pores; many thin clay films lining pores; strongly effervescent; abrupt, smooth boundary |
| Btk1 | 8–13 | Gray-orange to gray reddish-brown (5YR 6/4 dry, 5/4 moist) gravelly sandy clay loam; massive to medium weak crumb; many very fine interstitial pores; strongly effervescent; stage I carbonate occurs both disseminated and as 0.9- to 1.1-mm-thick coatings on bottoms of most pebbles; abrupt, smooth boundary |
| Btk2 | 13–21 | Strong reddish-brown to reddish-brown (5YR 5/8 dry, 5/6 moist) gravelly sandy clay loam; massive; loose, very sticky and very plastic; common very fine interstitial pores; common to many thin clay films on grains; strongly effervescent, stage I carbonate occurs both disseminated and as 0.2- to 0.4-mm-thick coatings on bottoms of pebbles; few medium discontinuous nodules of calcareous sand, occasional pockets of noncalcareous matrix; abrupt, smooth boundary |
| Btk3 | 21–30 | Dark reddish-brown to reddish-brown (5YR 4/6 dry, 5/6 moist) gravelly sandy clay loam; massive; soft, very sticky and very plastic; many very fine to fine interstitial pores; few thin clay films in some pores, less than 0.1 mm thick; strong to violently effervescent, stage II carbonate occurs both disseminated and as many fine to medium distinct nodules and as thick coatings up to 0.6 mm thick on bottoms of many pebbles; nodules are 5YR 8/1.5; abrupt, smooth boundary |
| Btk4 | 30–52 | Gray yellowish-orange to dark gray yellowish-orange (7.5YR 7/3 dry, 6/3 moist) sandy loamy gravel; massive; loose to soft, slightly sticky and slightly plastic; many very fine to fine interstitial pores; very few extremely thin clay films on grains strongly effervescent; carbonate occurs as disseminated and as 0.5- to 0.8-mm-stage II thick coatings on bottoms of pebbles; local engulfment of matrix by carbonate also, abrupt, irregular boundary |
| Bk1 | 52–56 | Light gray orange to brown (7.5YR 5/3 dry; 6/3 moist) loamy sandy gravel; massive; loose, very slightly sticky and slightly plastic; many very fine interstitial pores; strongly effervescent; stage II and III carbonate occurs both disseminated and as 0.5-mm coatings and pendants on bottoms of pebbles; bt band, 1.0 mm thick; abrupt, smooth boundary |
| IIKb | 56– | Carbonate cemented gravel |

Cobble weathering stages

B and C horizons: mafic plutonic, schist, and coarse-grained metamorphic rocks, 3–4; gneissic granitic, mafic metavolcanic rocks, 2–1; leucocratic, quartz-rich metamorphic rocks, 1–2; metadacites, 1

[a] Described by L. D. McFadden and W. B. Bull.

[b] See Table 2.3 for definitions.

[c] See Table 2.4 for definitions.

TABLE 2.10 Soil profile on late Pleistocene alluvium (Q2b) derived from schist, gneiss, and andesite of Whipple Mountains.[a]

*Location:* San Bernardino County, California; approximately 0.5 km south of Highway 62 in the NE 1/4 of the NE 1/4 of Sec. 12, T1N, R2E

*Physiographic position:* Pit and roadcut on moderately dissected varnished surface mapped as Q2b in age; slope 1° south; altitude 294 m

*Classification:* Typic haplargid

| Horizon[b] | Depth (cm) | Description |
|---|---|---|
| | 1–0 | Strongly varnished gravel of pavement; exposed clasts are black-brown (7.5YR 2/3), bases are strong orange-brown (7.5YR 6/6) and noncalcareous; many quartz-rich unvarnished rocks; most rocks in pavement are metavolcanic rocks weathered to incipient stage 3[c] |
| Avk | 0–2 | Light orange to gray-orange (5YR 8/2 dry, 6/3 moist) clayey silt; coarse to very coarse platy; soft, very sticky and very plastic; common to many microfine to fine inped vesicular pores; many thin films of clay-lined pores; strongly effervescent stage I carbonate; abrupt, wavy boundary |
| Btk1 | 2–24 | Reddish-brown or gray orange to reddish-brown (5YR 6/4 to 5/6 dry, 4/4 moist) gravelly to very gravelly sandy loam; massive; loose to soft, slightly sticky and slightly plastic; continuous, thin clay grain cutans; strongly effervescent stage II carbonate occurs disseminated and as segregated few, medium prominent nodules and veins, as 2- to 10-mm-thick coatings on bottoms of pebbles, or as rootlet-shaped coatings on tops and sides of pebbles; engulfing matrix; local pockets of non-effervescent matrix; abrupt, broken boundary |
| Btk2 | 24–45 | Light brownish-gray to gray reddish-brown (7.5YR 8/1 dry, 5YR 5/4 moist) gravelly loamy sand; massive; hard to very hard, slightly sticky and nonplastic; few very fine pores; few to common thin clay films on some grains where matrix not engulfed by carbonate; violently effervescent stage II and III carbonate occurs as up to 10-mm-thick coatings, as root-shaped coatings on tops and sides of pebbles, or engulfs matrix; abrupt, broken boundary |
| Btk3 | 45–67 | Gray-orange to gray reddish-brown (7.5YR 6/4 dry, 4YR 5/4 moist) gravelly sandy loam; weak coarse subangular blocky; soft to slightly hard, sticky and slightly plastic; many very fine to fine interstitial pores; few fine dendritic tubular pores; few to common thin clay films on some grains; strongly effervescent stage II carbonate occurs as up to 0.5-mm-thick coatings on bottoms of pebbles and disseminated; very abrupt, wavy boundary |
| K | 67–120 | Light brownish-gray to gray-brown (7/5YR 8/1 dry, 5/3 moist) loamy sandy gravel; massive; very hard to extremely hard, very slightly sticky and nonplastic; violently effervescent stage III carbonate engulfs matrix and occurs as multiple-layered coatings and pendants on bottoms of many pebbles; carbonate locally engulfs matrix of 7.5YR 6/3 color; abrupt, wavy boundary |
| Ckox | 120– | Gray orange to gray-brown (7.5YR 7/3 dry, 5/4 moist) gravelly loamy sand; massive; loose; slightly sticky and slightly plastic; strongly effervescent stage I carbonate occurs as thin, less than 0.3 mm thick coatings on bottoms of pebbles and disseminated in matrix |

Cobble weathering stages
Btk3 and Ckm horizons: mesocratic plutonic rocks, 4; mafic plutonic metamorphic and schistose rocks, 3–4; gneiss, 2; volcanic rocks, 1–2. Ck2 horizon: mesocratic plutonic rocks, 2; mafic schist, 2–3; gneiss, 1–2; volcanics, 1

[a] Described by L. D. McFadden.

[b] See Table 2.3 for definitions.

[c] See Table 2.4 for definitions.

FIGURE 2.18 Aerial view of Q2a surface 10 m above wash south of the Whipple Mountains.

TABLE 2.11 Soil profile on mid-Pleistocene alluvium (Q2a) derived from schist, gneiss, and andesite of Whipple Mountains.[a]

*Location:* San Bernardino County, California; 1.7 km north of Highway 62 in SW 1/4 of the SE 1/4 of Sec. 1, T1N, R24E

*Physiographic position:* Roadcut in dissected planar remnant of Q2a surface; slope 1.5° to south; altitude 274 m

*Classification:* Typic paleargid

| Horizon[b] | Depth (cm) | Description |
|---|---|---|
| | 2–0 | Strongly varnished gravel of pavement; exposed clasts are black-brown (5YR 2/3 to 7.4YR 2/3), bases are strong orange-brown (7.5YR 6/6) and occasionally possess coatings of carbonate up to 5 mm thick; metamorphic and volcanic cobbles weathered to stage 2[c] |
| Avk | 0–3 | Light orange to gray reddish-brown (5YR 8/2 dry, 5/4 moist) clayey silt; moderate coarse to very coarse platy; slightly hard, very sticky and very plastic; many microfine to fine, inped, vesicular pores; continuous thin clay films on pores; strongly effervescent stage I carbonate; abrupt, smooth boundary |

TABLE 2.11 (*Continued*)

| Horizon[b] | Depth (cm) | Description |
|---|---|---|
| Abvk | 3–8 | Gray-orange to orange (5YR 7/4 dry, 6/6 moist) sandy clay loam; weak medium-to-coarse subangular blocky; slightly hard, sticky and plastic; common very fine, inped, vesicular pores; strongly effervescent stage II and III carbonate occurs disseminated in matrix and as coatings on undersides of clasts up to 20 mm thick; abrupt, wavy boundary |
| Btk | 8–75 | Strong reddish-brown to reddish-brown (5YR 5/6 dry, 4/6 moist) clayey loamy gravel; massive breaking into medium crumb; soft to slightly hard, very sticky and very plastic; few very fine-to-fine interstitial pores; continuous moderately thick clay films on grains, in pores, and as bridges; strongly effervescent stage II and III carbonate occurs segregated as common medium-to-large prominent nodules and as coatings mostly on bottoms of large clasts up to 2 mm thick and as veinlets; zones of matrix (5YR 5/6) noncalcareous; clear, smooth boundary |
| Btk2 | 75–100 | Gray reddish-brown to dark reddish-brown (5YR 5/4 day, 4/4 moist) sandy loamy gravel; massive to medium crumb; soft, sticky and plastic; common thin-to-moderately-thick clay films on grains and as bridges; strongly effervescent stage II and III carbonate occurs as segregated common medium-to-coarse prominent nodules that locally engulf 5YR 5/4 matrix and as 1- to 2-mm-thick coatings on bottoms of large clasts; abrupt, smooth boundary |
| Btk3 | 100–116 | Gray reddish-brown (5YR 6/4 dry, 5/4 moist) sandy loamy gravel; massive; slightly hard to hard, sticky and plastic; common thin clay films on grains and as bridges; strongly to violently effervescent stage II and III carbonate occurs mostly segregated as common, distinct mottles and as thin, <1-mm-thick discontinuous coatings on bottoms of most large clasts and as thin rootlet-shaped coatings on tops of some clasts; abrupt, smooth boundary |
| Btk4 | 116–130 | Gray-orange (5YR 7/4 dry, 6/4 moist) sandy loamy gravel; massive; soft to slightly hard, sticky and plastic; many thin clay films on pores and grains; common bridges; strongly to violently effervescent stage II and III carbonate occurs as common distinct nodules and as thin, <0.2-mm-thick coatings on bottoms of large clasts and as common distinct nodules; abrupt, smooth boundary |
| Ckox1 | 130–185 | Gray-orange to brown (7.5YR 6/3 dry, 4/4 moist) loamy sandy gravel; massive; loose to soft, very slightly sticky and very slightly plastic; slightly effervescent stage II carbonate occurs disseminated in matrix and as thin, <0.5-mm-thick coatings on bottoms of large clasts; gradual, smooth boundary |
| Ckox2 | 185–195 | Gray-orange to brown (7/5YR 6/3 dry, 4/3 moist) loamy sandy gravel; massive; loose, very slightly sticky and nonplastic; very slightly effervescent to noneffervescent stage I carbonate occurs mostly as thin, <0.3-mm-thick discontinuous coatings on bottoms of some large clasts; abrupt, smooth boundary |
| IICb | 195– | Clayey gravel |

Cobble Weathering Stages

Bt horizons: leucocratic volcanic, 2–3; mafic plutonic and mafic schist, 3–4; gneiss, 3–2; mafic volcanic, 2; mafic metavolcanic, 4; mesocratic plutonic, 3–4. Ckox horizons: gneiss, 1–2; mafic metavolcanic and mesocratic plutonic, 2–3; mafic volcanic, 1–2

[a]Described by L. D. McFadden.  [b]See Table 2.3 for definitions.  [c]See Table 2.4 for definitions.

FIGURE 2.19 Weathered rubble from petrocalcic horizons and residual fragments of volcanic alluvium on Q1 surface southeast of the Mopah Range.

jacent to the Mopah Range. The pebble coatings and the cement for the gravel contain silica as well as calcium carbonate. Extremely thick carbonate coatings occur on only one side of individual weathered basalt cobbles, which suggests that carbonate illuviation on bottoms of cobbles may have largely engulfed the former Bt horizon. The soil-profile description in Table 2.12 for the Figure 2.19 site shows that the K plus Bkm horizons are thicker than 2 m and that a 0.5-m thick multiple laminar and pisolitic carbonate horizon is present above extremely hard carbonate-plugged gravels.

The soils (including their varnished desert pavements), sedimentology and stratigraphy, and topographic characteristics are sufficiently diagnostic to map the time-stratigraphic units of the desert piedmonts. Selected examples of parts of such maps are presented in the next section.

## 2.3 Mapping Desert Piedmonts

Mapping of alluvial geomorphic surfaces has many uses. Those interested in climatic geomorphology can use such maps to assess whether the regional distribution of surfaces is related to present or past patterns of regional airmass circulation and to evaluate the influences of climatic change on fluvial and pedogenic systems. Where climate has had widespread impact on fluvial systems, flights of roughly synchronous Quaternary surfaces provide a useful regional background for identifying anomalous local surfaces of possible tectonic origin. The ages of ruptured and unruptured surfaces identify stable areas as well as the times and places of faulting. Such information has been invaluable in assessing the tectonic stability of proposed nuclear generating stations or nuclear waste repositories (Bull, 1977b). Characteristics of alluvial geomorphic surfaces also greatly affect groundwater recharge, suitability of soils for agriculture and irrigation, and the search for and interpretation of archaeological sites. Studies of late Holocene surfaces can provide information about frequencies and magnitudes of major flood events on desert piedmonts and along streams (Costa, 1986).

Several types of maps can be made. The common approach used by geologists and hydrologists before 1970 was to map all Quaternary alluvium as one unit–Qal. Such generalization does not help engineers, planners, and farmers who work on desert piedmonts. Soil scientists provide highly

TABLE 2.12 Truncated soil profile on Q1 alluvium derived from basalt, andesite, and tuff of Mopah Range.[a]

*Location:* San Bernardino County, California; in roadcut on east side of U.S. Highway 95, 12.9 km north of State Highway 62, in the NE 1/4 of the SE 1/4 of Sec. 10, T2N, R22E

*Physiographic position:* Extensive piedmont surface; slope 1° to southeast; altitude 390 m

*Classification:* Petrocalcic paleorthid

| Horizon[b] | Depth (cm) | Description |
|---|---|---|
| | 28–0 | Rough pavement of boulders, cobbles, and laminara and brecciated petrocalcic fragments left as lag gravel from erosion of Btkm horizons; weathered basalt cobbles have 20- to 65-mm-thick stalactitic silicic carbonate coatings; varnish on tops of basalt cobbles is black (7.4YR 1.7/1); bases are gray-orange (7.5YR 6/4) |
| Avk | 0–4 | Light brown (7.4YR 8/3 dry, 5/4 moist) silt with petrocalcic and silicic petrocalcic fragments and cobble coatings; massive to very coarse platy; common fine, inped vesicular pores; slightly hard, slightly sticky and slightly plastic; violently effervescent; abrupt, smooth boundary |
| K1 | 4–12 | White (9/0) sandy stage V calcium carbonate; laminar; slightly hard at top to hard at base, nonsticky and nonplastic; violently effervescent; clear, wavy boundary |
| Kq2 | 12–53 | Light brown (7.5YR 8/3 dry, 6/3 moist) sandy matrix in bouldery gravel; stage VI carbonate is laminar, pisolitic; extremely hard, nonsticky and nonplastic; violently effervescent; clear, wavy boundary |
| K3 | 53–86 | Light brown (7.5YR 8/3 dry, 6/3 moist) to pinkish-white (7.5YR 8/2 dry, 7/2 moist) sandy gravel; massive, plugged; extremely hard, nonsticky and nonplastic; violently effervescent stage III and IV[c] carbonate; gradual wavy boundary |
| Bkm | 86–232+ | Light brown (7.5YR 8/3 dry, 6/3 moist) gravelly sand; massive, plugged, thick bedded; hard to very hard; lenses of soft to slightly hard sand, nonsticky and nonplastic; violently effervescent stage III carbonate |

[a] Described by W. B. Bull.

[b] See Table 2.3 for definitions.

[c] See Table 2.4 for definitions.

useful and detailed maps; their efforts, however, have been concentrated on loamy farmland soils except for special projects such as the Desert Soils–Geomorphology Project of the U. S. Soil Conservation Service at Las Cruces, New Mexico (Gile et al., 1981).

Some lower Colorado River piedmonts have been mapped by previous workers. Hamilton (1964) made an excellent detailed map of the alluvium between the Big Maria and Riverside mountains. Although he was concerned mainly with bedrock geology, he correctly identified and mapped key Quaternary units. His four categories of Quaternary alluvium correspond with the Q1 and Q2a, Q2b and Q2c, Q3, and Q4 of Table 2.2.

Piedmonts adjacent to the Whipple, Riverside, Big Maria, and Gila mountains were mapped as part of site evaluations for proposed nuclear generating stations in Vidal Valley west of Parker, California (Bull, 1974a,b), and south of Blythe, California (Bull, 1976a; Shlemon, 1978). The distinctive and extensive mappable units of the study region were defined on the basis of comparisons of topography, desert pavement, soil profiles, and stratigraphy and sedimentology of the alluvial geomorphic surfaces. Detailed maps of the immediate vicinity of a proposed nuclear generating station included all the surfaces identified in Table 2.2, and maps of the piedmont within 10 km of a proposed station identified each Q2 surface because they are critical for determining the surface-rupture history during the past 500 ky.

Regional reconnaissance maps can be made efficiently by mapping the four basic surfaces of Table 2.2 (Q1, Q2, Q3, and Q4), together with genetic types of deposits such as dune sand and playa deposits. Shlemon (1978) used the fourfold classification of piedmont geomorphic surfaces

(Bull, 1974b) for mapping regional Quaternary geology for the proposed nuclear generating station near Blythe. More detailed maps of smaller study areas have been made by Schenker (1977), Tucker (1979), McHargue (1981), Shih (1982), and Shih and Schowengerdt (1983). Christenson and Purcell (1985) correlated fan sequences in different parts of the Basin and Range Province.

The four basic classes of geomorphic surfaces can be mapped readily on orbital imagery and aerial photographs (Fig. 2.20). Q4 surfaces are unvarnished and have riparian vegetation. Gray tones and a plumose texture caused by the topographic pattern of bars and swales are diagnostic of the Q3 surfaces. Dissected but smooth black desert pavements that are virtually barren of veg-

FIGURE 2.20 Vertical aerial photograph of the Quaternary geomorphic surfaces northeast of the Gila Mountains. Note the plumose texture of the Q3b surfaces, the blackness of the Q2 surfaces, and the ridge-and-ravine topography of the Q1 surfaces. See Figure 2.24 for map.

114° 27′ 30″

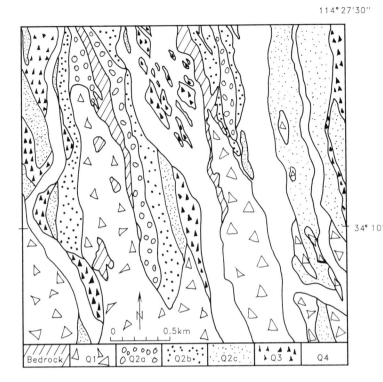

34° 10′

FIGURE 2.21 Map of geo-
morphic surfaces on the piedmont south of the Whipple
Mountains. From Dickey et al.,
1980.

etation identify the Q2 surfaces. Reddish Q2 argillic horizons stand out clearly on color photographs where the pavement is disturbed or for
grussy piedmont deposits derived from granitic
rocks. The ridge-and-ravine topographic pattern of
the dissected Q1 alluvium provides a distinctive
image (Figs. 2.10, 2.11D, 2.20). Careful field
checking that includes examination of many soil
profiles is necessary to map the subdivisions of
Q2 and Q3, particularly on piedmont deposits
derived from granitic mountains.

The four maps shown in Figures 2.21 through
2.24 are parts of much larger maps; they illustrate
typical features of the piedmonts. Downcutting by
streams tributary to the Colorado River has resulted in stream terraces with parallel longitudinal
profiles (Figs. 2.21, 2.24), instead of overlapping
climatic fans that are more typical of basins of
internal drainage. A diverse flight of stream terraces for geomorphic and soils genesis studies is

revealed to the left of center in Figure 2.21, and
a Q3b channel fan in the north-central part of the
map is being buried by Q4 deposits. The cross
sections and maps of Figure 2.22 reveal the influence of the nearby Colorado River. Holocene incision by the river has caused rapid downcutting
of the tributary streams. Only small mesas of
Pleistocene surfaces rise above the surrounding
flood of Holocene alluvium (Fig. 2.12D). The
Colorado River has migrated to the east of the Big
Maria Mountains (Fig. 2.23), but its former presence is revealed by low scarps high on the piedmont. Scarps of Q2a, Q2b, Q2c, and Q3 age
parallel the trend of the river valley, and the
pavements downslope from each are littered with
distinctive rock types deposited by the Colorado
River. The piedmont northeast of the Gila Mountains is graded to the Gila River about 50 km
upstream from its junction with the Colorado River.
Q1 deposits not only form a distinctive landform,

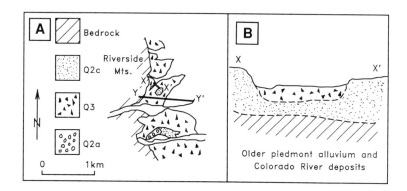

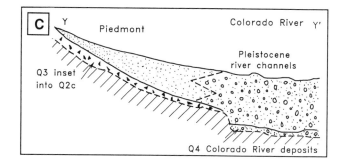

FIGURE 2.22 Relations of alluvial geomorphic surfaces along the east side of the Riverside Mountains, southeastern California, to past and present levels of the Colorado River. Dashed lines represent inferred contacts. A. Map. B. Section across the piedmont. C. Section down the piedmont.

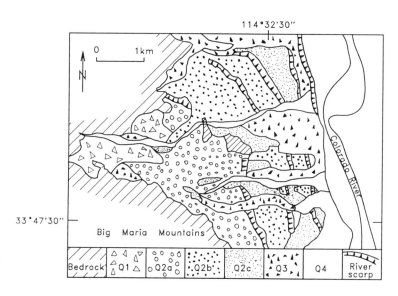

FIGURE 2.23 Map of geomorphic surfaces east of the Big Maria Mountains.

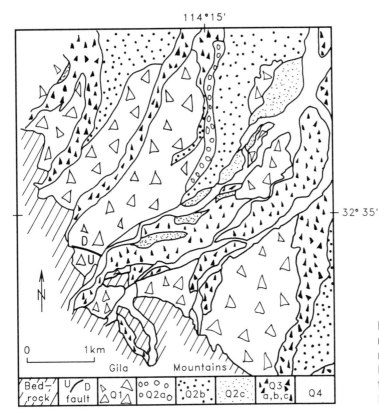

114°15'

32° 35'

0    1km

Gila    Mountains

| Bed‑rock | U/D fault | △Q1△ | o o Q2a o | Q2b | Q2c | Q3 a,b,c | Q4 |

FIGURE 2.24 Map of geo-morphic surfaces on the pied-mont northeast of the Gila Mountains. View of center of the mapped area is shown in Figure 2.10.

but also underlie all the younger deposits at depths of 1 to 5 m. Faults cut Q1 but not younger deposits.

As useful as maps may be, their full geologic potential is not achieved until the ages of the alluvial geomorphic surfaces are evaluated. Relative or numerical ages of mappable surfaces that are regional in extent provide the temporal aspect needed to define times of aggradation and degradation induced by climate change.

## 2.4 Dating Desert Piedmonts

### 2.4.1 NEW APPROACHES

A general lack of organic matter for radiocarbon dating in gravelly deposits and soils of hot deserts, particularly where mean annual precipitation is less than 100 mm, forces use of alternative dating methods. There are several new approaches for dating alluvial geomorphic surfaces in the important time span between 10 and 500 ka. The following sections discuss a variety of promising dating methods for desert piedmonts. Another method—thermoluminescence dating of fluvial and colluvial materials—appears to have great potential (Forman, 1989; Forman et al., 1988, 1989).

Ages of alluvial geomorphic surfaces of the lower Colorado River region have been estimated using several techniques. The results are summarized in Table 2.13. Some of the dating was done as part of the feasibility studies for proposed nuclear power generating stations. The headings of this section, and in Chapters 3, 4, and 5, reflect the excellent classification of dating methods presented by Colman, Pierce, and Birkeland (1987).

TABLE 2.13 Estimated numerical ages of alluvial geomorphic surfaces of lower Colorado River region.

| | | Estimated Age Range (ka) | Geomorphic Surface | Basis for Age Estimate |
|---|---|---|---|---|
| HOLOCENE | | 0 | Q4b | Loci of present streamflows |
| | Late | 0.1–2 | Q4a | Riparian trees not present in unvarnished abandoned channels |
| | ---------- 4 ka | | | |
| | | 2–4 | Q3c | Surficial cobbles old enough to have incipient coatings of light brown varnish |
| | Middle | 4–8 | Q3b | 4-ka age assigned because of brown varnish; 8-ka age based on $^{14}C$[a] dating of plant fossils in packrat middens |
| | ---------- 8 ka | | | |
| | Early | 8–12 | Q3a | Both ages based on $^{14}C$ dating of plant fossils in packrat middens |
| | ---------- 10 ka | | | |
| PLEISTOCENE | Late | 12–70 | Q2c | 12 ka age based on $^{14}C$ dating of plant fossils in packrat middens. One $^{230}Th/^{234}U$ age of 61 ka |
| | ---------- 130 ka | | | |
| | | 70–200 | Q2b | Seven $^{230}Th/^{234}U$[b] ages from one surface have mean of 82 ka. Seven ages from second surface have mean of 83 ka. 200-ka age based mainly on large amount of carbonate in Bk horizon compared with the Q2c geomorphic surface |
| | Middle | 400–730 | Q2a | Q2a soil profile in gravels overlying 700-ka-old Bishop Tuff on west side of Coxcomb Mountains. K/A age on 600 ka old basalt flow in Q2a alluvium of Rodman Mountains. Paleomagnetic samples have normal polarity |
| | ---------- 790 ka | | | |
| | | >1200 | Q1 | Oldest basalt flow at Toroweap in Grand Canyon has K/Ar[c] age of 1.16±0.2 Ma. Basalt is common in young terraces of Colorado River but is rare in Q1 terraces. Paleomagnetic samples have both normal and reversed polarities |
| | Early | | | |
| | ---------- 1640 ka | | | |

[a] Carbon-14.

[b] Thorium-230/uranium-234.

[c] Potassium–argon.

### 2.4.1.1 Numerical Ages

*2.4.1.1.1 Isotopic Age Estimates* Potassium–argon dating of volcanic materials was used to assess the ages of older geomorphic surfaces in several areas. A detailed set of ages spanning several million years was obtained as part of the studies of landscape change and soil-profile development in the Cima volcanic field of the eastern Mojave Desert (McFadden et al., 1984; Wells et al., 1985, 1987). Dating of basalt flows (Bull, 1974b) beneath Colorado River alluvium near Parker, Arizona, indicates that Q1 alluvium is younger than 3 Ma. A 600–ka basalt flow is interbedded in Q2a alluvium in the Rodman Mountains southeast of Barstow, California.

One isotopic method is thorium-230/uranium-234 ($^{230}$Th/$^{234}$U) dating of pedogenic carbonate. The initial attempts to use this isotopic clock to date calcic soils were at the proposed Vidal nuclear plant site on the piedmont of the Whipple Mountains (Ku et al., 1979). The minimum numerical ages agree with relative ages based on terrace height and are reproducible for seven-sample suites collected from different parts of alluvial fans derived from schist and andesite source rocks.

Samples for $^{230}$Th/$^{234}$U dating provide minimum ages even when collected from interior parts of dense pebble coatings from Bkm horizons. Thousands of years are needed to accumulate the carbonate in the soil after deposition of the gravel. Increments of young illuvial carbonate will have lower thorium/uranium ratios than will soil carbonate that has been isolated from pedogenic processes. Displacive growth by increments of new calcite is a continuing illuvial process in the formation of calcretes. Samples with macroscopic evidence of displacive growth were rejected for $^{230}$Th/$^{234}$U dating. The initial work at the Vidal site showed that with careful sample selection, carbonates from arid soils can provide minimum numerical ages for geomorphic surfaces when analyzed by $^{230}$Th/$^{234}$U isotopic dating.

The use of uranium-series isotopes to date soil carbonates has several advantages over the radiocarbon method. Radiocarbon dating of soil carbonate is fraught with difficulties of contamination by unknown amounts of carbon that are either younger or older than the age of the gravel. Owing to the relatively long half life of 75.2 ky of $^{230}$Th, the $^{230}$Th/$^{234}$U method presently has a potential dating span of about 350 ka. Thus the effect of contamination of old samples is much less than for radiocarbon dating. Furthermore, the method dates the time at which $CaCO_3$ precipitates from a solution because the major source of $^{230}$Th is the in-situ decay of uranium in the accumulated carbonates. This eliminates the parent -rock source problem for the radionuclides in gravel without limestone, which is not possible for radiocarbon dating of soil carbonates. With the advent of dating using mass spectrometric determination of thorium-230 (Edwards et al., 1987), both dating precision and dating span may be greatly improved and much smaller samples can be used.

The method involves two basic assumptions: (1) the thorium/uranium ratio is essentially zero when carbonate is precipitated in the soil and (2) the carbonate acts as a closed system subsequent to its formation. The first assumption is reasonable because uranyl carbonate complexes are quite soluble but thorium isotopes are highly insoluble.

The second assumption is difficult to verify and requires calibration by other dating methodologies. Let us consider three $^{230}$Th/$^{234}$U apparent ages of soil carbonates from Q1 geomorphic surfaces of the Whipple Mountains piedmont. Each is clearly older than the 350,000–year maximum age that can be attained by the $^{230}$Th/$^{234}$U method. A sample of porous, earthy carbonate from a depth of 0.2 m below the summit of an eroded ridgecrest yielded an apparent age of only $55 \pm 4$ ka. A second sample was an exceptionally dense 50-mm-thick pure cobble coating that probably was formed in a Btk or Bkm (petrocalcic) horizon but was collected from a planar degraded pavement. The innermost part of the carbonate coating was selected for dating and yielded an apparent age of $130 \pm 12$ ka. The third sample consisted of extremely hard nodules collected from a Ck horizon 0.8 m below the present surface and overlain by cemented silty sand with a low permeability. The central parts of the nodules yielded an apparent age of >350 ka. The apparent ages obtained for the first two samples show that it is easy to violate the second assumption by collecting material that is either too porous or that, although dense and pure, has been exposed to rain. Contamination was expected in the first sample, but the second sample appeared impermeable. The third sample had minimal contamination because it was both dense and removed from atmospheric and groundwater sources of contaminating uranium.

The inner parts of dense carbonate pebble coatings from Bk soil horizons were collected from two Q2b localities in Vidal Valley. At each locality seven samples were collected over a distance of about 3 km. One suite of samples provided ages that ranged from 69 to 94 ka and averaged 83 ka. The other suite provided ages that ranged from 73 to 93 ka and averaged 82 ka. The mean age for all 14 samples was 83 ka with a standard deviation of $\pm 9.5$ ky, which is only slightly more than the maximum laboratory measurement error for the individual ages; their range was $\pm 5$ to 9 ky. These results show that the dating methodology has a high degree of internal consistency, especially when one considers that the samples are from widely separated parts of two alluvial fans derived from schistose and andesitic drainage basins.

Isotopic dating of ubiquitous soil carbonate of arid regions has sufficient promise that Slate (1985, in press) compared $^{230}Th/^{234}U$ ages of pedogenic carbonate collected from basalt flows of the Pinacate volcanic field of northwestern Sonora, Mexico, with potassium–argon (K/Ar) ages of the same flows. The K/Ar ages ranged from 140 to 870 ka, but the $^{230}Th/^{234}U$ ages ranged only from 20 to 160 ka. The linear relation is

$$y = 5.3 + 0.16x \qquad (2.4)$$

where y is the $^{230}Th/^{234}U$ age and x is the K/Ar age. The constant of 5.3 ky is suggestive of the large time span between cooling of the lava flow and illuviation of sufficient carbonate to sample for dating. The coefficient of 0.16 clearly shows that accreted soil carbonate is not acting like a closed system even though the Gran Desierto is the driest region of North America. New increments of carbonate have been added to the soil system throughout the past 870 ky. Fractured lava flows appear to provide an excellent soil microenvironment for infiltration of abundant water from rare rainfall events. High soil permeability more than offsets an extremely arid late Quaternary climate. Slates' study shows that $^{230}Th/^{234}U$ soil ages are minima (which are still quite useful for

demonstrating tectonic stability of some engineering sites) and that the coefficient of Equation 2.6 is subject to lithologic controls that affect translocation of pedogenic $^{234}U$. The constants of Equation 2.6 can be expected to vary at each site within the Chapter 2 study region as a function of (1) amount, intensity, and temperature of rainfalls; (2) atmospheric and geologic sources of calcium carbonate: (3) infiltration capacity of surficial materials; and (4) transmissivity of water and partial pressure of $CO_2$ of subsurface materials. All five factors vary with time as well as with space. Dating of gravels with limestone clasts may introduce a parent-rock source for radionuclides. Sowers and colleagues' (1989) study in southern Nevada indicates that it may be possible to have $^{230}Th/^{234}U$ ages of soil carbonate that are too old because of $^{230}Th$ contamination from carbonate in limestone gravel. Thus it appears that isotopic dating of soil carbonate has to be done with special care, and may provide ages with large uncertainties.

Uranium-trend dating of Quaternary soils and deposits (Rosholt, 1980,1985; Rosholt et al., 1985) is an expensive, but valuable, new concept. In contrast to the $^{230}Th/^{234}U$ dating method, carbonate samples from each soil horizon are treated as an open system. For each sample, the concentrations of $^{238}U$, $^{234}U$, $^{232}Th$, and $^{230}Th$ are used to determine an isochron that can be used to estimate soil ages in the range from 4 to 900 ka.

### 2.4.1.2 Calibrated Ages

*2.4.1.2.1 Rock-Varnish Chemistry* Scanning electron microscopy by Dorn (1988) indicates that it takes a century for initial accretion of rock varnish. Most workers conclude that the visible coatings of rock varnish in the Mojave Desert require at least 2 ka to form (Blackwelder, 1948; Denny, 1965; Hooke & Lively, 1979; Hunt, 1962; Elvidge & Moore, 1979; Elvidge, 1982; Dorn & Oberlander, 1981). Vance Haynes (University of Arizona; oral communication, October 13, 1975) obtained $^{14}C$ ages for the Tule Springs site in southern Nevada that indicate that incipient varnish coatings started to form at about 4 ka. I described the colors of incipient varnish forming on chert nodules broken in the year 75 a.d. when the Romans built siege walls around the fortress of Massada in Israel. Thus, 2 ky seem to be needed to accrete visible amounts of varnish on stable lithologies of alluvial surfaces in several arid regions. Most Q3b gravel bars in the lower Colorado River region are darkly varnished, which suggests that they are older than 4 ka. These generalizations may not apply to some moist microenvironments and to certain lithologies such as quartz.

Some of the most exciting work being done in desert geomorphology concerns dating of varnished alluvial geomorphic surfaces. Aspects that appear to have particularly great potential include radiocarbon dating (Dorn, 1989a), cation-ratio dating, and correlations of sequences of microscopic laminations with different morphologies and compositions resulting from different climate-controlled modes of varnish accretion. Radiocarbon ages of minute amounts (<1%) of organic carbon in rock varnish from sites near Silver and Cronise lakes in the Mojave Desert were dated by the tandem accelerator mass spectrometer, which yielded ages of 1.4 to 16.8 ka $\pm$ 0.8 ka (Dorn et al., 1986). These ages agree with the estimated ages of pre- and post-lake-highstand (mainly Q3a) alluvium. When combined with cation-ratio dating of 167 surface artifacts from six sites and dating of petroglyphs (Whitley & Dorn, 1987,1988), they provide an important new line of evidence to suggest that the Mojave Desert was occupied by humans during the latest Pleistocene. Subsequently, radiocarbon dating of rock varnish has been applied to diverse geomorphic settings in the Basin and Range Province (Dorn et al., 1989).

Rock varnish has alternated between two modes of accretion during the Quaternary. Manganese-rich varnish with a botyroidal morphology forms during times of semiarid climate, and iron-rich to less manganese-rich varnish with a lammelate morphology forms during times of arid climates. Varnish is a stable sink for manganese over surprisingly long time spans (>2 Ma). It preserves

both the structural and chemical temporal variations as a micron-scale stratigraphy; these sequences can be used to provide relative age control and correlate alluvial fans and stream terraces of the region (Dorn, 1984a,b; Dorn et al., 1987a,b).

The layers formed during arid and semiarid times also record the variations in the types of organic carbon supplied to the accreting varnish by nearby plants. The abundance of $^{13}C$ relative to $^{12}C$ is significantly different between adjacent varnish layers (Dorn & DeNiro, 1985; Dorn et al., 1987c). Dorn and DeNiro sampled organic matter forming in arid and humid environments by scraping off the outermost 10-fm layer of varnish with a tungsten-carbide needle under 45X magnification. Carbon-isotope analyses of 15-mg samples of organic matter revealed strongly negative $\delta^{13}C$ values of about $-22$ per mil for varnish forming in the vicinity of semiarid to humid-type plants characterized by C-3 photosynthetic pathways. Varnish forming in the vicinity of arid-climate plants had less negative values of about $-15$ per mil indicative of C-4 photosynthetic pathways. The $\delta^{13}C$ notation used is defined as

$$\delta^{13}C = \frac{(^{13}C/^{12}C) \text{ sample} - (^{13}C/^{12}C) \text{ standard}}{(^{13}C/^{12}C) \text{ standard}} \times 1000/\text{mil} \qquad (2.5)$$

Thus carbon isotopes in rock varnish appear to have potential for recording late Quaternary fluctuations in plants and their associated climates (Dorn, 1988). Dorn and others tested their hypothesis in Death Valley where cation-ratio dating of varnished alluvial-fan surfaces agrees with the chronology of distinctive changes in flora recorded by packrat midden fossils (Wells & Woodcock, 1985). Organic matter extracted from rock varnish has isotopes that can be considered a closed system (neither additions nor deletions of carbon isotopes).

The climate-change-controlled sequence of surfaces on the Hanaupah Canyon (Fig. 2.25) and Johnson Canyon alluvial fans on the west side of Death Valley (Hunt & Mabey, 1966) are the same

FIGURE 2.25 Entrenched alluvial fan of Hanaupah Canyon derived from the Panamint Range on the west side of Death Valley. Large fan-shaped area above the salt pan is mainly Q3b surfaces with some active wash and Q3a surfaces. Q2c surfaces are to right of fanhead trench, and Q2b is to left of trench. High remnant of dissected Q2a alluvium occurs along the left edge of the fan.

as for the rest of the study region (Wells & McFadden, 1987). The carbon-isotope composition of the varnish, as sampled in 10-$\mu$m increments, shows a distinct layering (Table 2.14). Rock varnish is less than 10 $\mu$m thick on the Q3a fan segment, but thicknesses increase to more than 50 $\mu$m for the Q2a fan segment. The $\delta^{13}$C values of the outermost varnish at all eight sample localities range from 14.8 to 16.0 and are indicative of arid climates. The $\delta^{13}$C values of the underlying layer of rock varnish range from $-21.1$ to $-23.6$ at five of six sample localities. These distinctly larger negative values are indicative of accretion of varnish in the vicinity of semiarid climate plants. The Q2c locality on the Johnson Canyon fan also

has an underlying layer with a value of $-22$ per mil, but it occurs at a depth of 20 to 30 $\mu$m instead of at 10 to 20 $\mu$m. The value of $-17.5$ per mil for the sample from 10 to 20 $\mu$m is best interpreted as being a mixture of the overlying and underlying types of rock varnish. Four other mixed samples are underscored in Table 2.14.

As elsewhere in the study region, it is assumed that the large area of Holocene fan deposition (Fig. 2.25) resulted from a change to monsoonal and then interglacial arid conditions after the time of latest Pleistocene full-glacial semiarid climate. Varnish accumulated on the surficial rocks of the older fan segments during both semiarid and arid climates after stream-channel entrenchment iso-

TABLE 2.14 $\delta^{13}C$ values of organic matter in layers of varnish on Pleistocene and Holocene surfaces of Hanaupah Canyon and Johnson Canyon alluvial fans, Death Valley, California.

| | Depth Below Varnish Surface ($\mu m$) | Fan Surface[a] | | | |
|---|---|---|---|---|---|
| | | Q3a | Q2c | Q2b | Q2a |
| Hanaupah Canyon fan | 0–10 | **−15.1** | **−15.4** | **−15.1** | **−15.4** |
| | 10–20 | | −21.2 | −22.8 | −23.3 |
| | 20–30 | | | **−14.1** | −19.2 |
| | 30–40 | | | −18.3 | −23.4 |
| | 40–50 | | | −23.6 | **−16.1** |
| | 50–60 | | | | −23.2 |
| Johnson Canyon fan | 0–10 | −14.8 | **−15.4** | **−16.0** | **−14.6** |
| | 10–20 | | −17.5 | −21.8 | −22.1 |
| | 20–30 | | −22.2 | **−15.5** | −18.3 |
| | 30–40 | | | −18.5 | −23.4 |
| | 40–50 | | | −22.9 | **−15.7** |
| | 50–60 | | | | −21.1 |
| | 60–70 | | | | −22.2 |

[a] Bold numbers (**−15.4**) indicate values for varnish formed during arid climate mode. Plain numbers (−22.2) indicate semiarid mode. Underscored numbers (−18.3) are for samples that may consist of mixed materials representing both arid and semiarid modes.

*Source:* Dorn et al., 1987.

lated them from further deposition. Each fan segment was deposited as a result of a brief climate-change-induced sediment-yield increase in the Panamint Range watershed. The varnish from the desert pavement of the oldest surfaces provides the most complete isotopic record of major late Quaternary climatic changes. The layered varnish of the Q2a sample from both fans records two, and possibly three, times of drier climate presumably associated with the Q2b, Q2c, and Q3a aggradation events.

Cation-ratio dating is based on differences in leaching of potassium, calcium, and titanium from the rock varnish with time. Some of the basic assumptions of cation-ratio dating* include the following:

1. Varnish constituents are derived from atmospheric sources that have remained sufficiently constant during time spans of >5 to 10 ky to form a uniform layer of varnish.

2. Rock varnish serves as a cation-exchange complex whose capacity and properties are similar in all samples being considered.

*New dating methods evolve rapidly and tend to be controversial; cation-ratio dating is no exception. At the time that this book went to the printer, I became aware that Ronald Dorn's analytical technique did not accurately measure the concentration of Ti in the presence of Ba, which also occurs in rock varnish. Even so, the figures of chapter 2 based on proton induced x-ray emission methodology show fairly good internal consistency. Many other factors may also affect the results obtained from cation-ratio analyses of rock varnish, so it remains a technique to be used where nothing else will work. Continuing studies by several workers will better define the potential and problems of cation-ratio dating. You are referred to forthcoming articles on the subject, such as those by Bierman and Gillespie that should appear in the journal *Geology* in 1991.

3. K and Ca are more mobile than Ti in the varnish cation exchange complex.

4. The rates of leaching of varnish cations are similar for samples being considered, and rock varnish may be considered an open system with respect to leaching of mobile cations.

Rock-varnish samples should be collected from microenvironments that are similar in regard to aspect, microslope of the rock surface, amount of surface available to generate runoff above the rock varnish, height above dust-generating surfaces such as soil, and location relative to cracks and outcrops and plants. The samples are scrubbed in deionized water and Dorn detached the varnish from the underlying rock by scraping with a tungsten–carbide needle under 10 to 45 $\times$ stereoscopic magnification. This procedure produces samples that average the cation composition of the scraped interval. Dorn used particle-induced x-ray emission (PIXE) for his chemical analyses.

Flights of lake shorelines and alluvial-fan surfaces in the Death Valley region of southeastern California reveal systematic variations in the characteristics of rock varnish. The last highstand of 690 m at Searles Lake was at about 10.5 ka (Smith, 1979; Smith & Street-Perrott, 1983). Dorn collected varnish from five prehistoric shorelines below 690 m and on a desert pavement above the lake highstand. The Ca/Ti ratio was analyzed with x-ray fluorescence for 10 varnished cobbles at each site. This cation-ratio analysis shows a progressive increase in ratios from 2 to 9 with decreasing relative age (altitude) of shoreline sampled (Fig. 2.26A). Varnish on a Death Valley fan (Fig. 2.26B) has very high levels of Ti and correspondingly low cation ratios, but ratios of K+Ca/Ti decreased with increases in varnish darkness (Fig. 2.26C).

Cation ratios may be calibrated by samples from dated volcanic rocks. Dorn sampled the ubiquitous and excellent varnish on stable surfaces of 16 rhyolite domes and basalt flows in the Coso volcanic field 40 km northwest of Searles Lake that have K/Ar ages of 38 to >3000 ka (Dorn,

1983,1989b). The input cation ratio was estimated by collecting soil samples from 16 sites; for the clay-size fraction the average K+Ca/Ti ratio is 9.2. The general trend of the data indicates a progressive decrease in cation ratios to 3 Ma (Fig. 2.26D). The startling implication is that parts of volcanic outcrops remain uneroded while accreting varnish for more than a million years.

A similar study was made to determine the varnish cation-leaching curve for the Mojave River basin. Basalt flows from the Cima volcanic field were K/Ar dated, and post-14-ka ages were based on shoreline ages. The two plots of Figure 2.26D are fairly parallel, which suggests roughly similar rates of cation leaching. The ages represented by the cation ratios are markedly different, however. For example, leaching of varnish results in a K+Ca/Ti ratio of 4 after 18 ka at Cima and after 60 ka at Coso. Figure 2.26D suggests the possibility of a family of cation-ratio leaching curves for arid regions, which may be largely controlled by cation ratio inputs from soils and local climates. The difference in cation ratios suggests that each part of western North America requires its own calibration curve in order to date fans and terraces solely by cation-ratio analyses. This concept is illustrated by Dorn (1989b, Fig. 12). Thus cation-ratio dating may have constraints that are similar to those for $^{234}U/^{230}Th$ dating of soil carbonate (Section 2.4.1.1.1).

An example of a different cation-ratio curve is suggested by data from southwestern Arizona. Cation-ratio analyses were made for varnish samples collected from the five main alluvial geomorphic surfaces on the Gila Mountain piedmont shown in Figures 2.10 and 2.24, using scanning electron microscope-energy dispersive x-ray analysis. This procedure was described by Harrington and Whitney (1987). The age estimates are from Table 2.13. Jonathan Fuller noted that scans of varnish chips yielded more consistent cation ratios than spot analyses of chips or powdered samples. The Gila Mountains and Searles Lake sites have markedly different Ca/Ti ratios even for the samples with well-constrained mid-

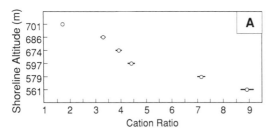

B

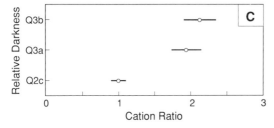

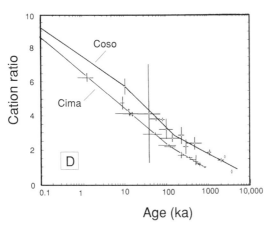

D

FIGURE 2.26 Cation ratios of rock varnish of the Mojave Desert, modified from Dorn (1983) and Dorn and colleagues (1986). A. Ratios of Ca/Ti in varnish from Searles Lake shorelines and a desert pavement at 701 m above the 10.5-ka highstand shoreline at 690 m. B. Five ages of debris flows on a 0.04-km$^2$ alluvial fan on the east side of Death Valley. Dorn analyzed varnish from the Q2c, Q3a, and Q3b surfaces. Varnish from the Q3c surface was too thin to sample, and the Q4 surface is aggrading. Faulting has ruptured the Q3c and older surfaces in the fanhead. See Table 2.12 for ages of units. C. K+Ca/Ti ratios of varnish from the surfaces shown in Figure 2.26B. D. Varnish leaching curves based mainly on K+Ca/Ti ratios of varnish from dated outcrops of the Coso and Cima volcanic fields. The horizontal and vertical bars indicate the age uncertainties and standard errors.

93

Holocene ages. The Gila Mountains ratios are twice as large as the Searles Lake ratios, and the rate of decrease of the cation ratios with time is twice that for the Coso and Cima volcanic fields (Fig. 2.26D). A lightly varnished rock from a ridgecrest in the highly dissected Q1 terrain also was analyzed, but the results were not included in the regression of samples from stable surfaces (Fig. 2.27). The thin varnish of the Q1 sample reflects the lack of surface stability and varnish degradation, but the cation ratio is quite low. Apparently, much of the remaining varnish is the innermost Q1 layer that was accreted when the surface was stable.

*2.4.1.2.2 Cobble Weathering-Rind Thicknesses* The thickness of weathering rinds on surficial basalt boulders was studied in western Arizona. Samples were collected from five different ages of faulted and unfaulted alluvial fans of Warm Springs Wash, which flows out of the Black Mountains of western Arizona and cuts through the Needles graben. Younger fan segments are inset into the older segments in much the same manner as for the Hanaupah Canyon fan (Fig. 2.25).

The rinds consist of light-colored zones of oxidation whose inner boundaries parallel the outer surfaces of the basalt boulders. Porter (1975a) was able to show that measurement of basalt weathering rinds of surficial boulders was highly effective for separating deposits of different relative or calibrated ages. Colman and Pierce (1981) and Colman (1982) measured over 7000 weathering rinds at 150 sites in the western United States. The mean annual precipitation of their study areas currently ranges from about 240 to 1300 mm. They collected basalt and andesite samples from B soil horizons. Colman was particularly interested in the mineralogic changes as a function of time and purposely sampled subsoil rocks that might be subject to solution but not to physical erosion that occurs in surficial lag gravels. Colman and Pierce identified many potential problems in using surficial weathering rinds for relative age control. Weathering rinds are softer than the parent

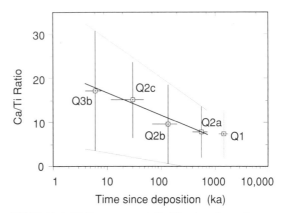

FIGURE 2.27 Decrease of Ca/Ti ratios in rock varnish with age for the surfaces of the Gila Mountains piedmont. Horizontal bars are the age ranges of the surfaces from Table 2.13 and the vertical bars are standard deviations. Analyses made by Jonathan Fuller, University of Arizona.

rock, so partial loss of rinds can occur. Other problems include the tendency for thicker rinds to occur beneath lichen-covered surfaces, spalling and oxidation caused by fires, disturbance of surficial stones by animals, and the difficulty of recognizing stones that are now at the surface but have been exhumed at some unknown time in the past.

Basalt weathering-rind thicknesses increase at rates that are functions of the kinetics of chemical reactions at the inner edge of the advancing rind. These reactions are partly controlled by the permeability of the underlying fresh basalt, types of eolian salts that affect pH, and the frequency and duration of wetting by rain and dew. Moisture is a dominant limiting variable in arid regions, so rates of increase of weathering-rind thickness may be one to two orders of magnitude less than for similar rock types in more humid climates. Rock varnish reduces influx of water to the weathering rind to an unknown degree; this factor may cause slower rates of rind formation.

Highly significant increases of surficial cobble-rind thicknesses with increasing age have been

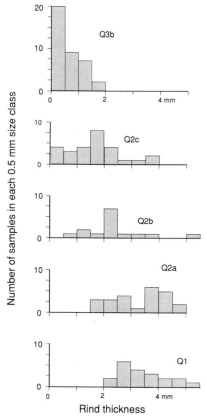

FIGURE 2.28 Variations of weathering-rind thicknesses for surficial basalt boulders of five ages of alluvial-fan deposits of Warm Spring Wash, Black Mountains, western Arizona. See Table 2.13 for ages of units.

considers that accretion of rock varnish should tend to inhibit rind erosion. Other potential difficulties such as lichen growth, fire, and animal disturbance seem rather unlikely for large boulders in the arid realm. Care was taken not to sample exhumed boulders, because rates of rind formation generally are different for surface and subsurface materials. Erosion of the Q1 surface has progressed to the stage where the petrocalcic soil horizon has been exhumed. Subsurface rinds were not sampled.

The results from Warm Springs Wash are sufficiently encouraging to warrant further evaluations of basalt weathering rinds in the southwestern deserts, particularly where rinds can be obtained from varnished outcrops that have been dated by K/Ar methods. Sample sizes were small (the largest being 38). Consideration of the potential problems just noted indicated that the most favorable sample localities would be the tops of the largest and densest boulders available. Such sampling is not realistic, however, because it can take half an hour to break a large basalt boulder with a 5-kg hammer. Fortunately, even the largest basalt boulders of hot deserts have a natural propensity for splitting in half. About 5 to 10 percent of the surficial boulders and cobbles have been split cleanly through the middle. These boulders of slightly vesicular to massive basalt provided an opportunity to collect samples of weathering rinds with a 4-kg sledge hammer.

Rind thicknesses were measured on smooth, convex, upper surfaces of boulders; undersides, corners, and depressions were avoided. Thicknesses were measured by using a 6X comparator containing a reticule graduated in 0.1–mm increments. Accuracy of measurements generally was ±0.2 mm, and thickness variations of a single rind commonly were ±0.3 mm. Each thickness measurement was assigned to a 0.5-mm-thick weathering-rind class.

Variations of weathering-rind thicknesses for the five ages of fans are compared in the set of bar graphs of Figure 2.28. Soil profiles were described for each of the five fans; soils and topo-

described by other workers in other areas. The most notable successes are those of Chinn (1981), Whitehouse et al. (1986), and Knuepfer (1984, 1988), who studied surficial greywacke cobbles on the South Island of New Zealand in a mean annual precipitation range of 1000 to 7000 mm (see Section 5.2.3.1). Mean annual rainfall at Warm Springs Wash is about 100 mm, which is only 40 percent of that for Colman and Pierce's driest site.

Attrition of rinds exposed at the surface might be sufficiently slow in some arid environments to make rind studies useful, particularly when one graphic characteristics were used to identify Q1, Q2a, Q2b, Q2c, and Q3b alluvial geomorphic surfaces. Each sample of weathering rinds shows a substantial range in thickness and each has a prominent modal thickness. For the youngest four surfaces, the modal thickness increases systematically with increasing age. All rind thicknesses are less than 2 mm for the Q3b and all exceed 2 mm for the Q1 population. The Q2a population is represented by a bimodal bar graph, and the bar graph for the Q1 suite of samples has a decrease in modal value when compared with Q2a. This apparent reversal in trend for the Q1 rind thicknesses may be the result of (1) erosion of the exposed rinds by granular disintegration or spalling of flakes, or both, followed by revarnishing of boulder surfaces; (2) exhuming of subsurface rocks during the past 200 to 300 ka that have subsequently developed weathering rinds and varnish; or (3) both processes.

Rates of weathering-rind formation on volcanic cobbles decrease exponentially with time and were described by logarithmic equations by Colman and Pierce (1981, Figures 17–19) for time spans up to 180 ka. Weathering-rind thickness versus time yields a significant semilogarithmic regression (Fig. 2.29). The relation seems to be consistent back to at least 500 ka. The intercept of the regression line with the zero thickness line suggests that at least 4 ka are needed to form incipient rinds.

The different modal values obtained for the segments of the Warm Springs Wash alluvial fan suggest the use of geomorphic relative ages described by ratios of weathering-rind thicknesses. For example, if the average rind thickness of the mode for Q3b (0.25 mm) is divided into the modal rind thickness for each of the other three samples, values of 7, 9, and 15 are obtained for Q2c/Q3b, Q2b/Q3b, and Q2a/Q3b rind-thickness ratios. Determination of similar ratios on other desert piedmonts might indicate similar ages of surfaces. One advantage of comparisons based on ratios is that the resulting dimensionless numbers may permit correlations even where rates of weathering-rind formation differ because of differences in surficial

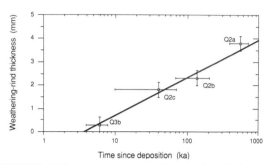

FIGURE 2.29 Increase in weathering-rind thickness for surficial basalt boulders for four ages of alluvial fans deposited by Warm Springs Wash, Black Mountains, western Arizona. Vertical bars indicate range of mode of rind thickness; horizontal bars are Table 2.13 age ranges.

wetting, salts, and dust between sites. In summary, it appears that the Warm Springs Wash data indicate a logarithmic model for weathering-rind development on surficial basalt boulders. The arid, hyperthermic climate and ubiquitous occurrence of rock varnish favor oxidizing conditions that promote rind growth and conditions of minimal rates of erosion of weathering rinds.

In marked contrast to rock varnish, weathering rinds appear to have a low sensitivity for recording fluctuations of late Quaternary climates for sites above a threshold moisture level. Knuepfer (1988) and other workers studying New Zealand weathering rinds on surficial cobbles (Section 5.2.3.1) found them useful over large areas because of minimal changes in rates of formation as a result of spatial variations in altitude, precipitation, and temperature.

*2.4.1.2.3 Sequence of Geomorphic Events* Another approach for dating soils and their associated alluvial landforms is to date climate-controlled changes in another subsystem of fluvial systems. For example, more than 1100 radiocarbon ages of plant fossils from packrat middens determine the times of major changes in plant communities of southwestern North America (Van Devender et al., 1987). This impressive data bank can be used to constrain assignments of maximum ages (dependent on response times; Section 2.5.4) for piedmont aggradation events that resulted from the hillslope sediment-yield increases associated with decreases of plant cover. For example, the 8- and 11-ka ages (Table 2.13) for the Q3b, Q3a, and Q2c geomorphic surfaces are based mainly on 100 radiocarbon ages of packrat middens in the desert mountain ranges (Fig. 2.6).

### 2.4.1.3 Correlated Ages

Correlated ages depend on age determinations on materials or events outside of the study region.

*2.4.1.3.1 The Astronomical Clock*    Variations in the earth's orbital parameters (Fig. 2.4) can be calculated with great precision for both the distant past and the future. Such potential as an ultimate age control has led to numerical ages being "fine tuned" against the *astronomical clock*. Chappell and Shackleton (1986) used $^{230}Th/^{234}U$ isotopic ages of fossil corals to obtain numerical ages of late Quaternary sea-level highstands, which were then fine tuned against the astronomical clock to give correlated ages of greater precision. Recent advances in K/Ar isotopic dating have led to revisions of the age of the Brunhes-Matuyama magnetic polarity reversal from 690 to 700 to 730 ka; all have substantial analytical uncertainties. R.G. Johnson (1982) used the astronomical clock to fine tune the age of the Brunhes-Matuyama magnetic reversal at 790 ± 5 ka. Thus the astronomical clock has considerable potential for refining age estimates of geomorphic events (Section 5.2.4).

*2.4.1.3.2 Paleomagnetism*    Samples were collected from sand pockets in gravel to evaluate the potential for dating the alluvium through studies of remnant magnetism characteristics. River terraces in the Yuma area that are younger than Q2a age all have normal polarity. Five samples from three localities were collected from the Q2a piedmont deposits in the Vidal area. All five samples had a normal polarity indicative of ages less than 790 ka.

*2.4.1.3.3 Tephra and Volcanic-Sediment Chronologies*
A Bishop Tuff locality was reported on the west side of the Coxcomb Mountains by Merriam and Bischoff in 1975. The Bishop Tuff is a widespread rhyolite ash fall in the western United States, and sanidine from pumice lapilli from three localities provided K/Ar ages of 0.7 Ma (Doell et al., 1966). A 2-m thick white lense of pure ash is exposed in a fanhead trench at the Coxcomb Mountain site. The Bkm horizon in gravels that cap the ash is similar in morphology and thickness to the Bkm horizons of Q2a age near the Colorado River. This <0.7-Ma age estimate for the Q2a is in agreement with the 0.6-Ma age for a basalt flow in Q2a gravels (Section 2.4.1.1.1)

The terraces of the Colorado River contain gravel derived from Quaternary volcanic rocks that provide a means of dating old piedmonts that were graded to former higher levels of the river. Volcanic mountain ranges near the river supply volcanic cobbles gradually to the river because of the slow rates and low competence of most hillslope processes in arid regions. A much different source of volcanic cobbles occurred where lava flowed into the river and provided an immediate and abundant source of volcanic gravel to be transported downstream.

Basaltic lava has flowed repeatedly into the Grand Canyon (McKee & Schenk, 1942) at Toroweap (fig. 2.30B). McKee et al. (1968) reported a K/Ar age of 1.16 ± 0.18 Ma for the basalt flow. W. K. Hamblin (Brigham Young University; written communication, October 23, 1975) estimated that a total of 1100 m of basalt flows has been eroded from a long reach of the canyon downstream from Toroweap during the past 1.2 my. It is presumed that large amounts of basaltic gravel survived transport from Toroweap and were deposited in the Q2 and Q3 river terraces.

The terraces of the lower Colorado River may be assigned to two general ages. Those terrace gravels that are essentially lacking basalt clasts are considered to be older than 1.2 Ma; those with basaltic gravel are considered to be younger than 1.2 Ma. The post-Q1 terraces of the Colorado

River contain gravel with more than 5 percent basalt clasts near Parker (Fig. 2.30A) but only about 1 percent basalt clasts north of Yuma. Volcanic clasts are abundant in all ages of terrace gravels downstream from the junction with the Gila River at Yuma. Upstream from this junction, most terrace gravels of Q1 age do not have basalt cobbles or the amounts are much less than 1 percent. The Q1 river terrace gravels northeast of the Big Maria Mountains contain less than 5 percent volcanic pebbles and cobbles, and andesite is a predominant volcanic lithology. Q2 river terraces at the same site have 10 to 25 percent volcanic gravel with abundant basalt.

The gravel lithologies indicate that most Q1 gravels were deposited previous to 1.2 Ma. This important correlated age constrains the Q2 terraces to being younger than 1.2 Ma.

### 2.4.1.4 Relative Ages

*2.4.1.4.1 Cobble Weathering Stages* Relative intensities of weathering of cobbles of different lithologies are useful for assigning relative ages to soils and alluvium. Mafic or coarse-textured cobbles weather more rapidly than do leucocratic or fine-textured cobbles. Hydrolytic alteration and expansion of biotite and weathering of calcium plagioclase cause changes in volume, which result in microfracturing and ultimately disintegration of rock into gravel or grus. Physical splitting of cobbles by accumulations of salts such as calcium carbonate, gypsum, and halite also is common in the lower Colorado River region. Features noted in the field included degrees of (1) preservation of abrasion surfaces and cobble shapes formed during stream transport; (2) abundance and depth of fracturing; and (3) intensity and depth of weathering as indicated by the sound emitted when the cobbles were struck by a hammer. Unweathered cobbles generally ring when struck, but strongly weathered cobbles emit a punky sound or crumble into monomineralic sand (grus). The four stages of cobble weathering outlined in Table 2.4 were noted for several cobble lithologies in desert pavements and B soil horizons. Cobble weathering is much more rapid in pavements than in subsurface horizons (Tables 2.8–2.11).

*2.4.1.4.2 Soil Profile Development* Indices are used to describe many phenomena, including relative development of soils. The soil-development index developed by Harden (1982) quantitatively measures degrees of soil-profile development by assigning numerical values for 4 to 10 soil properties for each horizon, and taking horizon thickness into account. These properties are clay films, texture plus wet consistence (total texture), rubification (color, hue, and chroma), structure, dry consistence, moist consistence, color value, and pH. Internally consistent indices can be calculated for most western United States soils chronosequences, with use of four parameters: texture, clay films, and dry consistence (all three emphasize amount and type of clay), and rubification.

The properties selected for computation of an index vary in their sensitivity to measure the effects of passing time. In the lower Colorado River, dry consistence increases rapidly in young soils and then stabilizes after about 50 ky. In marked contrast, films and bridges of clay around sand grains begin to form after a soil is 30- to 40-ky old. Then strength of development and abundance of clay films may increase for more than 1 my. Hue and chroma of soil color appear to increase progressively and can be used to class both young and old soils.

The soil-development index should be used with care because thresholds and feedback mechanisms of soil systems vary with soil microclimate. Vesicular horizons in deserts are not useful because they form too rapidly, and other soil characteristics change in a nonlinear manner (Figs. 2.15, 2.33). In humid regions, B-horizon soil rubification and A-horizon strength of structure commonly follow a pattern of initial increase followed by decrease. Nongravelly soils may have extremely variable rates of formation because of self-enhancing feedback mechanisms that occur as soil-profile development creates progressively poorer drainage (Dan & Yaalon, 1971; Section 3.2.1.2.2). Different

FIGURE 2.30 Identification and ages of Colorado River terraces. A. Surficial rocks on Q2a terraces southeast of the Whipple Mountains. Rounded quartzite cobbles identify the Colorado River as the stream that deposited the gravel. The rock hammer is on a boulder of locally derived basaltic andesite; the smooth vesicular boulder is presumed to be from the series of basaltic eruptions that flowed into the Grand Canyon at Toroweap. B. Vulcans Throne at Toroweap in the Grand Canyon has been the site of many fissure and cinder-cone eruptions that have poured large volumes of basalt into the canyon during the past 1.2 my.

modes and rates of soil formation change within short distances with differences in microclimate. The changes are functions of slope aspect and converging or diverging patterns of overland flow. The soil-development index empirical model works best for planar gravelly geomorphic surfaces. For more complex soils, such as those in cumulate colluvial settings, little may be gained by combining soil-profile characteristics into a single number.

The concept of a soil-development index centers about the pedogenic changes of parent materials that occur with time. Fortunately the soils of the lower Colorado River typically are thin, and streambank exposures commonly reveal good approximations of parent materials. Old soils in sub-humid to humid study areas (Chapters 4 and 5) typically are thick and Cox horizons may be greater than 10 m. For such soils B-horizon properties are compared with Cox-horizon properties, thereby producing indices that describe a lesser degree of contrast than that between horizon properties and true parent-material properties. Another way of estimating characteristics of parent material of soils in fluvial deposits is simply to assume that the parent material is similar to alluvium in adjacent stream channels. Three difficulties with this procedure are:

1. Most alluvium consists of sets of strata, each with different sedimentologic characteristics. During soil formation, stratification in the upper part of such deposits becomes mixed (bioturbated) and the resulting homogeneous deposit has characteristics that are a blend of several beds.
2. The proxy streambed alluvium may not be representative of those strata affected by soil-profile development because climatic changes can affect the lithology, size, and sorting of gravel and sand supplied to the streams from the hillslopes of a watershed.
3. The uppermost materials may be different from the underlying coarser grained materials because of deposition by floodplain rather than point-bar processes, or because the uppermost material

consists of loess or dust-impregnated gravel. Continuing additions of atmospheric dust may be thought of as progressively changing parent-material characteristics.

Field properties are normalized by quantifying them on a scale ranging from 0 to 1. The horizon index is obtained when the normalized properties are summed and divided by the number of properties. When the index for each horizon is multiplied by the thickness of that horizon, a summation of the horizon indices for the entire soil profile results in the profile-development index. Peter Birkeland divides the soil-profile development index by the maximum soil profile thickness in a study region in order to obtain normalized values (Birkeland et al., 1990, Fig. 2.1). This results in a weighted mean profile development index with values between 0 and 1 (maximum development) that are readily compared for a region.

Quantification of a horizon property (Xp), compared with the parent-material property (Xmin), involves several steps. For example, dry consistence may be assigned a 50-point range in five steps (Table 2.3 for abbreviations)

| Consistence | 0 | 10 | 20 | 30 | 40 | 50 points |
|---|---|---|---|---|---|---|
| | loose | soft | slightly hard | hard | very hard | extremely hard |

Maximum possible dry consistence (Xmax) is 50 points and gravelly parent materials are loose (0 points). To obtain the normalized value for dry consistence (Xpn), one divides the point value for the soil horizon by the maximum possible value (Xmax):

$$Xpn = \frac{Xp - Xmin}{Xmax} \qquad (2.6)$$

where Xmin = 0

If a Bt horizon has a very hard dry consistence, then:

$$Xpn = \frac{40 - 0}{50} = 0.80 \qquad (2.7)$$

In this manner a great variety of verbal descriptions of soil characteristics may be converted to numerical labels. However, because they represent diverse processes and rates of soil formation, these labels are difficult to use in rigorous statistical analyses.

Profile and maximum-horizon indices of soil development are shown in Figure 2.31 for the Whipple Mountains chronosequence (Table 2.7). The power function that describes both relations describes major decreases of rates of soil development with the passage of time (Harden, 1982; Harden & Taylor, 1983; Switzer et al., 1988).

Even though the approach is empirical, soil-development indices appear to be quite useful for estimating relative ages of piedmont surfaces of this arid to semiarid region. Index values are sufficiently similar, despite moderate differences in the climates, to allow the chronosequence of Holocene and late Pleistocene soils to be analyzed as a single population in the regressions of Figure 2.31. However, major differences in climate are reflected by major differences in regressions of soil-profile index and age (Fig. 6.1).

Maximum soil-horizon development indices are especially useful where erosion has partially truncated surface horizons (Chadwick et al., 1984; Hecker, 1985). Truncated horizons have had their thicknesses decreased by an unknown amount, so numerical values of horizon properties are minima. Where erosion is common, horizon indices for calcic horizons may be preferable to profile indices.

Q2 and Q3 soils on the Whipple Mountains piedmont have undergone minimal erosional truncation, so the Bt horizon was used as the maximally developed horizon for the Pleistocene surfaces. Some workers subdivide soil profiles more than others, so for consistency the maximum horizon index should use all parts of a major horizon such as a Bt or Bk horizon. These horizons are lacking in Holocene soils, so the Ck and Av horizons were used for the 7-ka and 0.2-ka soils, respectively. Figure 2.31 indicates that the values for maximum horizon index account for most of the values for the profile indices.

### 2.4.2 SUMMARY OF AGES

The preceding discussions outline a great variety of established and new techniques for dating chronosequences of stream terraces and alluvial fans on the desert piedmonts of the American southwest. The results of this work are summarized in Table 2.13 and 2.15, which show the age control presently available. Much more dating remains to be done to understand more fully the Quaternary history of the region.

The Q3a, Q3b, Q3c, Q4a, and Q4b geomorphic surfaces represent a continuum of deposition that has varied in space with a distinct, but brief, hiatus between the end of Q3a deposition and initiation of Q3b deposition. These surfaces may be regarded as the two main phases within the aggradation event that resulted from the Pleistocene–Holocene climatic changes.

Older aggradation events are separated by hiatuses. Of course the piedmont streams continued to flow throughout the Quaternary, but fluvial and pedogenic processes were only occasionally recorded as alluvial geomorphic surfaces. For much of the Quaternary, piedmont stream channels were incised below adjacent alluvial fans (Fig. 2.21) and conveyed their sediment load to trunk streams

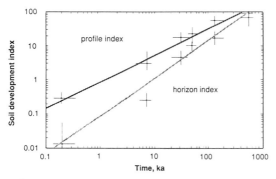

FIGURE 2.31  Soil profile and maximum horizon (Bt) indices for the Whipple Mountains chronosequences.

such as the Colorado River. A depositional hiatus of approximately 50 ky separates the Q2b and Q2c aggradation events.

Marked differences are readily apparent when heights above stream channels, degrees of dissection by streams originating on the surfaces, and soil-profile characteristics are compared for older surfaces (Tables 2.2, 2.7 and Fig. 2.9). A Q2a surface that is 10 m above the adjacent wash is shown in Figure 2.18, but Q2c and Q2b surfaces are only 1 to 2 m above the wash. This suggests that the Q2a is substantially older and that a long hiatus occurred between deposition of Q2a and Q2b deposits. Table 2.13 infers that prominent geomorphic surfaces have not been found for the estimated 0.2 my between general times of Q2a and Q2b aggradation, and for the estimated 0.5 my between general times of Q1 and Q2a aggradation. Although the assigned time gaps of Table 2.13 reflect obvious hiatuses, one should remember that old aggradation events had to be large depositional pulses to still be preserved as alluvial geomorphic surfaces. These pulses probably occurred during short intervals within the broad ranges of assigned times.

Deposits and terraces of desert fluvial systems record only a few of the climatic changes that have been defined by studies of marine fossils and sea-level fluctuations. The 18 sea-level highstands of the past 340 ky (Chappell, 1983) generally were associated with warmer-water fauna and with increases in northern hemisphere insolation, compared with the intervening lowstands. Thus marine subsystems seem much more sensitive to climatic changes than do the fluvial systems of the southwest deserts for a variety of reasons that are discussed in Section 6.4.3.

The Pleistocene–Holocene aggradation event (Q3a, Q3b) was much stronger than the preceding late Pleistocene aggradation event (Q2c), whose terrace deposits it tends to have buried. Apparently most Quaternary changes in airmass circulation had only minor effects on plant density and on water and sediment yield from arid and semiarid hillslopes. But several types of changes from full-

glacial to the Holocene interglacial climate had profound effects throughout the region.

Dating of alluvial geomorphic surfaces throughout a large region suggests that initiation of aggradation events coincided approximately with times of sea-level rise. The general assigned time spans of Table 2.13 acknowledge continued fluvial processes and allow for the possibility that aggradation surfaces for some fluvial systems may be diachronous. A set of ages from the deserts of southern Nevada and southern California (Table 2.15) indicates that climatic perturbations of the past 150 ky were strong enough to cause aggradation events at roughly 125, 55, and 10 ka. Two of these were during transitions from full-glacial to interglacial climates (Q2b and Q3), and one apparently was during a time of interstadial sea-level rise (Q2c).

A useful cross-check on the Pleistocene age assignments of the alluvial geomorphic surfaces of the study region is provided by late Quaternary lacustrine records. Stratigraphy, mineralogy, and 88 radiocarbon ages from the 100 km$^2$ Searles Lake salt pan in the northern Mojave Desert define general times of high and low lake levels (Smith, 1979). The presently arid Searles Lake basin received streamflow from the Sierra Nevada through a chain of lakes during times of larger effective precipitation. Searles Lake received such inflows only when all upstream lake basins overflowed. Chief dating uncertainties include apparent radiocarbon ages from samples of disseminated carbon that are 2 ka older than from adjacent samples of wood, and extrapolation of mud sedimentation rates for all lake beds older than 46 ka. Searles Lake lowstands are estimated to have occurred at about 150 to 130, 100 to 91, 53 to 40, and 10.5 to 0 ka; brief lowstands occurred at about 73, 64, and 27 ka. The Searles Lake record agrees fairly well with the marine record as interpreted by Chappell and Shackleton (1986), except for the major dry period ending at 130 ka (if an extrapolated sedimentation rate of 31 yr/cm were used instead of 33 yr/cm, the age would be 122 ka). Four major periods of lake desiccation occurred

TABLE 2.15 Estimated numerical ages for initiation of regional aggradation events in the deserts of southern Nevada and southern California.

| Aggradation Event | Place | Age (ka) | Dating Method | Reference |
|---|---|---|---|---|
| Q3b | Vidal, CA | >6 ± 1 | $^{230}$Th/$^{234}$U | Ku et al., 1979 |
| | Silver Lake Playa, CA | <8.7 | $^{14}$C | Wells et al., 1989 |
| Q3a | Silver Lake Playa, CA | >9.5 and <11.8 | $^{14}$C | Wells et al., 1989 |
| | Death Valley, CA | 11 | V/$^{14}$C [a] | Dorn, 1988 |
| Q2c | Cajon Pass, CA | 55 ± 8 | Calibrated fault slip [b] | Weldon and Sieh, 1985 |
| | Death Valley, CA | 55 | V/$^{14}$C [a] | Dorn, 1988 |
| | Vidal, CA | >61 ± 5 | $^{230}$Th/$^{234}$U | Ku et al., 1979 |
| | Yucca Mountain, NV | 55 ± 20 [c] | Uranium trend | Rosholt et al., 1985 |
| | Kyle Canyon, NV | 75 ± 20 [d] | $^{230}$Th/$^{234}$U | Sowers et al., 1989 |
| Q2b | Vidal, CA | >83 ± 10 | $^{230}$Th/$^{234}$U | Ku et al., 1979 |
| | Death Valley, CA | 125 | Cation ratio [e] | Dorn, 1988 |
| | Kyle Canyon, NV | 129 ± 6 [d] | $^{230}$Th/$^{234}$U | Sowers et al., 1989 |

[a] Corrected radiocarbon age of sample from basal layer of rock varnish using method described by Dorn et al. (1989). Age shown is oldest age of diachronous alluvial geomorphic surface on seven dated alluvial fans.

[b] Radiocarbon ages define the mean rate of horizontal displacement of the San Andreas fault, which can then be used to estimate ages of older offset fill terraces. See Section 4.4.2.

[c] Age shown is oldest age obtained for a diachronous? alluvial geomorphic surface.

[d] A parent-rock source problem for anomalous radionuclides may be present. $^{230}$Th/$^{234}$U age would be too old if thorium carbonate particles have been dissolved from limestone gravel clasts and translocated to sites of pedogenic carbonate accretion.

[e] Ratios of mobile versus immobile cations in rock varnish decrease with time (Dorn, 1983).

during the same interval in which the three major aggradation events (Q2b, Q2c, Q3) are believed to have occurred in the fluvial systems of the Mojave Desert. The three longest desiccation periods may have coincided with the three times of major piedmont aggradation. However, length of lake lowstand may not be indicative of strength of climatic perturbation affecting the desert hillslopes. The Searles Lake record appears to be intermediate in sensitivity—between the marine and fluvial systems—in regard to late Quaternary climatic changes caused by variations in the earth's orbital parameters.

The background information in the preceding sections describes characteristics of piedmont landforms and soils. Age control is fairly general but there is good internal consistency between a variety of dating techniques. It is readily apparent that the mappable units on the piedmonts of this large desert region span a million years instead of a few thousand years. It is equally clear that episodes of aggradation that spread new sheets of gravel across the desert piedmonts are infrequent events compared with the relatively common climatic variations suggested by marine records or by the astronomical clock.

## 2.5 Changes in Geomorphic Processes and Fluvial Landscapes Caused by Pleistocene–Holocene Climatic Change

Up to this point Chapter 2 has focused on many geomorphic and pedogenic variables of hills and streams that were affected by Pleistocene–Holocene climatic change. Now we are in a position to examine the behavior of specific subsystems. Surprising contrasts in the plants, alluvial sedimentology, and soil-profile development in glacial and interglacial times attest to major climate-controlled changes in the hot deserts of the American southwest. Paleoclimatic interpretations of these three subsystems are internally consistent.

### 2.5.1 ALLUVIAL GEOMORPHIC SURFACES

Massive aggradation of the valley floors and piedmonts seems to have been a rare geomorphic event, occurring only once every 50 to 300 ky. The Holocene aggradation event provides an unusual opportunity for insight into changing desert landscapes. The recency of this major climatic perturbation provides an ideal opportunity to date and examine the interactions of different components of desert fluvial systems. The contrasts between late Pleistocene (Q2c) and Holocene (Q3a and Q3b) outputs of watersheds are striking.

Latest Pleistocene to early Holocene alluvial geomorphic surfaces are products of distinct fluvial and pedogenic processes. They are representative of fluvial system outputs during major aggradation events. The sandy gravel of Q3 deposits is similar to the sediment of active washes; it represents source-area and transport conditions that provide substantial size reduction but little sorting during stream transport of gravelly detritus eroded from rocky hillslopes.

Late Pleistocene Q2c alluvial geomorphic surfaces were formed by much different fluvial and pedogenic processes than those of the Holocene. Pleistocene gravels also were deposited by braided ephemeral streams but they are markedly better sorted and finer grained than Holocene gravels. Sufficient soil and colluvium mantled the hillslopes to provide a moist microenvironment for

chemical breakdown of granitic and metamorphic rocks into grus and small rock fragments. The better sorting of the deposits and lesser rates of decrease of particle size in the downstream direction suggests more winnowing and less flashy streamflow than during the Holocene. These contrasts are graphically revealed in Figures 2.13 and 2.14. Scattered boulders in the late Pleistocene deposits indicate a large stream competence, but the general paucity of boulders suggests that the erosional subsystem supplied fewer boulders than during the Holocene. Although watershed lithology remained constant, the hillslope weathering products supplied to the streams changed with the changing climate.

#### 2.5.1.1 Stream Terraces

The piedmont geomorphic surfaces occur as terraces in valleys and as extensive sheets of alluvium deposited as thin alluvial fans (Figs. 2.21–2.24). The total thickness of climatic alluvial fans capping Tertiary deposits generally is 1 to 5 m thick, but increments of piedmont aggradation commonly are only 1 to 2 m thick (Fig. 2.17B).

All of the mountain ranges in the region have prominent Holocene fill terraces. Pleistocene terraces are not present. Most mountain valleys have prominent Q3a and Q3b terraces. Q3c terraces are much lower and are not as common; they should be considered as minor terraces. The regional extent of the Q3a–Q3b aggradation event and subsequent channel incisement suggests climatic controls of a regional nature. Valley-floor backfilling typically was less than 10 m, but more than 30 m of valley aggradation occurred in the Riverside Mountains and Mosaic Canyon in Death Valley.

Longitudinal profiles of Holocene fill terraces typically converge downstream. The Figure 2.32 watershed has an area of only 1.6 km$^2$, mean slope of 0.56, and relief ratio of 0.20. The Q3a terrace tread is 25 m above the stream channel at the start of the transect; it converges with the Q3b terrace tread and the stream channel. Sediment-yield increases from hillslopes have maximum impact on streams where stream power is least–in headwaters

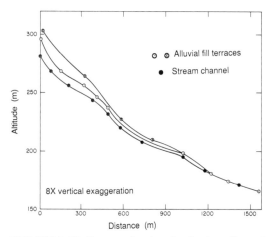

FIGURE 2.32 Converging longitudinal profiles of the present valley floor and the Q3a and Q3b Holocene fill terraces in the Riverside Mountains.

reaches. Maximum terrace height in Figure 2.32 is in the basin headwaters, which suggests that climatic changes on the hillslopes were responsible for the valley-floor aggradation event. The system has been tectonically quiet during the Holocene because the terraces sweep smoothly through reaches of potential earth deformation.

The fill terraces of the region record the following events:

1. Major Q3a valley-floor and piedmont aggradation increased stream gradients by 2 to more than 25 percent.

2. Stream-channel entrenchment removed much of the valley fill and entrenched the heads of the newly deposited climatic alluvial fans on the piedmonts.

3. Renewed aggradation resulted in inset valley fills in the mountains and burial of Q3a fans by Q3b deposits.

4. Renewed channel entrenchment created the Q3b terraces.

### 2.5.1.2 Effects of Climatic Change on Soils Genesis

Changes in late Quaternary soil-forming processes describe an important type of impact of climatic change in fluvial systems. Soils genesis was affected by changes in effective precipitation that were accompanied by changes in biota and influx of atmospheric dust and salts. Aridisols contain a mixed signature of present and past climates, because soil-profile formation is a cumulative process.

Typical Holocene pedogenic processes have consisted of varnishing or disintegration of surficial stones and illuviation of minor amounts of calcium carbonate on gravel clasts in the upper 5 to 20 cm of soils. Low intensity of subsurface hydrolytic weathering and shallow depths of carbonate illuviation both suggest that mid- and late-Holocene climates had an effective precipitation similar to that of the present arid, thermic to hyperthermic climate. Substantially greater depths of carbonate illuviation in the Q3a soils suggest a latest Pleistocene—early Holocene wetter or cooler climate, or both—a larger effective precipitation—compared with the present. Thus the characteristics of Holocene soils agree with the paleobotanical evidence (Figs. 2.6, 2.7) for maximal monsoonal rains during Q3a time. Even the Q3a carbonate illuviation depths were shallow compared with the Q2c soils that formed under much larger effective precipitation during times of much cooler full-glacial climates.

### 2.5.1.2.1 Influxes of Atmospheric Dust

Most stream-channel deposits and sediments beneath Holocene soil profiles are sandy gravel devoid of silt. Transformation of channels to bar-and-swale or smooth pavement topographies is aided greatly by heaving caused by wetting and drying of clay and silt in vesicular and argillic soil horizons. Nearly all the silt and clay in vesicular and argillic horizons is from atmospheric sources; negligible amounts are derived from incipient weathering of gravel.

Distant and local sources of atmospheric dust and salts represent an input into soils that may be regarded as one of continual change in parent materials from gravel to silty gravel. Sites such as the Cima volcanic field are downwind from Soda Lake bed, which is dry now but was wet during most of the Pleistocene when streamflow derived from the Transverse Ranges filled Pleistocene Lake Mojave (Enzel, 1989; Enzel et al., 1988; Wells et al., 1989). Cima soils receive occasional large influxes of atmospherically derived silt, clay and salts (McFadden et al., 1984,1986,1987; Wells et al., 1987), but soils downwind from more ephemeral lakes receive more frequent pulses of dust input. A major source of dust for much of the region is the lower Colorado River. Silt-charged overbank flows provided abundant silt, clay, and salts, as well as sand for dunes. The much-studied Whipple Mountains piedmont usually is upwind from the Colorado River, but occasional winds and rain from the southeast deposit silt and clay among the rocks of stable desert pavements. These are sites of minimal erosion where vesicular and argillic soil horizons can form. Occasional infiltration of water carries the silt into the layer beneath the pavement where it is filtered out by the subsurface deposits.

Changes in thickness of vesicular and argillic soil horizons for Whipple Mountains soils chronosequence are shown in Figure 2.33. Maximum thicknesses are used for the plot because of the susceptibility of these near-surface horizons to erosion. Air that is trapped during occasional times of infiltrating water expands when heated by the sun to form the vesicular structure (Evenari et al., 1974). Vesicular silt horizons develop rapidly and attain near maximum thicknesses in soils of the early Holocene geomorphic surfaces. This suggests that influx of atmospheric dust occurs mainly during times of interglacial climates and that vesicular-horizon development may be largely a Holocene pedogenic process.

Quaternary earth scientists think in terms of glacial and interglacial (or pluvial and interpluvial) climates, plant communities, and glacial and stream processes. The same general conceptual model—climate-controlled modes of system operation—should also be applied to temporal variations in soils genesis. Roger Morrison described episodes of rapid pedogenesis in the Lake Lahontan part of the Basin and Range Province in Nevada (Morrison, 1964; Morrison & Davis, 1984). Pleistocene soils seem to have formed during distinct, but widely separated, times of interglacial climates.

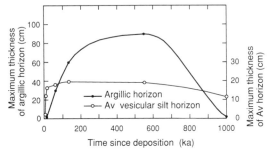

FIGURE 2.33 Variation of maximum thickness of the vesicular Av and argillic Bt soil horizons on the piedmont south of the Whipple Mountains.

Morrison believes that these geosols formed during times of warmer temperatures; pedogenesis at other times was minimal.

Recent workers have expanded on Morrison's concept of climatically controlled episodes of rapid soil formation by emphasizing the role of atmospheric dust (Yaalon & Ganor, 1973; Gerson & Amit, 1987; Reheis, 1987a,1990; Wells et al., 1987; McFadden et al., 1987; Chadwick & Davis, 1990). Chadwick notes that during times of major interglacial climates, such as the Holocene, permanent lakes evaporate and large amounts of silt, clay, and salts are blown from desiccated lake beds. Much of the atmospheric dust is deposited on the piedmonts and mountains that are downwind from the main sources, and the rest may be dispersed great distances. Major inputs of dust into stable piedmont surfaces occur episodically, but little transformation or translocation occurs until a subsequent climatic change increases the soil-leaching index. Thus arid Holocene soils typically have strongly developed Av horizons and weakly developed B horizons. The missing ingredient of effective soil moisture is supplied by the cooler or wetter climates of full-glacial times, which are much longer than the interglacial time spans. The combination of sufficient infiltration of water and abundant unweathered dust is conducive to episodes of rapid pedogenesis, which decline as unweathered dust that has accumulated in the uppermost soil horizons is weathered and translocated.

This fairly straightforward model explains some of the marked differences between soils of similar ages downwind from Pleistocene Lake Lahontan, and near Walker Lake in west-central Nevada which has persisted. Desiccation of 22,000 km$^2$ Lake Lahontan provided a great abundance of dust and salts to local desert piedmonts. Soils on the piedmonts adjacent to Walker Lake have minimal dust; even the Pleistocene soils on the upwind side are thin and weakly developed compared with those of piedmonts where abundant atmospheric dust has been an important pedogenic variable (Demsey, 1987).

Pre-Holocene episodes of dust production and enhanced pedogenesis can be dated where cumulate soils have formed on lava flows. More than 100 ka may occur between times of optimal pedogenesis in the Cima volcanic field of the eastern Mojave Desert (McFadden et al., 1984,1986; Wells et al., 1987).

*2.5.1.2.2 Development of Soil-Profile Horizons*  Differences in the late Quaternary soil-profile horizons of the region mainly reflect the passage of time and major climatic change. The late Pleistocene full-glacial climate appears to have been semiarid and mesic to thermic, as indicated by plants that grew with dated fossil pinyon pine collected from packrat middens. Q2c soils have argillic (Bt) horizons and the calcic (Bk) horizons generally start at depths of 25 to 35 cm, features that are indicative of a much larger leaching index than during the Holocene. Bt horizons indicate hydrolytic weathering of sufficient intensity to produce and translocate clay minerals and iron oxyhydroxides. Tops of Bk horizons at depths of less than 0.5 m reflect shallow depths of accumulation of atmospheric dust and salts that tend to lower infiltration capacity (Musick, 1975; McFadden & Tinsley, 1985; Wells et al., 1987). Q3a deposition between 11 and 8 ka coincided with the return of monsoon rains; pinyon had died out but juniper woodlands persisted. Q3a soils commonly lack Bt

horizons but have Bw and Bk horizons that extend to depths of 20 to 30 cm. Q3b deposition coincided with the initiation of arid, hyperthermic climates at about 8.5 to 7.8 ka that were typical of the remainder of the Holocene; coniferous plant communities changed to desert scrub. Q3b and Q3c soils commonly have weak Bk horizons at depths of 5 to 15 cm. These differences in development of soil-profile horizons reflect progressive decreases of effective precipitation.

The clay mineralogy of the Pleistocene soils contains paleoclimatic signatures indicative of well-drained sites and parent materials with intermediate to acid chemical composition (Barshad, 1966; Singer, 1980). Montmorillonite is common in soils of arid regions, and kaolinite, illite, and vermiculite generally are formed in greater abundance in semiarid regions. The limited chemical stability field for palygorskite is useful because it is common in all the Pleistocene soils. Palygorskite is stable only under conditions of alkalinity and high magnesium activity. The ubiquitous presence of palygorskite is an excellent indicator that the leaching index for Quaternary soils did not exceed that of a semiarid climate.

McFadden's (1982) analyses of samples collected from Q2c and Q3b pedons (described in Tables 2.9 and 2.8) provide an interesting comparison (Table 2.16) between latest Pleistocene and mid-Holocene soils. The total mass of illuviation products (in grams per square centimeter of pedon) reflects the cumulative effect of pedogenic processes with time. Total mass summations for the Q2c and the Q3b, respectively, are dithionite extractable Fe$_2$O$_3$, 0.18, 0.04; clay, 7.9, 1.5; CaCO$_3$, 6.6, 0.7; and organic carbon, 0.3, 0.06.

Latest Pleistocene (Q2c) soils show the effects of late Pleistocene and Holocene climates, but they developed their primary characteristics under a significantly wetter or cooler climate. The Bt horizon of Figure 2.17B (Table 2.9) has many medium prominent mottles of light yellow-orange calcareous silt. Illuviation of carbonate in an argillic horizon is indicative of a climatic polygenetic origin (Reheis, 1987a). The Q2c soil profile

TABLE 2.16 Laboratory data for mid-Holocene (Q3b) and latest Pleistocene (Q2c) pedons described in tables 2.8 and 2.9.[a]

| Horizon | Depth (cm) | Composition | | | Organic Carbon (%) | $Fe_2O_3$ (Dithionite) (%) | $CaCO_3$ (%) | Clay Compositions[b] | | | | | |
|---|---|---|---|---|---|---|---|---|---|---|---|---|---|
| | | Sand (%) | Silt (%) | Clay (%) | | | | Montmo- rillonite | Kaolinite | Illite/ Mica | Vermic- ulite | Chlorite | Palygor- skite |
| Holocene (Q3b) | | | | | | | | | | | | | |
| Avk | 0–4 | 46 | 32 | 33 | 1.2 | 1.1 | 7.6 | 3 | 2 | 2 | 1 | 1 | 0 |
| Ck | 4–14 | 83 | 13 | 4 | 0.4 | 0.8 | 9.5 | 3 | 2 | 1–2 | 1 | 2 | 1 |
| Cu | 14– | 89 | 9 | 2 | 0 | 0.8 | 7.0 | 3 | 2 | 2 | 1 | 2 | 1 |
| Latest Pleistocene (Q2c) | | | | | | | | | | | | | |
| Avk | 0–8 | 17 | 52 | 21 | 1.4 | 1.0 | 9.6 | 4 | 2 | 1 | 1 | 1 | 1 |
| Btk | 8–13 | 31 | 46 | 22 | 1.4 | 1.0 | 7.0 | 3–4 | 2 | 2 | 1 | 1 | 1 |
| Btk2 | 13–21 | 43 | 25 | 32 | 0 | 0.9 | 2.4 | 2–3 | 2 | 1–2 | 1–2 | 1 | 1–2 |
| Btk3 | 21–30 | 54 | 20 | 26 | 0 | 0.6 | 11.4 | 2–3 | 2 | 1 | 1 | 2–3 | 3 |
| Bwk | 30–52 | 65 | 21 | 14 | 0 | 0.5 | 9.7 | 2 | 2 | 1 | 1 | 2 | 3 |
| Bk | 52–56 | 82 | 12 | 6 | 0 | 0.4 | 3.6 | 3–4 | 2 | 1 | 1 | 2 | 2 |
| IICkm | 56– | | | | | | | | | | | | |

[a] Analyses by L. D. McFadden, 1982.

[b] 0, not detected; 1, trace (<10%); 2, small (20% ± 10%); 3, moderate (40% ± 10%); 4, abundant (60% ± 10%); 5, predominant (>70%).

with its Bt horizon is buried by Q3a alluvium about 300 m from the described pedon. Carbonate mottling is not present in a Bt horizon above a thin Bk horizon. Comparison of the unburied and buried soil profiles indicates that only minor additions of carbonate occurred during the Holocene. Holocene carbonate accumulated distinctly above the depth of late Pleistocene carbonate illuviation to produce the mottles in the Bt horizon of the unburied soil. McFadden's (1982) study of the arid Vidal Valley soils shows that the Bt horizons of the Q2c soils were leached of parent-material carbonate during the late Pleistocene. His calculated mass summations revealed only 1.0 to 1.4 g/cm$^2$ of CaCO$_3$ in the Bt horizon. This carbonate was illuviated during the Holocene as a result of a reduction in the leaching index from about 23 to 10 mm. Q3a soils with noncalcareous Bt horizons continue to form in semiarid parts of the Mojave Desert such as the Rodman Mountains, where the leaching index is about 36 mm.

Coarse prominent mottles, white carbonate nodules, and carbonate pebble coatings in the Btk horizons of the study region attest to the polygenetic nature of old soil profiles. The thickest carbonate pebble coatings occur in Btk horizons where isolated growth of the coatings occurs as pendants on bottoms of cobbles. Bk-horizon development is a combination of accretion of carbonate on cobbles and pebbles and massive plugging of matrix and interstices with carbonate. Carbonate illuviation probably occurs in Btk horizons during times of arid climate and in Bk horizons during times of semiarid climate because of shallower leaching depths. Although dense, thick carbonate pebble coatings occur in Btk horizons it is preferable to collect pebble coatings for $^{230}$Th/$^{234}$U dating from the Bk horizons. Btk horizons have been subject to alternating accumulation and leaching of soil carbonate as Quaternary climates alternated between arid and semiarid.

Much more time is needed to develop Bt horizons than Av horizons. Figure 2.33 suggests that Bt horizons have not formed during the arid thermic and hyperthermic climates of the Holocene but

that the rates of argillic horizon development were rapid during the semiarid mesic climate of the late Pleistocene. Although a smooth curve is shown for the cumulative increase in thickness of the Bt horizon with age, the rates probably varied from extremely slow during arid times to moderately rapid during semiarid times. Maximum thicknesses of argillic horizons may be about 1 to 1.5 m, because stability decreases as dissection becomes dominant. Degradation rates have exceeded Bt horizon development rates for most Q1 surfaces; argillic horizons have been eroded and continuing instability has prevented renewed formation.

*2.5.1.2.3 Translocation of Calcium Carbonate*    Rates of calcium carbonate accumulation in gravelly soils of the arid, hyperthermic parts of the study region are slow. More than 2 ky are needed to accumulate the first hints of thin, discontinuous pebble coatings. Discontinuous coatings occur in soils as old as about 8 to 11 ka, but some Q3a soils have continuous carbonate pebble coatings. Discontinuous coatings generally are thicker on the bottoms of pebbles, but soil carbonate occurs mainly on tops of pebbles and above depths of 15 cm in Ck horizons of Q3b soils on the piedmonts of the Aguila, Mohawk, and Gila mountains of southwestern Arizona.

Development of the continuous carbonate pebble coatings in the Colorado River region may have required much of the past 20 ky. The mass of carbonate needed to fill the interstices is much larger, and roughly 100 ky may be needed to cement gravels with pedogenic carbonate; and 400 ky may elapse before a capping laminar horizon starts to form. It appears that carbonate accumulates about three times more slowly in the soils of the lower Colorado River region than in the soils of the Rio Grande region (Table 2, Machette, 1985).

The abrupt contrast between the amounts and morphologies of illuviated carbonate of Holocene and late Pleistocene soils (Tables 2.8, 2.9) suggests that the late Pleistocene was a time of substantially more rapid and deeper illuviation of carbonate. Calcareous matrix materials and continuous pebble coatings are typical of late Pleistocene carbonate. The thickness of coatings and amounts of matrix carbonate increase substantially with increasing soil age (Fig. 2.34). Accretion of carbonate on subsurface pebbles and cobbles by infrequent wettings is slow and in the Bk horizon may be considered an irreversible pedogenic process (Yaalon, 1971). The morphology of the accreted calcite crystals appears to be influenced by climate (Chadwick et al., 1988). Overall rates of carbonate illuviation in the aridisols of Vidal Valley (McFadden, 1982; McFadden & Tinsley, 1985; Machette, 1985) and near Albuquerque, New Mexico (Machette, 1978,1985) remain constant, as is shown by their calculations of total masses of $CaCO_3$ in 1 $cm^2$ soil columns. The linear rate may reflect the importance of dust additions (Straknov, 1967; Reheis, 1990).

Rates of accretion of carbonate on gravel clasts in noncarbonate parent materials provide interesting comparisons between several areas in the southwestern United States. $^{230}Th/^{234}U$ ages were determined for the outer and inner portions of a pebble coating from the Bk horizon of a Q2b soil on the Whipple Mountains piedmont (Ku et al., 1979). The difference in the two ages indicates a mean maximum rate of carbonate accretion of about 0.11 mm/ky for carbonate that accumulated in both arid and semiarid climates. During arid climates such as the Holocene, the mean rate of carbonate accretion on pebbles is about 0.05 mm/ky. Dating of soils in the Tule Springs archaeological site in southern Nevada indicates a mean Holocene carbonate coating accretion rate of 0.07 to 0.13 mm/ky (Vance Haynes, University of Arizona; oral communication, October 6, 1975). Data from the Desert Soil-Geomorphology project of the U.S. Soil Conservation Service in southern New Mexico suggest average Holocene rates of about 0.2 mm/ky (Gile et al., 1979). Radiocarbon dating of the Holocene rate of carbonate accumulation on the bottom of a basalt boulder northeast of Flagstaff, Arizona, where annual precipitation

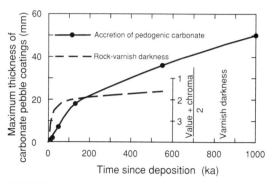

FIGURE 2.34 Changes in maximum thickness of carbonate pebble coatings for the geomorphic surfaces of the lower Colorado River region. (See Table 2.13 for age control.)

is about 450 mm, suggests a rate of 0.5 mm/ky (Sigalove, 1969). It seems that a 10-fold spatial variation of rates of carbonate accretion occurs in the gravelly soils of the southwestern United States. Marion (1989) notes that long-term accretion of total mass of pedogenic carbonate correlates well with modern values of mean annual precipitation.

The Holocene was the time of maximum illuviation rates of carbonate in much of western North America (Bachman & Machette, 1977; Machette, 1985; Reheis, 1987a,b). Four sites in New Mexico and Utah have average rates of pedogenic carbonate accumulation of 0.14 to 0.51 $g/cm^2/ky$ (Machette, 1985). Higher rates are associated primarily with Holocene increases of inputs of solid calcium carbonate in atmospheric dust and of $Ca^{++}$ dissolved in rainfall. Holocene accumulation rates in arid to semiarid New Mexico appear to be twice the semiarid to subhumid late Pleistocene rates. However, an opposite relation prevailed in the more arid, hyperthermic parts of the southwestern deserts.

Rates of accumulation of secondary carbonate in soils are mainly a function of influx of calcium ions and translocation of carbonate in surficial materials to deeper horizons (Fig. 2.35; Machette, 1985). The line separating the influx-limited and moisture-limited domains defines optimal conditions for soil-carbonate accumulation. Informed estimates of rates are based on stage of soil-carbonate morphology (Table 2.4) for soils of different late Quaternary ages. The effect of the Pleistocene–Holocene climatic change is dia-

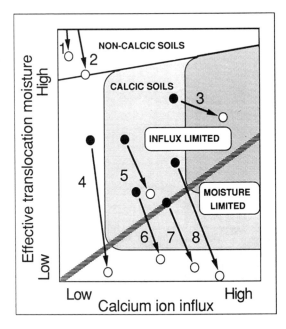

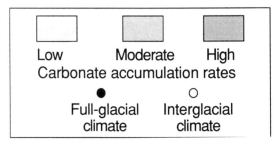

grammed with arrows between full-glacial and interglacial points. Estimates of Holocene moisture conditions are based mainly on present climates, for soils of the same general texture, infiltration capacity, and infiltration rate. Lengths of arrows and estimates of Pleistocene moisture conditions are based mainly on soil characteristics. These include the presence or absence of carbonate-free A or B horizons, depths to the top of the Pleistocene and the Holocene calcic horizons, and polygenetic soil-profile development indicated by features such as carbonate mottles in argillic horizons.

The arrows of Figure 2.35 refer to calcic-soil sites discussed in this book. Holocene accumulation rates increased in the Rio Grande Valley of New Mexico (arrow 3) but decreased in the lower Colorado River region (arrows 6, 7), which is in the moisture-limited domain. Decreased effective precipitation during the Holocene in the San Gabriel Mountains (arrows 1, 2) was insufficient to initiate development of calcic horizons. Extremely arid to arid regions (arrows 4, 8) have exceptionally thin Holocene soils with minimal carbonate; however, thick carbonate-free argillic horizons attest to much different Pleistocene climates. Some late Pleistocene soils on granitic pediments in the western Mojave Desert also have minimal carbonate (Boettinger and Southard, 1989). Schlesinger (1985) concluded, on the basis of $^{14}$C and $^{230}$Th/$^{234}$U ages of calcic horizons in the area southwest of the Coxcomb Mountains, that a major episode of carbonate illuviation occurred in the eastern Mojave Desert at about 20 ka. The Coxcomb site

FIGURE 2.35 Domains of influx-limited and moisture-limited accumulation of secondary soil calcium carbonate whose accumulation rates are largely dependent on calcium ion influx from atmospheric sources and the effectiveness of soil moisture in translocating carbonate. Arrows indicate general trends from times of full-glacial to interglacial climate. Key: 1 and 2, subhumid and semiarid parts of the San Gabriel Mountains (Chapter 4); 3, Las Cruces, New Mexico; 4, sites presently in the rain shadow of the Coast and Transverse Ranges of southern California such as the Anza Borrego part of the Salton Sea trough; 5, Rodman Mountains of the central Mojave Desert; 6, Whipple Mountains and Coxcomb Mountains piedmonts of the eastern Mojave Desert; 7, Gila Mountains of southwestern Arizona and the drier parts of Death Valley, California; 8, drier parts of the Negev-Sinai Desert (Chapter 3). Adapted from Figure 5 of Machette, 1985.

may be represented by arrow 6, but the Rodman Mountains site (McFadden, 1982; McFadden & Tinsley, 1985) lies wholly within the influx-limited domain and does not appear to have undergone major increase of carbonate-accumulation rate.

Recent studies of stable carbon and oxygen isotopes in pedogenic carbonates are providing new insight into late Quaternary plant communities and climates. Soil samples were collected along altitudinal transects where variations of present plants, isotopes, and climate can be analyzed (Cerling et al., 1989; Quade et al., 1989; Amundson et al., 1989a,b,c).

Modeling of the distribution of carbonate in soils is a powerful tool for evaluating the relative importance of several concurrently changing variables. One approach is to simulate soil-profile development as an open-system stack of 1-cm compartments with different values for temperature, available water-holding capacity, calcium carbonate content, and partial pressure of carbon dioxide (Rodgers, 1980; McFadden & Tinsley, 1985; Mayer et al., 1988). Mayer and others estimated values of available water-holding capacity on the basis of textural changes in soils of different ages that occur largely as a result of influxes of atmospheric dust and of soils genesis. Their measurements and model for the carbonate distribution in a 6-ka soil (Reheis et al., 1989) in Q3b piedmont gravels near Silver Lake (Fig. 2.1) are shown in Figure 2.36. The excellent correspondence between the modeled and actual distribution of soil carbonate indicates that inputs to the model are realistic, including the assumption that the present climate approximates mid- and late-Holocene climates.

In summary, weathering and soil-forming processes are extremely varied and differ greatly in their responses to changing precipitation and temperature. Late Quaternary soils also have been profoundly influenced by variations in inputs of atmospheric dust and salts. Soil-profile-development indices seem to reflect mainly the passage of time, until one compares them for greatly different climates (Fig. 6.1). Accumulation of pedogenic

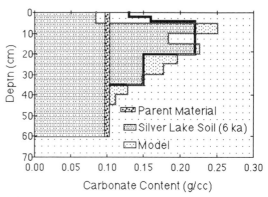

FIGURE 2.36 Modeled and actual distributions of calcium carbonate in a Q3b soil at Silver Lake. Areas between distributions and parent-material line represent total masses of pedogenic carbonate (g/cm$^2$); the total mass divided by the age of the soil is the mean carbonate flux (g [cm$^{-2}$] yr$^{-1}$). From Figure 2 of Mayer and colleagues (1988).

clay, iron oxides, and carbonate probably vary substantially with climatic variations, but the horizons that we sample represent summations of accumulation rates. Holocene and buried late Pleistocene soils are the least likely to be polygenetic; they may be the most representative of arid and semiarid modes of soil formation. Accretionary processes of rock varnish formation have great potential for recording and dating climatic histories with layered variations of iron and manganese abundance, leachable cation abundance, and carbon isotopes.

### 2.5.2 PROCESS-RESPONSE MODELS

#### 2.5.2.1 Biogeomorphic Response

Through profound influences on hillslope plant cover, geomorphic responses to climatic change affect processes on hills and in streams of humid and arid regions. Abrupt changes in late Quaternary climate caused changes in bedload transport rate and stream-channel pattern in Texas (Baker & Penteado-Orellana, 1977,1978). Knox (1972, 1983,1985,1987; Knox et al., 1978) analyzed the influence of climate-controlled changes in vegetation cover on sediment yield and aggradation and degradation of flood plains in southwestern Wisconsin.

Knox illustrated adjustments to climatic change in subhumid to humid regions with biogeomorphic response curves. Airmass circulation patterns may change abruptly at seasonal or longer intervals (Bryson & Wendland, 1967) and thus may be described by line A of Figure 2.37. For spans longer than lifetimes of plants, the density of plant cover protecting soil from erosion is described by line B. Knox's sediment-yield curve, line D, shows pronounced increases in sediment yields immediately after changes from semiarid to humid climates. Larger amounts of precipitation initially would fall on hillslopes with minimal vegetation cover to inhibit erosion. Subsequently, sediment yields would decrease. Conversely only a temporary reduction in sediment yield would occur at the onset of a semiarid period, after which sediment yields would increase rapidly as vegetation

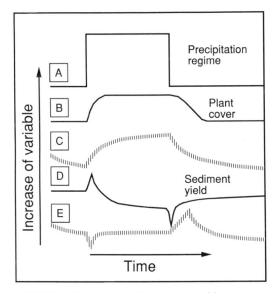

FIGURE 2.37 Models of biogeomorphic response of hillslope subsystems to abrupt changes in climate. Solid lines are for Wisconsin (from Knox, 1972), and dashed lines are for the rocky hillslopes of the southwestern deserts.

density decreased and soil became more exposed to erosion.

Biogeomorphic response curves for the southwestern deserts are opposite those of humid regions, as is shown by the dashed lines of Figure 2.37. A major difference between the two study regions is the availability of sediment to be eroded from hillslopes. Unlimited amounts of material appear to be available for erosion on the Wisconsin hillslopes. In the southwestern deserts the volume of hillslope material that can be eroded is limited and also changes with time as a climate-controlled response. Line C indicates that vegetation cover gradually decreases during arid times, whereas line B for a humid region is dominated by a horizontal line indicative of equilibrium conditions. Like the humid curve, the arid curve shows an immediate sharp decline in vegetation cover that is the result of an abrupt decrease in effective precipitation. Then vegetation cover continues to decrease because of the gradual net reduction of hillslope soil. The opposite is true when the climate changes from arid to semiarid. The rapid initial increase in vegetation cover that results from increased effective precipitation is followed by a gradual increase as weathering produces increasingly more soil to retain water for plants.

Changes in sediment yield—line E of Figure 2.37—reflect the interactions of desert vegetation and precipitation. When the climate changes from semiarid to arid, the concurrent decrease in vegetation cover results in a rapid increase in sediment yield. The sediment-yield maximum is attained quickly, after which the yield progressively declines as the area of hillslope colluvium decreases and outcrop area increases. The reverse occurs when the climate changes from arid to semiarid. The initial decline in sediment yield results from small areas of soil that remain from the preceding arid time and undergo a rapid increase in vegetation cover, but the much slower process of weathering of exposed outcrops into soil has yet to affect most of the slopes. With increasing time, weathering gradually converts rock to soil, which in turn is stabilized by increasing plant cover. Line

E approaches a steady-state condition, indicated by the horizontal dashed line that describes uniform rates of sediment yield, because all the outcrops have become covered with soil.

The preceding discussion of the complexity of different sediment-yield processes in arid and semiarid regions points out the futility of using general curves that relate sediment yield to climatic indices involving runoff or effective precipitation. The interactions between effective precipitation, plant cover, and sediment yield shown by the Langbein-Schumm sediment-yield curve (1958) portray logical increases and decreases of sediment yield with increasing effective precipitation; peak sediment yield is at about 300 to 400 mm of mean annual precipitation. These changes reflect accompanying changes in density of protective hillslope plant cover. An implicit assumption in using the Langbein-Schumm sediment-yield curve is that each data point represents a fluvial system that is in equilibrium. Lack of equilibrium may explain why some studies reveal sediment yields that are much different than the Langbein-Schumm relation (Douglas, 1967; Eybergen & Imeson, 1989).

Where hillslope equilibrium conditions do not exist one can expect sediment yields to be changed by self-enhancing feedback mechanisms even when climate remains constant. Another important consideration is whether valley floors are aggrading or degrading as a result of geomorphic responses to climatic change. Power-function equations relating sediment yield to drainage basin area may have larger negative exponents where part of the bedload is deposited on valley floors, and may have less negative exponents or even positive exponents (Church & Slaymaker, 1989) where streams obtain part of their load from alluvium deposited during earlier times of depletion of hillslope sediment reservoirs. In these ways, the heritage of past climatic change affects assessments of modern sediment-yield controls.

### 2.5.2.2 A Hillslope Process-Response Model
The marked contrasts between glacial and interglacial sediment yields probably were matched by an equally notable sequence of events on the hillslopes from which bedload was derived. Erosion rates were less than weathering rates during the late Pleistocene because plant cover was sufficiently dense to favor accumulation of colluvium. Mountain landscapes probably consisted of steep slopes with scattered bedrock outcrops rising above extensive sheets of colluvium and locally abundant talus. Valley floors had only a few meters of alluvial fill, or streams flowed on bedrock.

The change to drier or warmer climates during the latest Pleistocene decreased vegetation protection for hillslope colluvium that had accumulated during an extended period of semiarid climate. Hillslope sediment yields were greatly increased—partly because of increased rainfall intensities associated with the return of monsoon thunderstorms. Valleys became choked with alluvium in a manner first described by Huntington (1907,1914) in central Asia.

The aggradation event that was associated with the Pleistocene–Holocene climatic change consisted of two main pulses (Q3a and Q3b) with an intervening brief period of stream-channel incision. Recognition of these two pulses is possible only because they occurred recently. Aggradation events with ages of more than 100 ka generally are described as a single pulse, although they too may have been multiphase. The following model is one possible scenario for Q3 aggradation.

Geomorphic processes changed in hillslope subsystems as a result of Pleistocene–Holocene climatic change (Fig. 2.38). Decrease in precipitation or increase in temperature reduced the moisture available to support hillslope plants. The resulting reduction in vegetation density decreased infiltration rates and exposed more soil to raindrop erosion. Both runoff of water and concentration of entrained sediment increased for precipitation events of a given amount and intensity. Sources of coarser grained sediment became available where removal of colluvium exposed underlying joint blocks of bedrock. Increases in the amount and size of bedload increased resisting power so much that streamflow behavior was driven far to the aggra-

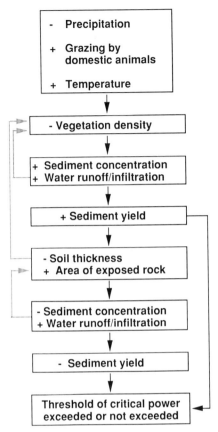

```
    - Precipitation

    + Grazing by
      domestic animals

    + Temperature
                │
                ▼
    - Vegetation density
                │
                ▼
    + Sediment concentration
    + Water runoff/infiltration
                │
                ▼
    + Sediment yield
                │
                ▼
    - Soil thickness
    + Area of exposed rock
                │
                ▼
    - Sediment concentration
    + Water runoff/infiltration
                │
                ▼
    - Sediment yield
                │
                ▼
    Threshold of critical power
    exceeded or not exceeded
```

FIGURE 2.38 Flow diagram showing increases (+) and decreases (−) of variables for rocky hillslope subsystems of the southwestern deserts. Self-enhancing feedback mechanisms are shown by dashed lines with arrows.

dational side of the threshold of critical power. This strong tendency for aggradation occurred despite concurrent increases in stream power caused by greater peak discharges associated with the return of monsoonal rains, and despite maximum valley aggradation in headwaters of drainage nets that increased streambed slopes (Fig. 2.32). Continuing decreases in soil thickness and concurrent increases in areas of exposed bedrock caused still more rapid runoff of water, but sediment concentration decreased as the sediment supply decreased (Fig. 2.38). The resulting decrease in sediment

yield was important because this decrease in resisting power occurred at a time when stream power had been increased by deposition of steep valley fills. The critical power threshold was crossed and erosion of the valley fill began.

Self-enhancing feedback mechanisms (Fig. 2.38) were important in perpetuating net removal of the hillslope soil mantle regardless of whether the valley floor was aggrading or degrading. Increased flashiness of runoff continued to decrease soil thickness and increase area of outcrops, both of which resulted in continued decreases in soil moisture and vegetation density. This model for the operation of the hillslope subsystem does not tend toward a steady state; indeed the stark mountainsides continue to become progressively more barren. These continuing hillslope responses to climatic change also drive the stream subsystem progressively farther to the erosional side of threshold conditions. The strength of the self-enhancing feedback mechanisms described in Figure 2.38 is important in determining the frequency of major aggradation events (Section 6.4.3).

Simple models may not be applicable for landscapes affected by climatic change. For example, stream-channel entrenchment commonly is ascribed to a change to either wetter or drier climate. The process-response model of Figure 2.38 indicates that a single climatic change resulted in both valley-floor alluviation and subsequent channel downcutting as self-enhancing feedback mechanisms changed stream power and resisting power. Although increases in gradient and peak streamflow magnitude increased stream power, changes in resisting power resulting from increase in bedload amount and size were even larger and occurred rapidly.

The Holocene fill terraces can be explained by the hillslope process-response model; they reflect climatic changes since 12 ka. Maximum valleyfloor aggradation occurred with deposition of Q3a fill terraces (Fig. 2.32) at about 11 to 9 ka, probably as a result of maximum availability of hillslope sediment that had been accumulating for about 50 ky. The paleobotanical record provided by packrat midden studies showed that another major climatic change was yet to come—the second phase of the Pleistocene–Holocene climatic transition.

The change to a markedly more arid and hyperthermic climate and locally to an extremely arid climate occurred at about 8.5 to 8 ka. This decrease in effective soil moisture triggered an acceleration of hillslope erosion that once again switched the stream subsystem from a degradational to an aggradational mode of operation. The resulting surge of sediment resulted in deposition of the Q3b valley fills and alluvial fans, which became the dominant Holocene piedmont surface in the region. Q3a deposits are not widespread on the piedmonts because they were buried by thin sheets of Q3b detritus.

The relative magnitudes of the effects of the climatic perturbations associated with the Q3a and Q3b aggradation events is indicated by the heights of the two fill terraces. In about two-thirds of the canyons the Q3a terrace stands above the Q3b terrace. Either the effects of the Q3b climatic impact were less, or the Q3a episode of hillslope stripping had removed much of the hillslope sediment reservoir. Q3b fill was more likely to bury Q3a fill in watersheds where lithologic controls favored thick hillslope colluvium (Section 3.2.2).

The sediment-yield increase associated with the Q3c deposition was much less than those of the Q3a and Q3b events. Small Q3c terraces are inset into valleys cut into Q3b piedmont deposits. Q3c aggradation should be considered as minor; several low aggradation or degradation terraces are typical. They either resulted from a minor climatic fluctuation or from complex responses (Section 1.7). Continuing depletion of the hillslope sediment reservoir has favored stream-channel downcutting (with concurrent decreases in gradient). Streams now are cutting into bedrock in most mountain and many piedmont reaches.

All of the past 12 ky were needed for the fluvial systems to adjust to the Pleistocene–Holocene climatic change. Climate-induced aggradation and degradation reflect the entire spectrum of rainfall-runoff events. In the next section we examine evidence for extreme events—the paleofloods of the late Quaternary.

### 2.5.3 EARLY HOLOCENE PALEOFLOODS

Most of the Holocene alluvium was deposited between 11 and 6 ka. The late Holocene has been characterized by removal of fills from valley floors by streamflows derived from rocky hillslopes, with progressively fewer remaining sources of fine-grained sediment.

Major flood events–with either water or debris-flow characteristics–occurred during the presumed latest Pleistocene–early Holocene maxima of monsoon thunderstorms. An index of sediment concentrations of Holocene streamflows is the proportion of water-laid and debris-flow deposits. A visible decrease in abundance of debris-flow beds occurs toward the top of Q3a Holocene valley fills in the Riverside Mountains and in northern Death Valley National Monument. Debris-flow beds are not common in most Q3b deposits. For metamorphic volcanic ranges such as the Whipple Mountains, water was the dominant mode of sediment transport throughout the Holocene. In the granitic Gila Mountains, debris flows were common during the early Holocene (Gerson et al., 1978). The debris flows built U-shaped channels and levees topped with boulders as long as 20 m, and alternated with water flows in aggrading valley floors. The early Holocene debris-flow levees have little remaining matrix. Exposed matrix has been raindrop splashed and washed away, leaving an open framework of weathered and brown to brownish black varnished boulders (Fig. 2.39B). Debris flows rarely occur now. Favorable conditions for initiation of debris flows such as steep slopes, sparse vegetation, and intense monsoon thunderstorms are present now, but abundant fine-grained hillslope colluvium—an essential ingredient for debris-flow matrix—has been removed from the hillslope sediment reservoirs. The Q3a debris flows are one of two types of deposits that record return of monsoon rains to the southwest deserts.

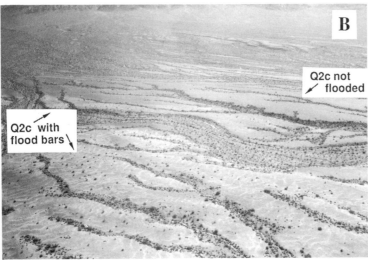

FIGURE 2.39 Types of flood deposits associated with the return of monsoon thunderstorms to the southwest deserts at the end of the Pleistocene. A. Bouldery debris-flow levee of Q3a age on a small alluvial fan derived from Sheep Mountain in the Gila Mountains. The 0.5- to 2-m granitic boulders are varnished and the exposed debris-flow matrix has been washed away. From Gerson et al., 1978, Figure 4.2. B. Oblique aerial view looking upslope on piedmont of the Eureka Mountains, southeastern California. Late Pleistocene (Q2) desert pavements in foreground are covered by sheetflow bedforms, but pavements of the same age that are closer to the mountains lack the flood-produced features. The upslope part of the piedmont consists mainly of Holocene deposits that have buried the Pleistocene surfaces. Photo courtesy of John Dohrenwend.

Exceptionally large streamflows during the early Holocene overtopped banks of partly aggraded stream channels inset into Q2 alluvial fans and spread out as major sheetfloods. Relict sheetflood bedforms (Fig. 2.39A) are most common down-slope from piedmont midslopes; upslope channels generally had sufficient capacity to contain these floods (Wells & Dohrenwend, 1985). The ripple-like features are narrow, widely spaced, trans-verse-to-slope bands of fine gravel and coarse sand with mean particle sizes of 2 to 8 mm.

Wells and Dohrenwend estimated paleoflood velocities of 0.3 to 0.6 m/sec and depths of 4 to 10 cm. Dunes and antidunes indicate that both upper and lower flow regimes were present. Shallow depths and low flow velocities may have minimized erosion of inundated Pleistocene surfaces. Pavement gravels were entrained, transported a short distance, and deposited as small-scale and large-scale bedforms whose clasts have degraded preflood rock varnish and weakly developed postflood varnish.

Ages of the youngest piedmont surfaces that were overtopped by the paleofloods constrain the maximum age of flooding. The youngest surfaces have Avk, Bw, and weak Bk soil horizons (McFadden et al., 1984) characteristic of Q3a soil development. Local sheetfloods of lesser magnitude have continued to the present (McGee, 1897).

Debris flows and paleofloods are interesting processes in the history of landscape change in the study region; they represent but one response to climatic change. Variables in each fluvial system had their own time lags of response to the Pleistocene–Holocene climatic change.

### 2.5.4 RESPONSE TIMES

Responses to late Quaternary climatic perturbations can be described with a simple model. The model of Figure 2.40 presumes a single abrupt decrease in effective precipitation at time B, which triggered responses in a variety of subsystems. A full-glacial semiarid climate dominated by winter precipitation occurred during time span A-B, which was characterized by equilibrium conditions

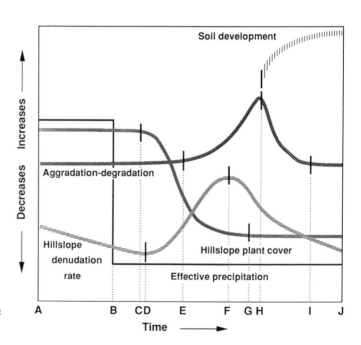

FIGURE 2.40 Time lags of responses in desert subsystems to the impact of the Pleistocene–Holocene climatic change.

in the dependent subsystems. An arid interglacial climate occurred during time span B-J, which was characterized by many subsystem changes. This is a simplification of the late Quaternary climatic changes of the study region, which were multiphase and complex—a sequence of full-glacial, transitional, monsoonal, and interglacial.

The vegetation, hillslope, piedmont, and soil subsystems had different time lags to the climatic perturbation. Plants reacted quickly, but had a long relaxation time (C-G) as thin hillslope soils were progressively stripped with concurrent increases in areas of exposed bedrock. During time A-B, denudation rates decreased slowly as soil thickness increased under good plant cover. Hillslope denudation rates had a short reaction time (C-D) because the perturbation for this process was decrease in plant cover. Self-enhancing feedbacks on these limited sediment supply slopes resulted in rapid increase (D-F) and decrease (F-J) of sediment yield.

Valley and piedmont aggradation and degradation lagged still more after the time of the stepwise climatic change. Static equilibrium conditions prevailed during time A-B, and then persisted until time E when rates and magnitudes of sediment-yield increase would no longer allow self-adjusting feedback mechanisms to maintain equilibrium streamflow conditions. Aggradation continued until the threshold of critical power was crossed at time H. The length of time span D-H was the time needed to move the climatically-induced slug of bedload down hillslopes and along valleys. During time I-J, the valley streams once again cut down to the same bedrock level as during time A-B. Attainment of the same base level of erosion indicates a lack of tectonically induced channel downcutting during all of time A-J. The fill terrace formed by channel entrenchment of the valley fill after time H provided a stable tread for soil-profile development; rapidly at first and then at progressively decreasing rates.

The scenario outlined in Figures 2.38 and 2.40 describes a transition between modes of fluvial-system operation. Our next step is to characterize the southwestern deserts as alternating between arid and semiarid climatic modes of operation.

## 2.6 Two Modes of Operation of Desert Fluvial Systems

The Quaternary has been a time of two modes of operation of geomorphic systems; for example, glacial and interglacial modes. Complex self-enhancing feedback mechanisms cause either a tendency for widespread glaciers or essentially no glaciers in the middle latitudes.

The concept of two climatically controlled modes of operation also applies to unglaciated fluvial systems that are now in arid (Chapter 2), semiarid to subhumid (Chapter 4), and humid (Chapter 5) regions. The sensitivity of fluvial systems to climatic changes varies greatly for systems with different mean annual precipitation and rainfall intensity, lithology (Chapter 3), and land-use patterns. The Holocene and late Pleistocene processes and landforms of the southwest deserts are representative of repeated alternations between the two types of climate-controlled system behavior during the Quaternary. Two modes of operation of the fluvial systems, which occurred during times of semiarid and arid climates, are described in table 2.17, although each aggradation event occurred during a time of climatic amelioration. The semiarid mode represents the more typical conditions of most of the Quaternary, whereas the arid mode represents brief excursions from the norm.

The Q3 and Q2c alluvial geomorphic surfaces represent the two modes of output of the desert fluvial systems. By definition, Q2c alluvium was deposited during times of semiarid climates of the late Wisconsinan glacial period. Desert pavements of the late Pleistocene geomorphic surfaces are smooth, tightly packed gravel (Fig. 2.12A,C) that do not reveal a hint of the former locations of the channels and gravel bars of the braided streams that deposited the gravel. However, Q2b and Q2a surfaces derived from the same drainage basins commonly have vestiges of former gravel bars that

TABLE 2.17 Characteristics of the two climatic modes of operation for desert fluvial systems.

| Mode | Characteristics |
|------|-----------------|
| Semiarid | Vegetated coluvium on hillslopes with less than one-third the slope area consisting of exposed outcrops |
| | Weathering of hillslope alluvium producing iron oxyhydroxides, clay, and calcium carbonate |
| | Weathering of schist and gneiss into fine gravel, sand, silt, and clay; weathering of granite into grus, silt, and clay |
| | Winnowing of the fine sediment during stream transport, and selective deposition |
| | Deposition of sheets of well-sorted gravel on piedmonts by braided streams from drainage basins underlain by schist or volcanic rocks; deposition of sheets of orange cross-bedded clayey sands from basins underlain by granitic rocks |
| | Formation of smooth desert pavements |
| | Accretion of manganese-rich rock varnish with a botryoidal morphology, and more negative $\delta^{13}C$ values |
| | Translocation of atmospheric dust and Av soil horizon silt and clay to B soil horizon |
| | Development of argillic soil horizons |
| | Calcium-carbonate illuviation in the depth range of about 25–100 cm |
| Arid | Sparsely vegetated colluvium on hillslopes with one-third to half the slope area consisting of exposed outcrops |
| | Extensive hydrolytic weathering and salt splitting of coarse-grained surficial rocks with granitic textures; varnishing of new outcrops, but minimal hydrolytic weathering, of schist, amphibolite, and volcanic rocks |
| | Transport by debris flows or flashy ephemeral streamflow with minimal winnowing of sand from bouldery sediment supplied from hillslopes |
| | Deposition of sheets of gray, poorly sorted sandy gravel on piedmonts by braided streams |
| | Formation of bar-and-swale pavements |
| | Accretion of iron-rich rock varnish with lamellate morphology and less negative $\delta^{13}C$ values |
| | Massive influx of atmospheric dust into surficial soils, rapid development of Av horizons |
| | No development of argillic soil horizons |
| | Calcium-carbonate illuviation in depth range of about 5–20 cm |

are similar to bars deposited during the Holocene Q3 aggradation event (Tables 2.13, 2.15). All Q2 soils have polygenetic Bt horizons because they have been through periods of semiarid and arid climate since their deposition. Q3 soils clearly were formed only under an arid mode.

The concept of two modes of operation for desert streams applies to a broad spectrum of flow magnitudes. During the early Holocene when hillslope sediment reservoirs were capable of high yields, small frequent flows infiltrated into streambeds and caused aggradation. Streamflow events with recurrence intervals of more than 100 years also caused aggradation throughout the stream system because of the large amount and size of bedload–stream power was much less than resisting power. During the late Holocene, when hillslope sediment reservoirs were depleted, small flows caused aggradation and flood flows caused degradation of the entire channel network–stream power was more than resisting power.

A model for the formation of stream terraces in

arid regions by extremely large flow events (superfloods) has been proposed by Schick (1974). The treads of some Q3c Holocene terraces in the lower Colorado River region may have been isolated by rapid stream-channel entrenchment during "superfloods." However, exceptionally large rainfall amounts and intensities during the early Holocene would cause only an extremely large debris flow that would aggrade the valley floor. Thus, availability of bedload is a major control on resisting power and on desert streamflow behavior.

Paleoclimatology studies have given us insight into the length and magnitude of the Pleistocene–Holocene climatic change and how it affected the southwestern deserts. Geomorphic and pedogenic variables were affected differently, but there is a good internal consistency between paleoclimatic insights from different subsystems. It is natural to equate the subject of paleoclimatology with paleobotany because of short reaction times of plants to climatic change and because of modern studies that provide abundant clues regarding the sensitivity of plants to precipitation and temperatures and their resistance to droughts and frosts.

Other subsystems also provide a distinctive record of climatic change. Hillslope denudation would seem to provide poor paleoclimatic information because hillslope erosion generally destroys evidence about prior topography. But Chap-

ter 3 will show that hillslope denudation is a valuable asset in the toolbox of paleoclimatic indicators. Aggradation-degradation responses of fluvial systems can help us better understand the nature of climatic changes, but we have yet to tap fully the wealth of paleohydrologic information contained in alluvial stratigraphic sections. The lack of sensitivity of soils to minor, short climatic fluctuations may be an advantage in providing an index of major climatic changes that occur every 100 ky.

The subject of landscape change in the southwestern deserts of North America is fascinating and complex. Multidisciplinary and diverse efforts by Quaternary scientists are continuing to reveal new and exciting pages of this important chapter of Earth's history. Many features of the hills, streams, and piedmonts, as well as the concepts discussed in this chapter, have strikingly similar parallels in other hot deserts of the world such as in the Middle East (Chapter 3). So far we have failed to discuss the relative importance of a major independent variable–lithology and structure. Lithologic control on landscape change is so important that the next chapter is devoted mainly to discussions of how different rock types influence geomorphic responses to climatic change in hot deserts.

# Lithologic Controls of Geomorphic Responses to Climatic Change on Desert Hillslopes

Dedicated to the innovative work and spirit of cooperation
in studies of deserts fostered by Ran Gerson

## 3.1 Introduction

The stream subsystems discussed in Chapter 2 are sensitive to climate-change-induced variations in stream power and resisting power. Fluvial landforms and deposits commonly record the effects of minor as well as major climatic fluctuations. A complete picture of climate-induced landscape change should also portray less sensitive landscape elements and geomorphic processes. Reaction times to climatic change for some hillslope processes are one or two orders of magnitude longer than those for streams.

The types of rocks that underlie hillslopes profoundly influence erosional and depositional processes in fluvial systems (Yatsu, 1966). Each rock chemistry and each crystal fabric has characteristic structural properties. Lithologic controls result in distinct erosional shapes of hills and streams, and exert a strong control on landscape changes induced by late Quaternary variations. Drainage basin lithology controls outputs of water and sediment through intermediate hillslope processes of runoff, weathering and soil formation, and plant growth. Earth materials as diverse as loessial silt,

quartzite, massive granite, and basalt flows will, for a given climate, have equally diverse stream-channel spacings, vegetation, and sediment yields.

The importance of lithologic controls on geomorphic processes can be illustrated with a single photograph. A dramatic example of self-enhancing feedback mechanisms that promote nonuniform erosion in hillslope materials is shown by erosion of bedded sedimentary rocks in Utah (Fig. 3.1). Where weakly consolidated materials overlie massive sandstone, any exposure of sandstone accelerates the erosion of overlying materials in the manner described by the process-response model shown in Figure 2.38. Lithologic contrasts are so important at the Figure 3.1 site that it is unlikely that the self-enhancing feedback mechanism can be reversed. Indeed, vast expanses of denuded dip-slopes of sandstone enhance the scenic grandeur of Dinosaur National Monument.

Discharge of water and yields of bedload sediment fluctuate rapidly; they are sensitive to minor climatic variations. Hillslope sediment reservoirs change from net depletion to net accumulation at much longer intervals. Climate-change-induced variations of rates of hillslope processes cause

FIGURE 3.1 Lithologic controls of self-enhancing feedback mechanisms on eroding hillslopes of bedded sedimentary rocks, Dinosaur National Monument, Utah. Exposure of outcrops of massive sandstone tends to accelerate stripping of the overlying weakly consolidated siltstones. Flashy runoff on sandstone erodes the margins of soft hillslope materials. The resulting thinner soils support less dense vegetation, which further accelerates erosion. Examples are at locations A. See Figure 2.38 for an appropriate process-response model.

desert streams to cross the threshold of critical power repeatedly with response times of 0.1 to 5 ky. Hillslopes appear much less sensitive to climatic change because changes of hillslope modes of operation in the southwestern deserts of North America occur only about every 50 ky. Studies of alternating mantling and stripping of hillslopes that are underlain by different rock types provide additional insight into landscape change.

We will consider mainly rock control of hillslope processes that are consequences of large and prolonged shifts in climate, not of minor variations. The study sites are small watersheds in the hot deserts of southwestern North America and the Middle East. Lithologic controls of geomorphic responses to climatic change in subhumid and humid watersheds are discussed in Chapters 4 and 5.

An ideal and extremely arid hyperthermic (see Table 2.1) drainage basin for such evaluations is the intensively instrumented small watershed of Nahal Yael in Israel adjacent to the Sinai Peninsula of Egypt (Fig. 3.2). Late Quaternary climatic changes in the Middle East differed from those of southwestern North America. This chapter begins with a summary of regional climatic changes in the Middle East and then examines climatically induced responses of geomorphic processes on

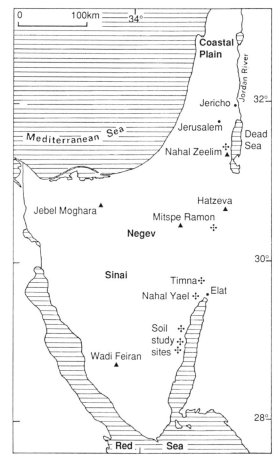

FIGURE 3.2 Location map of the region surrounding the Nahal Yael study area and locations of soil-profile study sites.

slopes in the Nahal Yael watershed. It concludes with a discussion of the sensitivity of rock types to climatic change.

## 3.2 Climatic Geomorphology of the Southwestern Dead Sea Rift Valley

### 3.2.1 CLIMATE

#### 3.2.1.1 Regional Climatology

Deserts in the Middle East occur at about the same latitudes as those in southwestern North America (Section 2.1.1), so airmass circulation patterns are similar. Subtropical anticyclones of slowly subsid-

ing air create a strong high-pressure dry belt over the northern Africa–Saudi Arabia region. The dry belt shifts to the south during the winters, but in contrast to western North America, few high mountains restrict passage of storms.

Instead the Mediterranean climate pattern extends far east of the Atlantic Ocean because winter storms derive additional moisture from the Mediterranean Sea. The highly seasonal Mediterranean climate is characterized by abundant winter rains along the northern and eastern margins of the sea followed by monotonous summer drought when dry belt airmass circulation returns.

To the south of the Mediterranean, dry belt airmass circulation predominates during most of the year resulting in a climatic transition zone only 300 to 400 km wide between subhumid and extremely arid conditions (Israel Meteorological Service, 1977). Coastal Lebanon and northern Israel receive 500 mm of rain per year, the northern Negev Desert about 200 mm, the southern Negev about 50 mm, and the lower altitudes of the Sinai Peninsula 10 to 30 mm per year.

A pronounced topographic rainshadow (Gerson, 1982b) is important in curtailing the eastward movement of Mediterranean sources of precipitation into the Dead Sea Rift Valley. The rainshadow is readily apparent to even the casual traveler; mean annual precipitation decreases from about 550 to 600 mm at Jerusalem to 100 mm at Jericho. Farther south, relief is less and the rainshadow is less pronounced. Mitspe Ramon receives 90 to 100 mm annual precipitation, which decreases to 45 mm at Hatzeva in the trough of the Rift Valley south of the Dead Sea.

Seasonal and long-term (Magaritz & Goodfriend, 1987) climatic changes are determined by interactions of airmass circulations of the monsoons, subtropical dry belt, and Atlantic Ocean storms and prevailing westerlies, as well as by the frequency and intensity of secondary winter storms that form over the eastern Mediterranean Sea. Incursions of moist air can produce major rainfall events. An example is the February 20, 1975, storm that caused extreme flooding in dry stream courses of the Sinai Peninsula and in the topographic rain shadow of the Rift Valley. Nahal Yael received 80 mm of rain—almost triple the mean annual rainfall. Streamflow did not issue from Nahal Yael during the next 6 years. This unusual storm was caused by northward movement of a monsoonal airmass (normally a summer type of rainfall) into an area that is usually dominated by dry belt and Mediterranean climate patterns.

### 3.2.1.2 Paleoclimatology

Insight about the past climates of the large region comes from studies of lakes, snails, soils, pollen, and human occupation sites. Data from the Sinai Peninsula, Negev Desert, and Rift Valley provide particularly useful clues about the general times and types of climatic changes in the small watershed of Nahal Yael. (Readers seeking a summary rather than a discussion should skip ahead to Section 3.2.1.2.5.)

Similar trends in late Quaternary climates occurred in the large region that extends from India to the central Sahara (Fig. 3.3). Postulated types and times of climatic change are based on many studies of pollen, foraminifera and land snails, soil profiles, lake sediments and highstands (Street & Grove, 1979), archaeological remains, isotopic compositions of modern and prehistoric waters, and cores from the eastern Mediterranean Sea. The latest Pleistocene (see Table 1.2 for definition of Quaternary temporal terms) full-glacial climate of 25 to 15 ka was dry (Butzer, 1975), as were the past 3 to 5 ky. The onset of wetter climates at $13 \pm 2$ ka peaked at $7 \pm 2$ ka at nearly all sites.

Wetter periods were associated with increases in monsoonal rain in North Africa (Wickens, 1975, 1982) and in southwest India. In North Africa, areas that now are extremely arid were wet enough in the early to mid-Holocene for sustenance of semiarid land snails, plants, ostriches and for hydrolytic pedogenesis (Mehringer, 1982; Haynes, 1982, 1987; Lézine et al., 1990; Pachur & Braun, 1980; Pachur & Roper, 1984; Ritchie et al., 1985; Haynes & Mead, 1987; Brookes, 1989).

The sequence of regional climatic change summarized in Figure 3.3 contrasts with the sequence in other regions because the wettest time during the late Quaternary occurred between 10 and 5 ka. The Pleistocene ended abruptly in Greenland as sea level rose at 10 ka (Dansgaard et al., 1969). Thermic, arid conditions have prevailed in much of southwestern North America since 8 ka.

*3.2.1.2.1 Effects of Changing Orbital Parameters* The dry climate of the "last glacial maximum would be related to diversion of the monsoons from the Indian and Arabian subcontinents" (Gat & Magaritz, 1980). The colder and drier climates associated with full-glacial conditions may have resulted from cooling and 130-m lowering of sea level (Lamb, 1977); this resulted in weaker winter storm cyclogenesis in the eastern Mediterranean Sea. Weak monsoonal airflow during cool full-glacial times may have been a function of low reception of solar radiation in the northern hemisphere because of (1) an unfavorable combination of the earth's orbital parameters and (2) a high albedo of central Asian ice sheets and snowfields (Prell, 1984).

Conversely, wet periods were related to increases in moisture from monsoonal or Mediterranean sources, or from both. Monsoonal incursions strengthen with increases in low-level atmospheric pressure gradients between land and ocean (Kutzbach & Guetter, 1984). Just as cold winters with above-average snowpacks delay summer heating of continental areas, extended terrestrial ice sheets during the Pleistocene would weaken and delay summer monsoonal rainfall.

The times of optimal orbital parameters for strong northern hemisphere summer monsoons were assessed by Kutzbach and Guetter (1984), using the NCAR (National Center for Atmospheric Research) Climate Community model. Strengthening of monsoonal circulation would be favored by increases in seasonal temperature contrasts. At 9 ka, land temperatures apparently were 1.8 to 2.0°C warmer in July and 1.5°C cooler in January. The magnitude of climate change induced by such variations in solar input is also a function of size and latitude of landmasses. Thus the effects of the

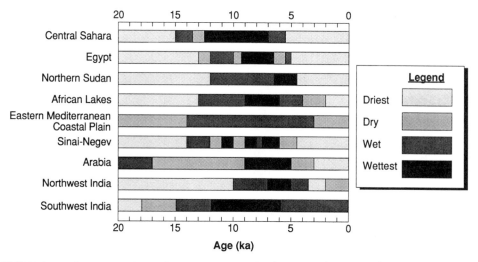

FIGURE 3.3 Summary of late Quaternary climatic change in the region from North Africa to India during the past 20 ky. Results based primarily on lacustrine, pollen, pedogenic, and archaeological studies.

modeled changes would be more for the African–Indian monsoon than for the monsoon of southwestern North America. The combined precessional and eccentricity effects maximize the modeled monsoon at about 9 ka. The amount of northwest Indian monsoonal rainfall appears to have been triple its present amount during the early Holocene but variable since then (Bryson & Swain, 1981). Monsoonal rainfall in southwest India appears to have been more consistent (Van Campo, 1986); a summer rainfall maximum at 11 ka was close to the northern hemisphere maximum of summer insolation (Fig. 2.4). Prell (1984), in his study of summer monsoon-induced upwellings of the Arabian Sea, found a 3- to 5-ky time lag between times of maximum solar radiation input and strength of summer monsoons as measured by oxygen isotopes in a planktonic foraminifer.

Much of the Arabian Peninsula lies south of winter storm tracks, so variations in monsoonal rainfall are an important factor in climatic change. Lacustrine deposits in the Rubse-Khali area indicate a prolonged wet period between 36 and 17 ka followed by a dry period (lack of lacustrine deposition) until about 9 ka. A second period of lacustrine deposition occurred between 9 and 6 ka, climaxing at about 6 ka (McClure, 1978).

*3.2.1.2.2 Changing Soils and Climates* In Israel soils provide valuable, but general, data about past climates (Fig. 3.2). Coastal plain soils are complex. Leaching in this area of highly seasonal winter precipitation and influx of atmospheric dust blown from distant deserts gradually transform calcareous eolianite parent material deposited as beaches and dunes into red noncalcic soils (Dan & Yaalon, 1971). With continuing additions of atmospheric dust, oxidized and well-drained soils become reduced and water logged—in soils terms, alfisols become gleyed and change to vertisols. Such extreme changes in physicochemical properties are further complicated by position of a pedon in a hillslope catena. This makes it difficult to separate the effects of reduced rates of unsaturated flow through soils from the effects of climatic change on sequential pedogenesis during the middle and late Pleistocene.

Coastal plain soils provide clues about climatic changes during the past 25 ka. Decreased leaching of calcium carbonate in eolianite soils allowed calciorthids to form during dryer climates of the last glacial maximum (personal communication from Bakler & Kaufman, 1978, cited by Gat & Magaritz, 1980). It is important to note that this decrease in effective precipitation occurred despite the cooler temperatures. By 14 ka the climate had become sufficiently wet to again favor leaching of soil carbonate, and eventually red noncalcic soils (xeralfs) formed (Dan & Yaalon, 1971). Further south in the Negev Desert this was also a time of more rapid pedogenesis and calcic horizons formed (Goodfriend & Magaritz, 1988). Sand dunes that buried the red soils attest to decreases in stabilizing vegetation as the climate became drier at about 3 ka.

Comparisons of Pleistocene and late Holocene soil profiles in the presently arid Dead Sea Rift Valley suggest major changes in effective precipitation for soil leaching. Higher and older terraces (Fig. 3.2) have 20- to 40-cm-thick bright reddish-brown argillic horizons that are underlain by well developed (stage III of Table 2.6) calcic horizons that are more than 50 cm thick (Dan, 1981; Dan et al., 1981,1982). The presence of argillic horizons and much greater depths to the tops of the calcic horizons indicate comparatively much greater soil leaching than is occurring at present.

These strongly developed soil profiles are considered to (1) be indicative of formation under a semiarid climate and (2) require more time to form than the 5 to 7 ky of semiarid climate that is estimated to have occurred during the past 25 ky. These alluvial geomorphic surfaces were deposited sufficiently before the ~25- to 15-ka period of full-glacial arid climate to allow ample time for soil formation.

Monsoonal rains are considered important for the development of typic haplargids in the eastern Sinai Peninsula and Dead Sea Rift Valley because Mediterranean sources of moisture are subject to extreme rainshadow effects along the west side of the Rift Valley and rarely penetrate as far as the eastern Sinai. Monsoonal rains from the south can spread easily into the Rift Valley, as indicated by the rare storm of 1975.

Soil profiles on late Holocene alluvial terraces south of the Dead Sea (Fig. 3.2) have desert pavements with brown rock varnish and discontinuous carbonate coatings on gravel clasts at depths of only 2 to 10 cm (see Fig. 3.17B). Minimal leaching of these torriorthents indicates that present climates are representative of late Holocene climates. Deposition and soil-profile development on these alluvial fans and stream terraces occurred after the last prolonged wet period, with sufficient soil leaching to form argillic horizons above deep calcic horizons.

Streams, such as Nahal Zeelim (Bowman, 1974,1978,1988; Begin, 1975), that flow east from the Judean hills to the Dead Sea formed spectacular flights of degradation terraces and eroded and buried the shorelines of receding full-glacial Lake Lisan. The youngest alluvium into which the highest lake strandline is cut has a thin argillic soil horizon (Amit & Gerson, 1986). None of the 14 inset fluvial terraces of post full-glacial age along Nahal Zeelim west of the Dead Sea have argillic horizons. Their desert pavements are lightly varnished and salic horizons are progressively weaker with decrease in terrace age. Pedogenic horizonation of calcium carbonate, gypsum, and halite with increasing depth attest to persistent arid conditions.

Evidence for a temporary slight increase in moisture consists of lichen growth patterns (Danin et al., 1982; Danin, 1986) etched into carbonate boulders on Holocene terraces; the climate is presently too arid for lichens but algae grow in fissures of boulders where rainfall is 100 mm/yr (Ran Gerson, Hebrew University of Jerusalem, oral communication, August 21, 1980). The soils and lichens provide strong evidence for general lack of rain from monsoonal sources and highly effective rainshadow blockage of Mediterranean moisture sources during all of the past 18 ky in the Dead Sea area and to a lesser extent in much of the central and southern Rift Valley.

*3.2.1.2.3 Lakes as Sensitive Barometers of Climatic Change* Lake levels, surface areas (Benson &

Paillet, 1989), and associated deposits provide exceptionally sensitive records of climatic change in arid regions, where they are sinks for the outputs of water and sediment from large watersheds (Street & Grove, 1979). It is difficult to determine relative importances of climatically induced changes in temperature, precipitation, and windiness on past lake levels and surface areas, streamflows, vegetation, and soil-profile development (Brakenridge, 1978). Investigators of Quaternary climatic change generally are faced with situations of equifinality (similar effects caused by several different combinations of variables), so it is prudent to avoid labeling temperature or precipitation change as the sole cause of changes in pedogenic and geomorphic processes.

The hydrologic regime of the Dead Sea closed basin (Fig. 3.2) is controlled by the rate of evaporation from the lake surface and runoff from its drainage area, which is mainly 200 to 300 km to the north in subhumid to humid mountains (Begin et al., 1985). Prior higher lake levels are indicative of wetter or cooler climates in the watershed and wetter or cooler and less windy conditions at the Dead Sea. Strandlines of Pleistocene Lake Lisan are as much as 220 m above the present level of the Dead Sea (Neev & Emery, 1967; Neev & Hall, 1977; Bowman, 1971,1974). $^{14}$C and $^{230}$Th/$^{234}$U dating indicates that the lacustrine Lisan Formation was deposited between 70 and 60 ka and 18 and 15 ka (Vogel & Waterbolk, 1972; Kaufman, 1971).

Paleoclimatic interpretations of Pleistocene levels of the Dead Sea are complicated because the distant headwaters receives abundant winter precipitation. However, a lesser rise in lake levels at 9 to 7 ka occurred during the climatic warming associated with onset of the Holocene. The level of the Dead Sea was 120 m above its present level even at 6.7 ka (Goodfriend et al., 1986). Such rises suggest that increases in headwaters runoff were more than enough to offset concurrent increases in temperature.

Deposition of bouldery alluvium on the deposits of retreating Lake Lisan between 18 and 11 ka may be interpreted in several ways. Begin et al. (1974) used the fluvial boulder berms as evidence for a rainy period in the Judean hills. Alternative explanations include stripping of hillslope colluvium as a result of change to more arid conditions and return of intense monsoonal rains.

Oxygen-isotope composition of carbonates precipitated in Lake Hula in the upper Jordan River valley indicate that temperatures have warmed since a cooler period during >16 to 10 ka (Stiller, 1979). The warmest period was between about 6.6 and 4.3 ka.

*3.2.1.2.4 Paleoclimatic Clues from the Negev and Sinai Deserts* Lakes and glaciers may have been present in the past. Glacial cirques and U-shaped valleys in mountains of the Sinai Peninsula more than 2000 m high indicate cold late Pleistocene temperatures (Gvirtzman, 1976). Studies at Jebel Moghara indicate lacustrine expansion in the Negev at 14 to 12 ka (Goldberg, 1977; Bar Yosef & Phillips, 1977).

Groups of nomadic hunters and even small pastoral communities were able to subsist in the Negev and northern Sinai during wetter times. Goldberg and Bar Yosef (1982) used radiocarbon dating to identify periods that were sufficiently wet for human settlement. A dry period from 20 to 14 ka was followed by a 2-ky wet period (calendric ages of approximately 23.8 to 16.6 ka using the corrections of radiocarbon ages described by Bard et al. [1990]). Drier conditions returned until 9 ka, except for a slightly wetter spell at 11 to 10 ka. A major Holocene wet period persisted between 9 and 6 ka, except for a brief dry spell between 8 and 7.5 ka. Wetter conditions also allowed pastoral communities to exist from about 9.5 to 4.5 ka along a now-extinct river system in northern Sudan, which now receives 25 mm mean annual precipitation (Pachur & Kropelin, 1987).

Land snails from the Negev Desert provide valuable information about Holocene changes in rainfall because their shells can be analyzed for radiocarbon ages and for carbon isotopes ($^{13}$C/$^{12}$C) that reveal the types of plants they ate (Goodfriend, 1987,1988, in press). Rodents collected and ate snails in their middens and like the plants in the packrat middens of the American Southwest (Section 2.1.2.3) the paleoclimatic information from the snail shells reflects only the immediate vicinity of the middens. Goodfriend's surveys of modern plant and snail communities show that C4 plants, which are enriched in $^{13}C$ (Teeri, 1979), typically grow where mean annual precipitation is less than 230 mm (Vogel et al., 1986). Comparisons of modern and mid-Holocene $\delta^{13}C$ values (defined in Equation 2.7) suggest that mean annual precipitation between 6.5 and 30 ka was about twice that of the present.

Fine-grained 10-m-high fill terraces occur along the presently bouldery Wadi Feiran in southwestern Sinai (Issar & Eckstein, 1969; Nir, 1970). Although they are of discontinuous ephemeral stream origin instead of lacustrine origin, the fine-grained deposits indicate a major climatically induced change in stream behavior. Goldberg ascribed the fill-terrace deposits to a wet period that began in the latest Pleistocene, and the soil profiles suggest late Holocene stream-channel entrenchment. Weathering formed extensive grus mantles on granitic hillslopes during a presumably long period of semiarid climate in the headwaters of Wadi Feiran (Gerson & Yair, 1975; Gerson, 1982b). Then hillslope erosion aggraded the valleys with water-laid and debris-flow deposits. After the climate became drier or warmer, the dominant processes became physical weathering on rocky hillslopes and entrenchment of valley fill.

*3.2.1.2.5 Summary of Late Pleistocene Climatic Change* The diverse and abundant types of evidence for late Quaternary climatic change in the regions can be used to outline a scenario of changing paleoclimates. Interactions of converging and shifting airmasses still are poorly understood, and it is easy to fall into the trap of oversimplifying feedback mechanisms and consequences of climatic change. The scenario of late Quaternary climatic change summarized in Table 3.1 and

Figure 3.4 is merely a set of informed estimates that are internally consistent with the facts presently available. Only trends comparing precipitation amounts and temperatures with the previous climate are noted; speculations are not attempted regarding the magnitudes of change of precipitation and temperature or changes in rainfall intensity.

The interpretation of possible trends in late Quaternary climate is noted for three areas—Nahal Yael, the Dead Sea, and the coastal plain (Fig. 3.2). Each area has distinctive geomorphic and pedogenic characteristics that are primarily the result of its location with respect to airmass circulation boundaries. It is assumed that the abrupt north-to-south gradient to more arid climates was present but that its latitudinal position changed several times. Seasonality of precipitation in Table 3.1 is indicated by the dominant airmass; monsoon for mainly summer, and Mediterranean for mainly winter. Trends of climatic change are noted for amounts of precipitation and temperature.

Periods of semiarid climate occurred between 70 and 25 ka, and thick argillic horizons are assumed to be at least this old. This late Pleistocene period of occasional semiarid climate ended with the approach of full-glacial climates; dry belt circulation and monsoonal airmasses retreated to the south.

The highest shoreline of Pleistocene Lake Lisan occurred at about $18 \pm 3$ ka because of two factors. Markedly lower lake temperatures may have greatly reduced evaporation losses while precipitation increases in distant high headwaters to the north maintained large streamflows through the Jordan River valley. Then the very dry and cool climate that prevailed from India to the Sahara during the last glacial maximum at about 25 to 18 ka (calendric ages of approximately 27.5 to 21.5 ka using the corrections of radiocarbon ages described by Bard et al. [1990]) was replaced by a thermic arid climate and lake levels declined.

Local climates fluctuated from extremely arid to semiarid during the past 14 ky. The periods

TABLE 3.1 Possible trends in late Quaternary climate and climate-controlled processes for three areas in the Middle East.

| Time (ka) | Area | Change in Dominant Geomorphic and Pedogenic Processes | Dominant Airmass Circulation | Climatic Change Relative to Preceding Period[a] | |
|---|---|---|---|---|---|
| | | | | Precipitation | Temperature |
| >25 | Nahal Yael–E. Sinai | Slope aggradation; haplargids form on alluvium | Monsoon, Mediterranean | + | |
| 23 | Nahal Yael, Dead Sea | Slope degradation and valley-floor aggradation | Dry belt, Mediterranean | − | − |
| 18 | Nahal Yael | Slope and valley-floor degradation | Dry belt | − | − |
| | Dead Sea | Highest shoreline, slope degradation | Dry belt, Mediterranean | − | − |
| | Coastal plain | Calciorthids form | Dry belt, Mediterranean | − | − |
| 14–6 | Nahal Yael | Slope aggradation | Monsoon, Mediterranean | + | + |
| | Dead Sea | Sediment deposited on receding shorelines, then temporary rise in lake level at 7 ka | Dry belt, Mediterranean | = | + |
| | Coastal plain | Formation of alfisols as calciorthids are leached | Mediterranean | + | + |
| 3–0 | Nahal Yael | Slope degradation and valley-floor aggradation | Dry belt, Mediterranean | − | + |
| | Dead Sea | Lake-level decline | Dry belt, Mediterranean | − | + |
| | Coastal plain | Sand dunes form | Dry belt, Mediterranean | − | + |

[a] +, increase; −, decrease; =, no significant change.

between 14 and 12 ka and between 9 and 6 ka were much wetter and warmer than climates during full-glacial times. The 14- to 12-ka wetter period probably was not monsoonal, or it should have been detected by paleoenvironmental studies on the Arabian Peninsula. Instead increased rain in the Negev and northern Sinai may have come mainly from Mediterranean Sea and Atlantic Ocean sources of winter precipitation as those bodies of water rose and became warmer.

Boundaries between dry belt and monsoonal airmasses appear to have shifted northward more frequently between about 10 and 5 ka, bringing convective storms at least as far north as Sinai Peninsula, Nahal Yael, and the southern Negev. Return of especially abundant monsoonal rains from 9 to 6 ka favored lacustrine expansion, deeper soil leaching, and increase of pastoral communities. Increase of monsoonal rainfall during the early and middle Holocene fits the model of climate control by the earth's orbital parameters fairly well. But the fit is even better if one accepts the

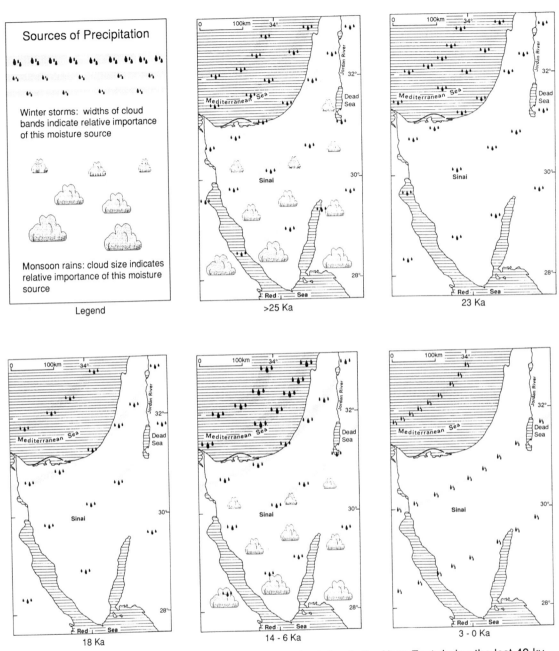

FIGURE 3.4 Approximate trends in dominant airmass circulation in the Near East during the last 40 ky.

reaction time of 3 to 5 ky postulated by Prell (1984).

Holocene shorelines of Làke Lisan presumably retreated mainly because of increased temperatures and decrease of headwaters precipitation. Horizonation of calcite, gypsum, and halite splitting the chert clasts in the gravelly soil profiles on postglacial stream terraces attests to (1) the severity of the rainshadow effects for Mediterranean moisture sources and (2) a general lack of abundant monsoon rains this far north.

Increased importance of dry belt airmasses since about 5 to 3 ka heralded the onset of the present arid to extremely arid, and thermic to hyperthermic, climate that is summarized in Figure 3.5 for Elat, about 4 km east of Nahal Yael. Grazing of animals and human settlements in the Sinai and southern Negev declined in the late Holocene. The level of the Dead Sea declined rapidly, while dune sand capped latest Pleistocene–Holocene alfisols on the coastal plain.

The general pattern of paleoclimatic change outlines the climatic heritage that caused profound changes in geomorphic processes in the Nahal Yael study area. Knowledge of shifting late Quaternary airmass circulations provides a useful background for investigations of how lithologic controls affect hillslope processes during changes to markedly drier or warmer climates.

### 3.2.2 LITHOLOGIC CONTROLS OF HILLSLOPE PROCESSES IN NAHAL YAEL

The extremely arid fluvial system of Nahal Yael (Fig. 3.2) has been exceptionally sensitive to late Quaternary climatic change (Bull & Schick, 1979). It is ideal for detailed evaluations of the importance of lithology in changing outputs of geomorphic processes on hillslopes when the climate becomes drier and hotter. Hillslopes of extremely arid regions consist of bedrock outcrops with ephemeral thin mantles of colluvial materials. In extremely arid regions, a 40-mm decrease of mean annual precipitation, or a 6°C increase in temperature, may halve the effective precipitation for plants and weathering. Brief variations in climate affect desert streams, and changes in hillslope sediment reservoirs may occur when a climatic change is sufficiently long.

The study area consists of the adjacent small drainage basins of Nahal Yael and Nahal Naomi (Fig. 3.6). Observations of precipitation, streamflow, sediment transport, erosion, and deposition have been made in Nahal Yael since 1965 (Schick, 1970, 1977, 1980; Sharon, 1972; Schick & Sharon, 1974; Salmon & Schick, 1980). The hillslopes are almost devoid of plants but a sparse growth of trees, bushes, and grass persists along the stream channels.

#### 3.2.2.1 Rock Types

Lithologic control of geomorphic responses to climatic change is best studied in drainage basins underlain by a variety of rock types. Diverse metamorphosed Pre-Cambrian rocks of the Elat Massif (Fig. 3.6) underlie the study area. A sheared complex of equigranular metabasites underlie the headwaters of Nahal Yael; biotite-plagioclase amphibolite, hornblende-plagioclase schist, and gabbroic rocks are common. The central part of the Nahal Yael and the headwaters of Nahal Naomi

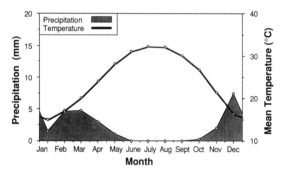

FIGURE 3.5 Monthly variations of precipitation and temperature at an altitude of 30 m at Elat, Israel, which is strongly seasonal extremely arid and strongly seasonal hyperthermic. Data from Israel Meteorological Service, 1967, 1983.

drainage basins are underlain by metasedimentary rocks, chiefly pelitic schists. Biotite-quartz schist, muscovite-biotite schist, and biotite-plagioclase schist are common. The downstream parts of both basins are underlain by the coarse-grained, jointed and sheared, pink porphyroclastic granitic gneiss-to-gneissic granite, which was intruded into the metasedimentary and mafic rocks. Dike swarms cut across all the rock units. Common dike lithologies are feldspar porphyry, lamprophyre, quartz porphyry, and pegmatite.

The two drainage basins have different proportions of underlying rock types. The Nahal Yael basin upstream from the fan apex is 0.54 km$^2$ and is underlain by metabasite rocks (19%), metase-

dimentary rocks (49%), gneissic granite (13%), and dike rocks (19%). The 0.07–km$^2$ Nahal Naomi basin is underlain by metasedimentary rocks (17%), gneissic granite (70%), and dike rocks (13%). Both basins are rugged. Nahal Yael has a relief of 180 m and Nahal Naomi 135 m. Sites of figures and tables for Nahal Yael are shown in Figure 3.6.

### 3.2.2.2 Changes in Hillslope Processes

The regional climatic variations described in Section 3.2.1.2 affected rates of geomorphic processes by changing rates of chemical weathering and erodibility of weathering products through changes in temperature, soil leaching, and vege-

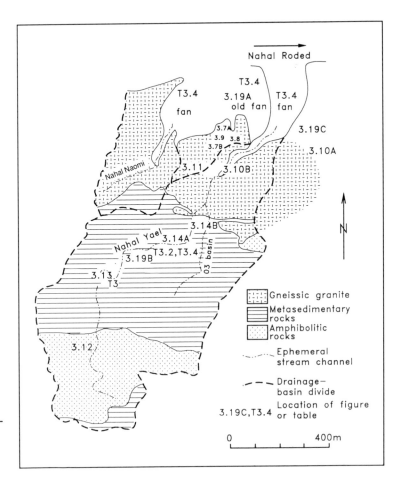

FIGURE 3.6 Geology and location of study sites in the Nahal Yael research area. Geology chiefly from Shimron, 1974.

tation density. Erosion rates also were affected by changes in amount and intensity of rainfall. Rainfall–runoff relations for hillslopes underlain by each rock type changed in different ways as temperature and precipitation changed. It is assumed that late Quaternary climates were warm enough to promote chemical weathering but that adequate moisture for rapid chemical weathering was present only briefly during time spans of 2 to 8 ky.

*3.2.2.2.1 Gneissic Granite* The most striking feature of the bare slopes of coarse-grained gneissic granite is the marked contrast in smoothness of outcrops (Figs. 3.7, 3.11). Rough, craggy exposures are coated with rock varnish, are little weathered, and ring or emit a solid sound when struck with a hammer. Smooth exposures occur adjacent to rough outcrops and there is no visible change in petrologic fabric, mineralogy, or fracture density (Fig. 3.8). The smooth slopes lack rock varnish, are crumbly, and emit a dull or punky sound when struck with a hammer. The two types of gneissic granite outcrops are the result of different weathering microenvironments of the same rock type.

Rough outcrops occur on topographic highs, on hillslopes sloping more than 25 degrees, and in streambeds. Joint and shear fractures promote differential weathering of the gneissic granite where exposed to raindrop impact, sheetflow, and streamflow. Chemical weathering is slow in such a microenvironment, and outcrops retain rock varnish (left side of Fig. 3.11) except where abraded by streamflow.

Smooth outcrops occur in topographic hollows sloping less than about 25 degrees and under rare remnants of colluvium. The coarse-grained texture of the gneissic granite favors rapid weathering to smooth surfaces where a cover of colluvium provides a moist microenvironment after prolonged rains.

Outcrop roughness and rock varnish were surveyed along a transect (A–A' of Figs. 3.8 and 3.9) with both rough and smooth outcrops. Frac-

ture density (shears and joints) is the same for both types of outcrops.

Roughness and degree of varnishing are markedly different for rough and smooth outcrops. Outcrops along the line of transect were assigned classes of relative roughness values, the roughest given a value of 5 and the smoothest a value of 1. Intermediate roughness classes were assigned visually. For example, the right side of Figure 3.11 has class 2 roughness and the left side has class 4. Rougher outcrops occur on the steeper parts of the ridgecrest of Figures 3.7 and 3.9. Outcrop roughness increases between distances of 33 to 44 m because streamflow along the first-order rill and the trunk stream has tended to erode preferentially along fractures.

Rock-varnish color was described for the darkest outcrop along each part of the transect. Outcrops are brownish-black (5YR 2/2) at the ridgecrest but the varnish color value increases—becomes lighter—progressively downslope. Typical colors are as follows: for a value of 3 (5YR 3/4), dark reddish-brown; and for a value of 7 (7.5YR 7/3), dull orange. The smooth outcrop between distances of 14 and 18 m is darker than the lower smooth outcrops, which suggests longer exposure to varnishing processes. Between distances of 20 and 44 m the gneissic granite is so weathered that disintegration rates exceed varnishing rates.

Although the hillslopes now are bare, the weathering contrasts just noted are clear evidence that large portions of the slopes previously were mantled with grussy colluvium. Grus has been eroded, leaving only scattered pieces of dike rocks on smooth outcrops of gneissic granite.

Thicknesses of colluvium that have been removed by erosion were determined by measurement of heights to nearby 0.1- to 2-m$^2$ outcrops of rough dark gneissic granite. These thickness measurements probably are accurate to within +20 percent. The lower part of the slope does not have any rough outcrops because even the small ridge between two first-order streams (such as the swale in the center foreground of Figure 3.7A) had been completely covered by colluvium. Measurements

FIGURE 3.7 North-facing hill-slopes of gneissic granite with contrasts in smoothness of outcrops. A. View along transect (indicated by patterned line) of Figures 3.8 and 3.9. B. Hilltop. Note the comparatively gentler slopes of the smooth outcrops. Cobble-size blocks littering the smooth slope are from dikes, which are equally common in the hilltop rocks. Hat is for scale in the left-center foreground.

of heights to adjacent bare ridgecrests provided a means of making rough estimates ($\pm 50\%$) of minimum thicknesses of prior colluvium in the swale; they are shown by dashed lines in Figure 3.8. The former mantle of grussy colluvium on the hillslope was thin but it thickened downslope. The smoothest outcrops (distances 25–33 m in Fig. 3.8) occur downslope from extensive areas of class 2 roughness; this suggests that they have been uncovered more recently than the class 2

outcrops and that progressive stripping of colluvium was completed first on the midslopes and then on the footslopes.

The different weathering modes of the gneissic granite allow mapping of the former extent of the colluvial cover. The weathering contrasts of smooth and rough outcrops show clearly on aerial photographs and indicate that about two-thirds of the granitic hillslopes had a mantle of colluvium that has since been stripped. Hillslope characteristics are shown in Figure 3.8, and a map of stripped hillslopes is shown in Figure 3.9. Most of the area of rough granitic outcrops is along a low southwest-trending ridgecrest. Isolated outcrops of rough granite also occur as small hills that rise above the adjacent smooth slopes. The boundaries between the areas of smooth and rough gneissic granite

may parallel or cross joints and shear zones, which are common in both rough and smooth outcrops.

Much of the gneissic granite is sufficiently massive to favor *tafoni* (cavernous weathering). Tafoni form under conditions of subaerial exposure as a result of variations in case hardening that affect rates of chemical weathering (Jennings, 1968; Winkler, 1975) and by flaking induced by salt splitting (Bradley et al., 1978). Moist subsurface microenvironments eliminate the differences in weatherability caused by salt accumulations or case hardening, prevent formation of new tafoni, and promote destruction of buried tafoni. In Figure 3.10A only the upper part of the gneissic granite hillslope has abundant varnished outcrops and tafoni, which suggests the presence of a former colluvial mantle on the footslope. Remnants of

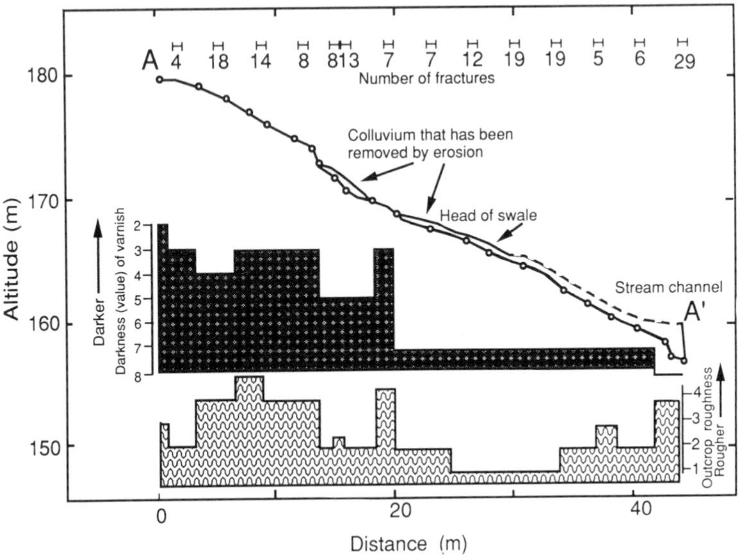

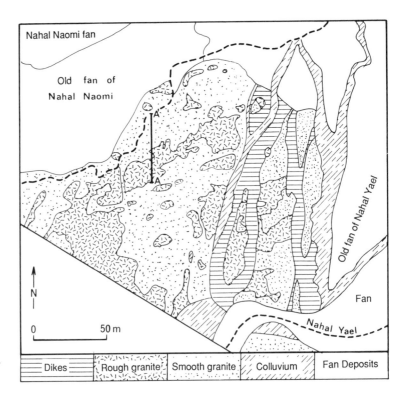

FIGURE 3.8 Former thicknesses of colluvium, outcrop roughness, rock varnish, and fracture density in the gneissic granite. The location of transect A–A' is shown on Figure 3.9.

FIGURE 3.9 Map of gneissic granite hillslopes that have been stripped of colluvium (the areas of smooth granite). Transect A–A' is shown in Figures 3.7 and 3.8.

exhumed and partially destroyed tafoni in Figure 3.10B are exposed where streamflow has removed 6 m of Holocene alluvium.

At least two periods of subcolluvial weathering resulted in the contrasting roughly planar hillslope surfaces shown in Figure 3.11. Smooth unvarnished outcrops on the right have characteristics similar to those described for the lower part of the slopes illustrated in Figures 3.7A, 3.8, and 3.9 and are assigned to the most recent episode of hillslope stripping. The younger, smooth type of rock surface has a broad scalloped boundary where it joins the older, slightly higher but much rougher and darker varnished planar surface shown on the left side of Figure 3.11. This older rock-weathering surface probably owes its planar nature to an expansion of the hillslope sediment reservoir; it abuts craggy, steep slopes that have never been mantled with colluvium. The smooth surface lies about a meter below the rough surface, which implies either (1) deep weathering of the area of presently smooth gneissic granite during the most recent episode of mantling or (2) several episodes of mantling that reoccupied the lower, but not the upper, of the two planar surfaces.

The gneissic granite hillslopes of Nahal Yael have been amazingly sensitive to the effects of changing climates. The visual impression of change is enhanced because the total lack of vegetation on the stark slopes leaves nothing hidden from view. The stripping is so profound that it would seem unlikely the process could be reversed. Reversals have occurred repeatedly, however. Plants first recolonize footslopes; this favors progressively expanding mantles of grussy colluvium. The sensitivity of these slopes to changing climate has been enhanced by the contrast between dry subaerial and moist subsoil weathering rates, the presumed major contrast in density of hillslope plant cover of semiarid and extremely arid climates, and the small forces needed to entrain and transport the monomineralic detritus (grus) resulting from decay of the gneissic granite. The thin nature of the hillslope mantles contributed to the rapidity of

erosional stripping despite infrequent rainfalls during the late Holocene.

*3.2.2.2.2 Amphibolite and Metabasites* In contrast to the uniform petrologic and structural properties of the gneissic granite, extreme shearing of the mafic-rock complex in the headwaters of Nahal Yael has resulted in diverse adjacent lithologies. Mapping of the former extent of colluvial mantles would be difficult, because lithologic and structural differences are partly responsible for areal variations of outcrop roughness and rock varnish.

Stripping of prior colluvium and alluvium is apparent where hillslopes are underlain by uniform amphibolite. Massive mafic metaplutonic rocks weather much like gneissic granite except that weathering rates are more rapid, former colluvium was thicker, grus is finer grained, and exhumed outcrops seem to varnish more readily. The slope shown in Figure 3.12 consists of biotite-plagioclase amphibolite. Fresh rock in the stream channel is olive-gray (10YR 5/2), but even a few meters above the channel the rock is varnished dull brown (7.5YR 5/4) to dark brown (7.5YR 3/4). Punky outcrops are moderately smooth and locally show concentric spheroidal layers indicative of subsoil weathering of corestones of quadrangular joint blocks. In contrast the upper part of the hill has rough outcrops varnished brownish black (7.5YR 2/2). The change in amphibolite weathering characteristics coincides with the upper limit of a former debris slope, a remnant of which is preserved downstream from the view of Figure 3.12. The debris slope is graded to the tread of a fill terrace. Valley-floor aggradation occurred either before or at the same time as hillslope colluviation.

In most ways the responses of the granitic-fabric amphibolite were similar to those of the granitic-fabric gneiss. For a much different response to the same climatic changes one needs only to examine nearby hillslopes of schist.

*3.2.2.2.3 Pelitic Schists* Slopes underlain by schist show a third type of weathering response to climatic change. Intergrown small crystals of meta-

FIGURE 3.10 Tafoni in gneissic granite. A. Contrasts in varnishing and abundance of tafoni on the upper and lower parts of a hillslope suggest prior burial of the more gently sloping footslope. B. Buried and partially destroyed tafoni that have been exhumed as a result of erosion of fill-terrace deposits. A 15-cm scale is shown in the center.

morphic fabrics, abundant joints, and strongly developed foliation parallel to cleavage planes of oriented micas favor weathering to angular gravel instead of grus. In contrast to hillslopes of gneissic granite that are nearly completely stripped of colluvium, schist slopes range from stripped of colluvium to largely mantled by colluvium. Dissected remnants of formerly extensive wedges of colluvium interfinger with alluvium in valley floors to which they were graded.

The colluvial wedge shown in Figure 3.13 consists of interbedded well-sorted sheet and gully

talus and poorly sorted colluvium. Bed thicknesses average 0.4 m. Most of the schist clasts are 30 to 80 mm in size; maximum clast size in the beds typically is 90 to 210 mm. The variety of gravel sizes and matrix types suggest accumulation by rockfalls, water transport, and debris flows. Debris flows were common during times of hillslope colluviation (Gerson, 1982a), in contrast to absence of debris flows on the present bare bedrock hills.

The wedge was deposited on a low fill terrace, which has dull yellow (2.5YR 6/4) extremely micaceous sand beds that contrast with the grayish-

brown (10YR 5/2) sandy gravel of younger fill terraces. A 14-cm-thick paleosol remnant above the micaceous sands has argillic and calcic horizons suggestive of formation in a semiarid climate.

Locally, the bedrock colluvium contact may be about the same as before the latest episode of hillslope stripping on nearby gneissic granite slopes, as is suggested by the dark varnish and stable surface of the debris slope on the right side of Figure 3.13A. Jagged outcrops predominate whether schist is buried or exposed, so no change in outcrop roughness is apparent on slopes that have been stripped of colluvium. Thus in regard to weathering and erosion contrasts, slopes underlain by pelitic schists are much less sensitive to climatic change than are slopes underlain by gneissic gran-

FIGURE 3.11 Smooth and planar outcrops of gneissic granite. The hat in the center is on contact between rough and smooth granite; a 15-cm white scale is in the center foreground to left of contact.

ite. Colluvium that extends to the stream-channel bedrock floor, and which is interbedded with water-laid micaceous sands, indicates that colluvial wedges accumulated during an aggradation event that involved both valley floors and adjacent hillslopes. Stream-channel entrenchment allowed formation of a thin argillic soil profile (typic haplargid) on the terrace tread. Deposition of footslope colluvium buried the alluvium. Winnowing of silt and sand and varnishing of surficial rocks have continued on stable colluvial slopes.

The outputs of water and bedload from the hillslopes of Nahal Yael affected the streams, as is recorded in the stream terraces.

### 3.2.2.3 Climate-Change-Induced Variations in Sources, Erosion, Transport, and Deposition of Sediment

*3.2.2.3.1 Changes in Hillslope Processes*   The Nahal Yael research watershed is excellent for climatic geomorphology studies because the magnitudes and lengths of late Quaternary climatic changes were sufficient to change rates and magnitudes of geomorphic processes on hillslopes underlain by different lithologies. Erosional processes associated with each hillslope rock type provided different inputs of sediment and water to the streams. The Nahal Yael study underscores the importance of hillslope lithologic controls on geomorphic pro-

FIGURE 3.12 East-facing hillslope of amphibolite that has been stripped of colluvium. Note concentric weathering structures at lower right. The hat is shown for scale in the center foreground.

cesses after a change to a markedly drier or warmer climate.

Weathering processes underwent extreme changes during the late Quaternary (Table 3.2). The late Holocene has been extremely arid to arid, whereas semiarid conditions appear to have characterized latest Pleistocene and early to mid-Holocene times (Section 3.2.1.2). The moderately to strongly developed soil profiles required more than 10 ky of semiarid climate to form. They are assumed to have formed during both Pleistocene and the Pleistocene–Holocene transition semiarid climates

FIGURE 3.13 Colluvium derived from metasedimentary rocks that has been incised by Nahal Yael. A. Small remaining source area for wedge of colluvium and surficial sheet talus. A person is shown for scale in the lower left corner. B. Colluvium exposed by stream erosion. The horizontal beds in the lower right are prior stream deposits of Nahal Yael.

144

characterized by vastly more effective precipitation than at present. Thus the presence of a well-developed argillic horizon was the main reason for assigning a Pleistocene age to the old alluvial fan at the mouth of Nahal Yael.

Alluvial-fan deposits derived from Nahal Naomi record changes in weathering of granitic hillslope materials. Holocene sediments contained progressively less clay and iron oxyhydroxides as the hillslope sediment reservoir was stripped. Late Holocene fan deposits are gray to dull yellow-orange, loose, sandy gravel. Reddish-brown, hard, clayey gravel underlies brown to reddish-brown mid-Holocene deposits, and initially suggests an exceptionally thick argillic horizon. However, soil-profile horizons and clay cutans are absent and fragments of fresh biotite schist are present. The material is water-laid clayey grus that was deposited during or shortly after an extended period of moderately intense chemical weathering of granitic hillslope colluvium and bedrock. Remnants of weathered gneissic granite footslopes that have been exposed recently by stream erosion record the intensity of weathering during the most recent time of semiarid, thermic climate. The weathered gneissic granite (weathering stage IV of Table 2.6)

noted in Table 3.2 consists of interlocking crystals that resemble an outcrop whose surface has been cleaned by raindrop splash; when cut with a shovel, bright reddish-brown silty grus is revealed. Apparently, the reddish-brown, clayey colluvial soils overlying stage IV weathered bedrock were widespread on hillslopes prior to the mid-Holocene stripping event. Cumulative particle-size distribution curves (Fig. 3.14) are similar for weathered gneissic granite and fan deposits derived mainly from it. Fine gravel in the fan deposits reflects the small area of metasedimentary rocks that underlie the headwaters of 17 percent of the watershed.

Amounts and intensities of rain needed to initiate overland flow on hillslopes underlain by different materials were evaluated by infiltration tests (Salmon & Schick, 1980; Yair & Lavee, 1976,1982). The amounts of rain needed to initiate runoff during simulated rainfalls of 20 mm/hr were as follows: unjointed to sparsely jointed gneissic granite, 0.6 mm; amphibolite with closely spaced joints, 0.8 to 1.2 mm; schist with closely spaced joints perpendicular to the slope, 1.3 mm; and colluvium with 85 to 95 percent stone cover, 2.9 to 3.2 mm. A 10-mm rainfall should be enough to generate sufficient runoff to erode grus adjacent

TABLE 3.2 Characteristics of materials weathered primarily under arid (A) and semiarid (S) conditions[a] during late Pleistocene (P), latest Pleistocene and early to mid-Holocene (P&H), and mid- and late Holocene (H).

| Location | Material, Age, and Mode of Weathering | Lithology | Color | Wet Consistence |
|---|---|---|---|---|
| Nahal Yael fill terrace | Ck soil horizon (A,H) | Sandy gravel | Light yellow-orange (7.5YR 7/4) | Nonsticky, nonplastic |
| Nahal Yael fill terrace | Btb soil horizon (S,P) | Silty gravel | Bright brown (7/5YR 6/8) | Slightly sticky, plastic |
| Nahal Yael alluvial fan | Bt soil horizon (S,P) | Gravelly sandy clay | Bright reddish-brown (5YR 5/8) | Sticky, plastic to very plastic |
| Nahal Naomi hillslope | Weathered granitic gneiss (S,P&H) | Silty grus | Dull reddish-brown (5YR 5/4) | Slightly sticky, plastic |
| Nahal Naomi alluvial fan | Water-laid deposits (A,H)[b] | Sandy gravel | Dull yellow-orange (10YR 7/3) | Nonsticky, nonplastic |
| Nahal Naomi alluvial fan | Water-laid deposits (S,P&H) | Clayey, grussy gravel | Reddish-brown (5YR 4/8) | Sticky, plastic |

[a] Arid mode simply means insufficient water to form and translocate clay minerals and iron oxyhydroxides to form a Bw or Bt soil horizon, and includes both strongly seasonal arid and extremly arid hyperthermic climates. Semiarid mode means that sufficient water was available for these soil-leaching processes.

[b] The color noted here is for late Holocene deposits. Typical colors of earlier Holocene fan deposits are redder (5YR 5/4, 5/6, 6/4, and 6/6).

to outcrops of gneissic granite, thereby increasing the area of bedrock that contributes to runoff in subsequent rainfall events.

The low entrainment shear stresses for grus particles has favored complete stripping of colluvium from gneissic granite hillslopes sooner than from slopes of blocky schist colluvium. Angular blocks of varnished dike rocks remain as a lag gravel on smooth weathered granite (Figs. 3.7, 3.11). This ease of transport of hillslope weathering products makes granitic rocks highly sensitive to increases in temperature and decreases in precipitation.

Reestablishment of colluvial mantles on gneissic granite hillslopes that are less than 25 degrees involves the processes of weathering of gneissic granite and hillslope aggradation. Reversal of aggradation or degradation is associated with long response times because on hillslopes both processes are maintained by self-enhancing feedback mechanisms. Formation of colluvium on bare, smooth granitic slopes requires the stabilizing presence of vegetation to create self-enhancing feedback mechanisms favorable for progressive aggradation rather than progressive degradation of grus. The climate during times of grus accumulation would have to persist for a long time and would also have to be consistently wet or cool, or both, to eliminate severe droughts that prevent reestablishment of stabilizing hillslope vegetation. Thus hillslope stripping episodes and especially hillslope aggradation events result from large and prolonged shifts in climate, not from minor climatic variations lasting less than 1 ky. Such hillslope studies complement studies of streams, which are much more responsive to small, frequent changes in climate.

The metaamphibolitic rocks in the headwaters of Nahal Yael react to climatic change in approximately the same manner as the gneissic granite;

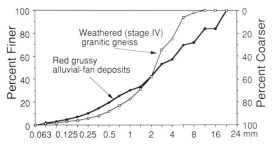

FIGURE 3.14 Cumulative particle-size distributions of samples of weathered granitic gneiss and of alluvial-fan deposits from the Nahal Naomi basin. Analyses courtesy of Asher P. Schick.

both rocks have a granitic fabric. Amphibolite outcrops are deeply weathered, and sandy colluvium mantled with blocks of dike rocks has not been completely stripped from some slopes. The amounts of rain needed to initiate runoff are one and one-half to two times that needed on slopes of gneissic granite.

Angular blocks of metamorphic and dike rocks are especially abundant on colluvial slopes of metasedimentary rocks. Both the large amount of rainfall required to initiate runoff and the large discharges needed to move angular blocks produced by weathering of schist and quartzite have resulted in less stripping of the hillslope sediment reservoir than for gneissic granite and amphibolite slopes. Colluvium is rapidly stripped from granitic hillslopes but there may be insufficient time between periods of hillslope mantling to erode colluvium completely from metasedimentary hillslopes.

Progressive burial of small schist outcrops may inhibit erosion of adjacent colluvium, because decreases in the area of exposed bedrock decrease runoff to erode colluvium. Some small ridges (Fig. 3.13A) have been sufficiently buried that the remaining outcrop area generates less than the minimum discharge needed to entrain colluvium cobbles even at times of no vegetation cover. Only a cataclysmic runoff event will promote the crossing of this stability threshold. Thus the reaction time

needed to initiate stripping of colluvium after change to an extremely arid hyperthermic climate is subject to several lithologic controls that include (1) the capacity of outcrops of different areas to generate runoff, (2) infiltration/runoff ratios of the colluvium derived from each rock type, and (3) resistance to erosion of colluviums with different particle sizes and cohesiveness.

Stripping of hillslope sediment reservoirs should decrease the frequency of debris flows. Although debris-flow deposits comprise less than 10 percent of fan and terrace deposits, they represent a process that may have been more common during times of semiarid climates than at present. Debris flows occurred on the few remaining colluviated hillslopes in the Elat area and in the eastern Sinai during the unusual February 20, 1975, storm (Gerson, 1981), but water is the mode of fluvial transport along the active stream channels.

*3.2.2.3.2 Changes in Gravel Lithologies* Descriptions of overall particle-size distribution and of gravel lithologies in alluvium of different ages provided comparisons of fluvial system outputs for the reaches upstream from sample sites in the headwaters and alluvial-fan reaches of Nahal Yael. The reach upstream from the headwaters site (Fig. 3.13B) drains the upstream third of the watershed, which is underlain by some mafic-complex rocks (40%), metasedimentary rocks (40%), and dike rocks (20%). The 1975 gravelly sand has only one-tenth as much material finer than 0.25 mm, but has more gravel than the highly micaceous Pleistocene sand (Fig. 3.15). About 24 percent and 6 percent gravel are present in the two samples, respectively. The climatically controlled suites of gravel lithologies (Fig. 3.16) are assumed to represent outputs of semiarid and arid weathering modes on hillslopes upstream from the sample sites.

A size range of 16 to 64 mm was selected for the gravel lithology counts of 200 and 228 clasts because the Pleistocene alluvium contains few clasts larger than 64 mm. The proportions of schist and pegmatite pebbles are the same in Pleistocene and modern alluvium, but the amounts of other dike rocks and amphibolite pebbles are markedly different (Fig. 3.16). Dike-rock pebbles are chiefly red quartz porphyry, which is the rock type in Nahal Yael most resistant to chemical weathering. Amphibolite is the least resistant. The markedly lower amphibolite content of gravel from the terrace deposit shows that the Pleistocene deposits were derived from hillslopes subject to stronger chemical weathering than during the late Holocene. The terrace deposits contain many clasts of acidic dike rocks compared with their areal abundance on the source hillslopes, because they are resistant to chemical weathering. The larger proportion of dike rocks in the Pleistocene stream deposits than in the present deposits suggests that schist and pegmatite, as well as amphibolite, were weathered more during Pleistocene semiarid than during late Holocene extremely arid climates.

The greater chemical weathering on the hillslopes that supplied the detritus for the Pleistocene terrace deposits also explains the fourfold lesser amounts of gravel and the extremely micaceous nature of the deposits compared with the modern stream deposits. The mica was weathered from biotite-hornblende-plagioclase amphibolite and mica schist, presumably in a moist subsoil microenvironment during a period of semiarid climate. This example of climatic controls on abundance of gravel serves as a reminder that not all increases of gravel content in ancient stratigraphic records are indicative of tectonic uplift of source areas or of increases of stream power.

Lithologies and sizes of gravel in the Pleistocene and modern alluvial fans of Nahal Yael (Fig. 3.6) represent bedload outputs from the entire watershed. Cobbles were described along randomly selected transects. Although the drainage basin is underlain by only 13 percent gneissic granite, 30 percent of the cobbles on the active fan consist of gneissic granite. Gneissic granite hillslopes occur just upstream from the fan apex. The surface of the Pleistocene fan has only 4 percent gneissic granite cobbles; they are weathered but appear to have undergone minimal reduction in size. Appar-

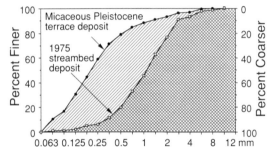

FIGURE 3.15 Cumulative particle-size distribution of samples of late Pleistocene and modern stream-channel deposits from the Nahal Yael basin. Analyses courtesy of Asher P. Schick.

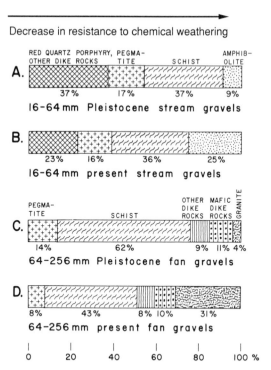

FIGURE 3.16 Climate-controlled changes in lithologies of gravel clasts weathered and transported from the hillslopes of the Nahal Yael basin. A and B are from the headwaters reach. C and D are from the alluvial fan reach.

ently fewer granitic cobbles were produced by Nahal Yael when the Pleistocene fan was deposited because of greater mantling of hillslopes by grus. Colluvial mantles on gneissic granite hillslopes resulted in more monomineralic weathering products and fewer sources of coarse gravel.

Ratios were used to describe increases of gravel lithologies relative to an index lithology that is resistant to chemical weathering. Clasts of red quartz porphyry are abundant in the 16- to 64-mm size range analyzed at the headwaters site; this lithology is most resistant to chemical weathering. Quartz porphyry does not weather to cobbles and boulders so it was not used as the index lithology for the alluvial-fan samples. Pegmatite is the most resistant lithology in the 64- to 256-mm size range. Although much coarser grained than the gneissic granite, the pegmatite has a lower biotite/muscovite ratio, which may account for its greater resistance to chemical weathering. The other dike rocks

category at the fan site consists mainly of metalamprophyres.

The ratios used as relative weathering indices (Table 3.3) are functions of the hillslope areas underlain by each lithology relative to the area underlain by the index lithology. These areas are assumed to be constant with time. The percent increase of ratios for a given lithology is a relative weathering index that describes outputs of semiarid and arid weathering modes on hillslopes upstream from the sample sites. Relative weathering indices of 350 and 1200 for amphibolite and gneissic granite, respectively, underscore the magnitude of climate-controlled changes in hillslope weathering environments for sensitive lithologies.

The relative weathering indices of Table 3.3 also provide a general ranking of sensitivity of different hillslope lithologies to climatic change. The ranking emphasizes weathering but also includes the processes of erosion, transportation, and

TABLE 3.3 Relative weathering indices of sediment yielded from Nahal Yael hillslopes during arid and semiarid weathering modes.[a]

| | Weathering Mode | | Relative Weathering Index |
| Lithology | Semiarid | Arid | (% increase from semiarid to arid) |
|---|---|---|---|
| Headwaters site | | | |
| Quartz prophyry[b] | 1.00 | 1.00 | 0 (comparative lithology) |
| Pegmatite | 0.46 | 0.70 | 52 |
| Schist | 1.00 | 1.57 | 57 |
| Amphibolite | 0.24 | 1.09 | 350 |
| Alluvial-fan site | | | |
| Pegmatite[c] | 1.00 | 1.00 | 0 (comparative lithology) |
| Schist | 4.43 | 5.38 | 21 |
| Other dike rocks | 0.64 | 1.00 | 56 |
| Mafic dike rocks | 0.79 | 1.25 | 58 |
| Gneissic granite | 0.29 | 3.88 | 1200 |

[a]Data consist of Figure 3.16 percentages of abundance of gravel clasts of different lithologies with varying susceptibilities to chemical weathering that are expressed as ratios relative to the most resistant lithology.

[b]Red quartz porphyry is regarded as the most chemically stable lithology in the 16–64-mm size range and is assigned a ratio of 1.00.

[c]Pegmatite is regarded as the most chemically stable lithology in the 64–256-mm size range and is assigned a ratio of 1.00.

FIGURE 3.17 Strath and fill terraces of Nahal Yael. A. Remnants of a strath are shown on the right side of the photo. Fill terrace capped with locally derived alluvium and colluvium is shown to the left of Nahal Yael and at the tributary stream junction in the center. Photo by Asher P. Schick. B. Shallow depth of carbonate illuviation on schist boulder on early Holocene terrace. Ground level was at 0 cm. Foliation of schist is apparent above and below a 6-cm band of calcium carbonate accumulation.

deposition; all determined the characteristics of the alluvium that was sampled. In order of increasing sensitivity to climatic change the Nahal Yael lithologies are quartz porphyry, pegmatite and schist, metalamprophyre and mafic dike rocks, and metaamphibolite and gneissic granite. Lithologic controls of sensitivity to late Quaternary climatic changes played an important role in aggrading valley floors and piedmonts.

*3.2.2.3.3 Fill Terraces* An alluvial fill along a valley integrates the effects of discharge of water and of bedload from all parts of the watershed. In multilithologic drainage basins, different reaction and relaxation times that are associated with each rock type, and microclimatic settings, cause infinite variety of fill terrace characteristics.

Maximum backfilling in headwaters reaches shows that hillslopes are the source of the perturbation responsible for valley-floor aggradation (Fig. 2.32). This implies that climate change increased yields of gravelly detritus from hillslopes or re-

duced the stream power needed to transport bedload, or both. The thickest terrace deposits are 0.5 to 1.0 km downstream from the headwaters divide of Nahal Yael. Burial of Pleistocene terraces indicates that the strongest late Quaternary aggradation event occurred during the middle Holocene.

Sensitivities of hillslope lithologies to climatic change also are reflected by the volumes of fill terraces and alluvial-fan deposits for a given drainage basin area. Nahal Naomi (70% sensitive gneissic granite) has only one-fifth the basin area of Nahal Yael (30% sensitive gneissic granite and amphibolite) but has equally thick fill terraces. Deposition of Pleistocene alluvial fans by both fluvial systems probably occurred during a pre-Holocene hillslope stripping event. Late Holocene fan alluviation in Nahal Yael more than filled the trench cut into the Pleistocene fan, but less than 1 m of gneissic granite-rich alluvium lapped onto part of the fan apex. The granitic slopes of the Nahal Naomi basin supplied sufficient debris to bury the Pleistocene fan completely. It seems that

Nahal Naomi was more sensitive to climatic changes than Nahal Yael, and that the Holocene climatic perturbation was stronger than the preceding late Pleistocene climatic perturbation.

Strath as well as fill terraces are present in Nahal Yael (Fig. 3.17A) and Nahal Naomi. The beveled bedrock of strath surfaces are capped by the gravels of fill terraces. Straths formed during times of equilibrium and are graded to the top of a waterfall, which acts as a local base-level. Poststrath processes of valley-floor aggradation and reincision of stream channels caused the fill terraces. They reflect changes in the operation of the fluvial systems that caused the streams to cross the threshold of critical power.

The terraces are not the result of tectonic activ-

TABLE 3.4  Soils and fill terrace deposits of Nahal Yael.

| Horizon | Depth (cm) | Description[a] |
|---------|-----------|----------------|
|  | +2 to +8 | Loosely packed desert pavement; varnish varies with rock type; quartz-biotite schist is gray (7.5YR 5/1), top is grayish-brown (7.5YR 4/2), and bottom is orange (7.5YR 6/6) |
| Avk | 0–2 | Light yellow-orange (7.5YR 8/4) silt; vesicular; slightly hard, slightly sticky, plastic, abrupt wavy boundary |
| Bk | 2–6 | Light yellow-orange (7.5YR 7/4) sandy gravel; massive; slightly hard, slightly sticky to nonsticky, nonplastic; clear wavy boundary; thin (<0.1 mm) incipient discontinuous carbonate films on gravel particles |
| Cu | 6–61 | Brownish-gray (10YR 6/1) to dull yellow-orange (10YR 7/3) silty gravel, and sand; bed thickness 5 to 21 cm, most 6 to 12 cm; loose to slightly hard |
| IICu[b] | 61–180 | Brownish-gray (10YR 6/1) to dull yellow-orange (10YR 7/3) sandy gravel, gravel, and sand; bed thickness 3 to 17 cm, most 6 to 12 cm; loose to slightly hard |
| IIIBtb | 186–217 | Bright brown (7.5YR 5/8) to orange (7.5YR 6/8) silty gravel; massive; hard to very hard, slightly sticky to sticky, plastic; clear wavy boundary |
| IIIBkb | 186–217 | Dull yellow-orange (10YR 7/4) sandy gravel; massive; hard to very hard, non-sticky, nonplastic; gradual, wavy boundary; 0.5- to 3-mm-thick continuous carbonate coatings on bottoms of pebbles and cobbles; horizon extends horizontally more than 4.5 m under active channel where it is buried by 10–45 cm loose gray sand and gravelly sand |
| IIICk | 217–405+ | Yellowish-gray (2.5Y 5/3) and grayish-yellow (2.5Y 5/2) to dull yellowish-brown (10YR 5/3) sand and sandy gravel; beds 2 to 25 cm thick, most 8 to 12 cm |

[a] All colors are for dry materials.

[b] Cobble weathering stages: I and II—schist, stage 1; amphibolite, stages 1 and 2. III—schist, stage 1; amphibolite, stage 2 (one stage 4).

ity. Neither the Holocene terraces along the stream channels in the mountains nor the Pleistocene alluvial fan at the mountain front is ruptured by faulting or deformed by folding. Furthermore, base-level fall at the mountain front would not be transmitted upstream past the waterfall. Both strath and fill terraces resulted from adjustments in the fluvial system to late Quaternary climatic changes.

Four meters of deposits and soil profiles were described (Table 3.4) for an alluvial fill that records two aggradation events. The surficial profile is typical of a late Holocene soil in either the Dead Sea Rift Valley or the Mojave Desert. A lightly varnished desert pavement is underlain by an incipient soil profile (a torriorthent) with no hint of a Bw horizon. A weak Bk calcic horizon at 4 to 6 cm has discontinuous thin carbonate films (stage I).

Minimal carbonate illuviation is illustrated dramatically by carbonate coatings on partially buried boulders. The schist boulder shown in Figure 3.17B has a brownish-black varnish (7.5YR 2/2), but the continuous to discontinuous carbonate coating extends only about 6 cm below the surface. Fresh grayish-green schist is present below 6 cm. Rare prolonged rains, such as 80 mm on February 20, 1975, infiltrate 0.3 to 0.8 m into the terrace deposits, but the shallow depth of carbonate illuviation indicates that infiltration typically is less than 1 decimeter (dm).

Two Holocene episodes of valley-floor aggradation occurred in Nahal Yael. Two Holocene terraces occur along the 03 tributary (Fig. 3.6) in an inset relation. Burial of the age II deposits (Table 3.4) represents an overlapping relation. The age II deposits are siltier and more poorly sorted than the overlying age I deposits, suggesting either a decrease in availability of silt or an increase in winnowing of silt.

The buried soil 1.8 m below the fill terrace tread formed on Pleistocene deposits (age III of Table 3.4). The A horizon and part of the Bt horizon were truncated by stream erosion. In contrast to 10-cm-thick Holocene soil profiles, the buried soil was more than 37-cm-thick. An argillic horizon and a 31-cm-thick calcic horizon with 3-mm-thick carbonate coatings would seem to require an interval of increased effective precipitation longer than the 5-ky early Holocene semiarid period. It appears that during the late Pleistocene an episode of moderate valley aggradation occurred; subsequent channel entrenchment allowed soil profiles to develop in a climate with much greater effective precipitation than at present.

Preservation of terraces varies with hillslope rock type. Terrace remnants are most common in reaches of schist hillslopes where they are protected by a cover of cobbly colluvium (Fig. 3.13B).

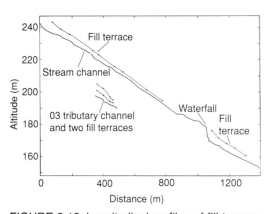

FIGURE 3.18 Longitudinal profiles of fill terraces and stream channels of Nahal Yael. Stream profiles courtesy of Ran Gerson.

Few terrace remnants are present in reaches of granitic hillslopes where runoff from a given rainfall event is larger and grussy deposits are more susceptible to erosion than for remnants in reaches of schist hillslopes.

Longitudinal profiles of Holocene fill terraces converge with the stream channel to the top of the waterfall (Fig. 3.18). Holocene aggradation did not completely bury the waterfall and a plunge pool remained at the foot of the falls. Channel slope and cross-sectional characteristics downstream from the plunge pool tend toward equilibrium conditions for efficient conveyance of water and sediment. Gradients became steeper during aggradation and gentler during degradation (Section 2.5.1.1). Both terraces in the largest tributary also converge downstream. Progressively steeper gradients during times of aggradation were one way of increasing stream power when channel pattern and hydraulic adjustments were insufficient to permit attainment of equilibrium conditions.

By now the reader might think that all aspects of geomorphic responses to climatic change have been discussed for Nahal Yael. One unusual feature remains to be pointed out–relict trails that have been preserved because of the present extreme aridity.

### 3.2.2.3 Climatic Secrets Suggested by Trails
Variations in abundance of trails on geomorphic surfaces of different ages provide clues about possible climate-controlled changes in vegetation. The trails are gently sloping benches a few decimeters wide that are clear of cobbles and boulders and commonly are anastomosing. Origins such as solifluction seem implausible for trails that cross contacts between resistant rocks, traverse fans that slope as little as 4.5 degrees, and for which no surface or subsurface evidence of mass movements can be found. Dense networks of trails occur on the Pleistocene fan (Fig. 3.19A), Pleistocene colluvium derived from metasedimentary rocks (Fig. 3.19B), and quartz porphyry dikes in gneissic granite (Fig. 3.19C). Trails are much less common on late Holocene alluvial and hillslope surfaces.

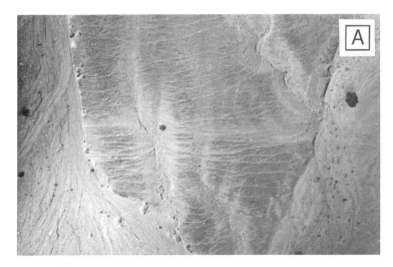

FIGURE 3.19 Trails on surfaces that are relict from a time of wetter climate. A. Pleistocene alluvial fan of Nahal Yael. The dirt road is shown in the upper left for scale. Photo courtesy of Asher P. Schick. B. Colluvium on hillslopes underlain by metasedimentary rocks. Photo courtesy of Asher P. Schick. C. Quartz porphyry dike in the Gneissic granite.

Most trails are not used because the hillslopes and alluvial surfaces are practically devoid of vegetation. Many trails do not continue onto surfaces presently undergoing erosion or deposition. Only three or four trails of a dense network cross active gullies incising the toe of the Pleistocene fan. Trails on schistose colluvium (Fig. 3.19B) end abruptly at a small valley. Trails are preserved on quartz porphyry dikes (Fig. 3.19C) but are absent on adjacent slopes where granitic-gneiss colluvium has been stripped. The discontinuous, unused trails identify landscape elements that have undergone minimal change since onset of the present extremely arid climate. A few ibex and gazelle feed on sparse riparian plants but apparently use only a few of the trails between streamcourses.

Extended droughts have prevented permanent increases in density of hillslope vegetation and in populations of grazing animals. Evidence of the effect of increased rainfall occurred following 80 mm of rain (almost triple the mean annual rainfall). Within 3 weeks the north-facing slope shown in Figure 3.19B was green with newly sprouted annual plants. This rare burst of plant growth was not repeated during the next 15 years. It seems that conditions have not favored extensive use of the trail network since the period of semiarid climate that ended at roughly 5 ka.

### 3.2.3 TRIANGULAR TALUS FACETS

Our discussion of the Nahal Yael study area compared rock controls of hillslope processes that responded to climatic change. Next we examine alternation of storage and removal in hillslope sediment reservoirs and their lithologic controls. Intervals between times of major stripping events on hillslope and major aggradation events on piedmonts may be similar.

Variations of the earth's orbital parameters are the ultimate cyclic control for the timing of late Quaternary climatic changes. The latest Pleistocene–early Holocene return of monsoonal rains to the Middle East and lower Colorado River regions were partly the result of short-term precessional variations of the earth's orbit (Fig. 2.4). Variations in the obliquity of the earth's orbit have a periodicity of roughly 100 ky and seem to be associated with longer term climatic changes. When combined with shorter term precessional effects, major interglacial climates such as the Holocene may occur.

Where does one look in the landscapes around us for changes that occur with a frequency of about 100 ky? Hillslope rather than stream subsystems should be studied because of comparatively large masses of materials to be eroded relative to the small frequency and magnitude of most erosional processes. Tectonically active humid hillslopes underlain by soft rocks erode too fast to allow partial preservation of 100–ky episodes of stripping of hillslope sediment reservoirs. Hillslopes that evolve slowly relative to the frequency of climatic changes are more likely to show evidence only of those paleoclimatic events of sufficiently long duration to have markedly changed geomorphic processes and landforms. Arid hillslopes underlain by resistant rocks are good sites for preservation of evidence for long-term responses to major climatic change. Infrequent rainfall–runoff and streamflow events of the arid realm dictate slow landscape changes.

In his studies of the extremely arid landforms of the Dead Sea Rift Valley, Gerson (1982a) described talus relics that seem to record major but not minor climatic fluctuations. These triangular relics were eroded from formerly continuous sheets of colluvium and talus that accumulated under dominantly semiarid climatic conditions in much the same manner as for hillslope sediment reservoirs in Nahal Yael. The present arid climate is responsible for the opposite mode of operation— erosion of stored hillslope colluvium.

Rates and styles of cliff retreat and talus accumulation are subject to simple, but important, lithologic controls. Triangular talus facets form most readily where resistant cliff-forming strata are underlain by weak slope-forming strata (Koons,

FIGURE 3.20 Triangular talus facets, Dead Sea Rift Valley. A. Multiple facets near Timmna where dolomitic limestone overlies more easily eroded sandstone. Facets are numbered from oldest to young-est. B. Multiple facets in the central Negev Desert where flint-rich beds overlie chalk. Facets are num-bered from oldest to youngest.

1955). Examples include hillslopes where resistant limestone and dolomite overlie soft sandstone, and where flint overlies chalk (Fig. 3.20). Talus relicts can develop on massive igneous and metamorphic rocks (Fig. 4.19). Gerson studied such relicts in the Dead Sea Rift Valley and in southwestern Arizona. Triangular talus facets are not likely to occur where thick, soft rocks lie above hard rocks.

The concept of two climate-controlled modes of operation of fluvial systems (Table 2.17) includes the modes of escarpment retreat in arid regions. Continuous 1- to 5-m-thick aprons of talus and colluvium accumulate slowly during times of semiarid climates. The talus aprons are rapidly dissected by permanent gullies that cut into the underlying softer rocks during times of arid cli-mate. Gully erosion separates triangular remnants of talus-mantled bedrock from their source cliffs. Once removed from their cliff-derived sources of water and gravel, the stable surfaces of the talus remnants change very slowly and have persistence times of several hundred ky. Minor climatic vari-ations have insufficient durations to switch the mode of operation from gullying to accumulation of new sheets of talus, or vice versa. Talus re-moval, and especially talus accumulation, are slow processes that occur in response to major and prolonged shifts in late Quaternary climate. Alternating modes of geomorphic process may result in several generations of detached talus relicts at progressively greater distances from the retreating source cliff (Figs. 3.20, 3.21).

Micron-scale stratigraphy of rock varnish on chert and flint records the sequence of major climatic shifts. Manganese-rich varnish forms during relatively semiarid climates only to be covered with more iron-rich varnish during subsequent arid climates. Rock varnish from the multiple talus relicts of the Dead Sea Rift Valley has the same number of manganese-rich layers (Dorn, 1983) as the number of semiarid periods needed to form the several generations of talus sheets postulated by Gerson (1982a).

The time needed to form several generations of triangular talus facets has yet to be firmly dated, but must be long. Accretion of alternating layers

of rock varnish also may require several hundred ky (Section 2.4.1.2.1). The aggradation phase of each remnant may require at least 25 to 50 ky of dominantly semiarid climate (Gerson, 1982a). Incisement and stripping during arid climates may occur only during times of major interglacials such as about 130 to 115 and 15 to 0 ka. The eight relatively minor episodes of interglacial climate recorded by sea-level highstands between 103 and 30 ka (Chappell, 1983; Chappell & Shackleton, 1986) probably tended to decrease rates of accumulation of talus sheets temporarily, but generally lacked sufficient duration to cause hillslopes to change modes of operation. Substantially more age control is needed from a variety of sites to determine whether one or several hillslope mantling events occurred between 115 and 15 ka. Such geomorphic studies seem to have considerable potential for paleoclimatic insight into major shifts of Quaternary climates in arid regions.

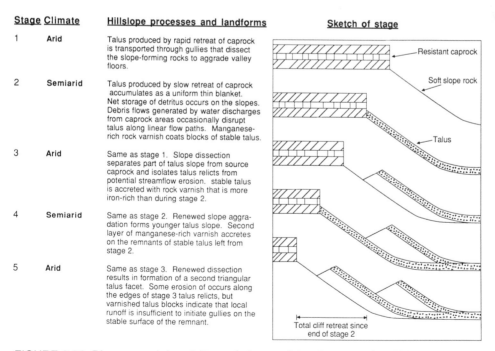

| Stage | Climate | Hillslope processes and landforms |
|-------|---------|-----------------------------------|
| 1 | Arid | Talus produced by rapid retreat of caprock is transported through gullies that dissect the slope-forming rocks to aggrade valley floors. |
| 2 | Semiarid | Talus produced by slow retreat of caprock accumulates as a uniform thin blanket. Net storage of detritus occurs on the slopes. Debris flows generated by water discharges from caprock areas occasionally disrupt talus along linear flow paths. Manganese-rich rock varnish coats blocks of stable talus. |
| 3 | Arid | Same as stage 1. Slope dissection separates part of talus slope from source caprock and isolates talus relicts from potential streamflow erosion. stable talus is accreted with rock varnish that is more iron-rich than during stage 2. |
| 4 | Semiarid | Same as stage 2. Renewed slope aggradation forms younger talus slope. Second layer of manganese-rich varnish accretes on the remnants of stable talus left from stage 2. |
| 5 | Arid | Same as stage 3. Renewed dissection results in formation of a second triangular talus facet. Some erosion of occurs along the edges of stage 3 talus relicts, but varnished talus blocks indicate that local runoff is insufficient to initiate gullies on the stable surface of the remnant. |

FIGURE 3.21 Diagram and descriptions of stages of development of multiple generations of triangular talus facets, where arid and semiarid climates alternate at 100-ky intervals.

## 3.3 Lithologic Controls of Sensitivity of Geomorphic Processes to Climatic Change

The study of geomorphic processes associated with the rock types of Nahal Yael followed the theme that hillslope lithology plays an important role in determining sensitivity of geomorphic processes to climatic change. Thus it is appropriate now to summarize the important aspects of rock control on geomorphic responses to climatic changes.

Rates of rock weathering, plant growth, and erosion on hillslopes undergo minor or major changes for a given late Quaternary climatic change. Magnitudes of the resulting geomorphic changes are in large part a function of rock types. Lithologic controls on sensitivity of geomorphic processes to climatic change is evaluated by examining hillslope inputs to streams, and outputs of bedload from drainage basins. Data from thermic to hyperthermic, extremely arid to semiarid drainage basins in the southwestern United States, northern Mexico, Israel, and the Sinai Peninsula are used to define basic concepts that can be adapted to evaluate rock control of geomorphic processes in climatic settings that range from humid and tropical to periglacial. Locations of most of the study sites are shown in Figures 2.1 and 3.2.

### 3.3.1 AMOUNTS AND TYPES OF HILLSLOPE SEDIMENT YIELD

Weathering of rocks is influenced by hillslope microclimates, vegetation, and lithologic properties such as mineralogy, microscopic fabric, and macroscopic petrologic structures. Intergranular openings, bedding planes, joints, fractures, foliations, and shear planes control movements of weathering solutions through rocks. Weathering microclimates on hillslopes change markedly where mantling colluvial materials are eroded from bedrock. Colluvium-mantled bedrock remains moist for weeks or months after being wetted by infiltrating rainfall or snowmelt. Thus a change from buried to bare outcrops that dry rapidly after rainfalls may decrease weathering rates by several

orders of magnitude. The magnitude of the change in weathering rates is small for some rock types and large for others. The latter can be regarded as relatively more sensitive to burial or exposure caused by climatic change than the former. Rock types are ranked in order of decreasing sensitivity to change from semiarid to arid climate in Table 3.5.

Hillslopes underlain by sensitive rock types undergo large changes in weathering rates, sediment yields, and abundance of boulders transported by streams as a result of a climatic change. Granitic rocks, such as granite, quartz monzonite, and diorite, may be regarded as being sensitive to climatic change because they weather rapidly in moist microclimates and slowly in dry microclimates (Budel, 1957; Wahrhaftig, 1965). Hydrolytic weathering and oxidation cause biotite to expand and fracture adjacent feldspar and quartz crystals; porosity and permeability increase, and rock strength decreases. Such disaggregation of of granitic rocks into grus is further enhanced by crystallization of salts derived from rain and dust on many arid hillslopes (Goudie, 1978; Gerson et al., 1985). Geomorphic processes on hillslopes underlain by micaceous granitic rocks in Nahal Yael–gneissic granite and amphibolite–were very sensitive to climatic change (Sections 3.2.2.2, 3.2.2.3.2). Sediment-yield changes were extreme, and changes in amounts of gravel supplied to streams were much larger than for the less sensitive rocks of the drainage basin.

Hillslopes underlain by sensitive rock types undergo large changes in weathering rates, sediment yields, and abundance of boulders transported by streams as a result of a climatic change. Granitic rocks, such as granite, quartz monzonite, and diorite, may be regarded as being sensitive to climatic change because they weather rapidly in moist microclimates and slowly in dry microclimates (Budel, 1957; Wahrhaftig, 1965). Hydrolytic weathering and oxidation cause bioltite to expand and fracture adjacent feldspar and quartz crystals; porosity and permeability increase, and rock strength decreases. Such disaggregation of

TABLE 3.5 Ranking of sensitivity of different rock types to change from semiarid to arid climate in hot deserts.

| | Hillslope Subsystem | | Fluvial-System Output |
|---|---|---|---|
| Decreasing Contrast of Subaerial and Subsoil Weathering Rates | Decreasing Contrast in Density of Hillslope Vegetation Cover | Increasing Shear Stress Needed to Erode Hillslope Weathering Products | Decreasing Thickness of Holocene Fill Terraces in 0.5–10-km² Drainage Basins |
| Coarse-grained, biotitic, porphyritic granite to diorite | Thin-bedded limestone and dolomite | Coarse-grained granitic rocks, mudstone and shale, soft calcareous sandstone | Coarse-grained biotitic, porphyritic granite to diorite, thin-bedded limestone and marble |
| Gneissic granite, soft calcareous sandstone | Coarse-grained granitic rocks, gneissic granite | Thin-bedded limestone and dolomite, fine-grained granitic rocks, amphibolite | Gneissic granite, soft volcanic tuffaceous rocks |
| Fine-grained granitic rocks, thin-bedded limestone, basalt, and shale | Biotite schist, calcareous sandstone, basalt | Biotite schist, quartz-biotite gneiss, calcareous sandstone, shale | Fine-grained granitic rocks, gneiss soft calcareous sandstone, shale |
| Quartz-biotite schist, andesite, and dolomite | Andesite, quartzite, welded tuff | Quartzite, andesite | Biotite-plagioclase schist, phyllite, metaamphibolite |
| Welded tuff, quartzite | Quartzite, welded tuff | Basalt, welded tuff | Basalt, andesite, quartz-biotite schist |

*(Left margin, rotated: Sensitivity to climatic change, with upward arrow)*

granitic rocks into grus is further enhanced by crystallization of salts derived from rain and dust on many arid hillslopes (Goudie, 1978; Gerson et al., 1985). Geomorphic processes on hillslopes underlain by micaceous granitic rocks in Nahal Yael—gneissic granite and amphilbolite—were very sensitive to climatic change (Sections 3.2.2.2, 3.2.2.3.2). Sediment-yield changes were extreme, and changes in amounts of gravel supplied to streams were much larger than for the less sensitive rocks of the drainage basin.

Hillslopes underlain by insensitive rock types undergo minor changes in geomorphic processes as a result of a climatic change. Basalt may be regarded as an insensitive lithology, even though its mafic composition favors rapid chemical decomposition. Although vesicular basalt has a high porosity, its permeability is low because blocks

between cooling joints of lava flows generally have few fractures. An overall low surface area for percolating solutions to cause hydrolytic alterations results in weathering of basalt blocks that proceeds only as rapidly as weathering rinds move into fresh rock. Basalt weathers rapidly in humid climates and slowly in arid climates. A 170-ka basalt flow in the Cima volcanic field (Fig. 2.1) is virtually unweathered (Wells et al., 1984,1987). Weathering characteristics of primary flow structures on K/Ar-dated basalt flows in northern Mexico and southwestern Arizona indicate that more than 5 Ma are needed to weather 1 dm of surficial basalt.

Chemical composition, crystal size, fabric, and macrostructures affect contrasts in subaerial and subsoil weathering rates of granitic (Budel, 1957; Wahrhaftig, 1965; Yaalon, 1971), carbonate (Gerson & Yair, 1975), and many other rocks. Chemically inert quartzite weathers slowly in both subaerial and subsoil microclimates. The importance of crystal size is illustrated by fine-grained welded tuff, or basalt, which weathers more slowly than chemically similar coarse-grained granite, or gabbro. Biotite is especially unstable in subsoil weathering environments; the presence of several percent of this mica makes plutonic lithologies much more sensitive to changes in weathering microclimates. Metamorphic rocks commonly have a fabric of tightly knit intergrown crystals that make them more resistant to weathering than chemically similar rocks with isotropic fabrics.

The marked contrast between subaerial and subsoil weathering of granitic rocks is obvious when recently exhumed bedrock is compared with adjacent outcrops that have been exposed to continuous subaerial weathering. Subsoil hydrolytic weathering is greatest along protruding irregularities that are subject to several converging weathering fronts. Even 0.5 m of alluvium (Fig. 3.22) or colluvium (Figs. 3.7, 3.11) cover acts as a smooth mantle that prevents flowing water from eroding preferentially along joints and shears to create the rough outcrops that are typical of exposed rock. Instead, subsoil weathering tends to transmit the smooth surface of the gently sloping alluvium mantling the pediment to the underlying quartz monzonite. Occasional stripping of the mantling materials allows fluvial processes to remove weathered rock, thereby completing a process that Mabbutt (1966) described as mantle-controlled planation of pediments.

Rock controls of density of vegetation cover during times of changing climate are complex and involve plant nutrients, moisture retention, and drainage. General appraisal of this parameter and how it varies with change from a semiarid to an arid climate was evaluated in two ways. Density and types of vegetation on arid and semiarid hillslopes in several parts of Arizona were compared qualitatively for different rock types. The second approach was to use sites in Arizona, California, and Nevada to compare densities of plant cover on hot, south-facing slopes and cool, north-facing

slopes underlain by the same lithology. Interestingly, carbonate rocks showed the most variation (Fig. 3.23B). Apparently, fine-grained, nutrient-rich soils are weathered from carbonate rocks in semiarid and humid climates but not in arid and extremely arid climates. Coarse-grained granitic rocks ranked high in Table 3.5 because granitic hillslopes typically support a good growth of grasses in semiarid but not in arid climates. Schist, sandstone, and shale had intermediate densities of plant cover and quartzite had the least.

The variety of Paleozoic and Mesozoic sedimentary rocks in the semiarid Swisshelm Mountains of southeastern Arizona provide many examples of microclimatic controls on density and type of plant cover (Fig. 3.23B). This Chihuahuan Desert plant community is a juniper-oak-pinyon pine-grass woodland with patches of manzanita, ocotillo, acacia, creosote bush, and mesquite. Sandpaper bush is common in many microclimates, and *agave* and *opuntia* succulents flourish on the more arid slopes. Extreme contrasts in density of plant cover on north- and south-facing hillslopes occur on the limestone slopes at the upper left. Minimal contrast occurs between north- and south-facing hillslopes underlain by quartzite (small area at right) and dolomitic sandstone near the range front. Intermediate plant density contrasts occur on shale, sandstone, and volcanic rocks.

Both volume and size of sediment stored in hillslope sediment reservoirs are included in the general category of shear stresses needed to erode hillslope weathering products. Erodibility of hillslope weathering products is largely a function of abundance of boulders. Basalt and welded tuff are classed as insensitive rocks in Table 3.5 because unfractured blocks between cooling joints weather slowly to fine materials in arid climates. Large forces are required to move the resulting hillslope lag gravel of large blocks downslope into streams. At the other extreme is sensitive coarse-grained biotite granite, which weathers readily to monomineralic decay products (grus) that are readily transported downslope by small flows generated by minor rainfall events (Section 3.2.2.2.1).

FIGURE 3.22 Contrasts in subaerial and subsoil weathering of quartz monzonite, exhumed pediment, Coxcomb Mountains. Continuously exposed outcrops are rough and are solid when struck with a hammer; they have rock varnish and tafoni (cavernous weathering features). The pediment has been exhumed from beneath a thin mantle of grussy alluvium during the late Holocene. The pediment exposures are smooth and punky when struck with a hammer; they lack rock varnish and tafoni.

Contrasts between hillslope weathering rates, vegetation cover, and sizes of weathered detritus vary greatly with rock type. These factors are important controls on variations in sediment yielded from drainage basins underlain by different lithologies.

### 3.3.2 AMOUNTS AND TYPES OF DRAINAGE BASIN OUTPUT

Thicknesses of alluvial fills along valley floors is a parameter that integrates the hillslope processes described in the first three columns of Table 3.5, and includes the stream processes of transport and deposition. Thus this aspect of lithologic control of sensitivity of geomorphic processes to climatic change concerns outputs of fluvial systems, not just hillslope subsystems. It is not surprising that basins underlain by coarse-grained, micaceous, granitic rocks have thick fill terraces and that basins underlain by quartzite have thin fill terraces. Thin-bedded carbonate rocks rank surprisingly high (Fig. 3.24A); some basins are as responsive to climatic change as micaceous quartz monzonite.

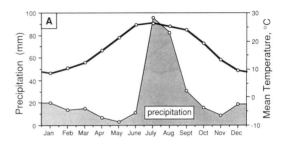

FIGURE 3.23 Climatic and lithologic controls on plant communities in southeastern Arizona. A. Monthly variations of precipitation and temperature at Tombstone, Arizona; altitude, 1405 m. Data from U.S. Department of Commerce (1985). B. View toward west of the central Swisshelm Mountains in the southeastern corner of Arizona. Altitude of foreground valley is 1400 to 1500 m; ridgecrest is 2000 to 2300 m . The contrast in vegetation density on north- and south-facing slopes is in part controlled by lithology. Contrasts are greatest for limestone and least for quartzite. These spatial contrasts appear to be the same as for the sensitivity of hillslopes underlain by different rock types that have undergone climatic change during the late Quaternary. Key: ls, limestone; dol ss, dolomitic sandstone; Q2, Pleistocene fan; Q3, Holocene fan; sh, shale; ss, sandstone; qtzite, quartzite.

Other rocks vary in their influence on geomorphic processes. Amphibolite with a metamorphic fabric behaves more like schist (Fig. 3.24B), but where the amphibolite has a granitic fabric (Nahal Yael), its geomorphic response to climatic change is similar to that of coarse-grained granite (Section 3.2.2.2.2).

Characteristics of piedmont deposits derived from drainage basins underlain by sensitive lithologies are much different from those derived from insen-

sitive lithologies. Large differences in size and abundance of boulders and in sorting of alluvium characterize the arid and semiarid output modes of drainage basins that are sensitive to climatic change (Figs. 2.12A, 2.13; Tables 2.17, 3.3). Piedmont deposits derived from drainage basins underlain by insensitive lithologies have characteristics that are virtually the same for both arid and semiarid modes of fluvial system operation. Late Pleistocene alluvium derived from basaltic

FIGURE 3.24 Contrasts in Holocene fill-terrace heights for drainage basins of similar size that are underlain by different rock types. A. Forty- to fifty-meter-high fill terrace in Mosaic Canyon, Death Valley; basin is underlain mainly by limestone and volcanic rocks. B. Two and one-half-meter-high fill terrace, Whipple Mountains; basin is underlain by schist and andesitic agglomerate. Person is pointing at alluvium-bedrock contact.

watersheds of the Dead Sea Rift Valley and the Mojave Desert is bouldery and has the same bar-and-swale topography as the Holocene alluvium derived from the same source areas. This lack of contrast between arid and semiarid output modes for basaltic watersheds implies minimal changes—low sensitivity—to late Quaternary climatic changes.

The response times of fluvial systems to cross-ing the threshold of critical power after the Pleistocene–Holocene climatic change in the Mojave Desert also was a function of rock type. Pre-Cambrian schistose rocks occur at the mountain front and marble and schist occur at the crest of the Marble Mountains (Fig. 3.25). Late Holocene (Q3c) detritus has practically buried the smooth black-pavement Pleistocene (Q2c) surfaces at the

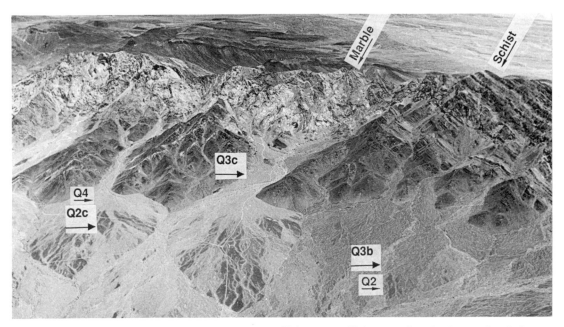

FIGURE 3.25 Response times of fluvial systems to Pleistocene–Holocene climatic change for drainage basins underlain by marble and schist (left) and by schist (right), southwest side of the Marble Mountains. The relaxation time (elapsed time between the climatic perturbation and the crossing of the threshold of critical power as each system switched from aggradation to degradation) was all of the Holocene for the marble basins and only until mid-Holocene for the schist basins. Dirt roads are shown for scale. See Table 2.13 for ages of alluvium.

apexes of fans whose streams head in the range crest and have backfilled the broad valley floors upstream from the fans. The late Holocene aggradation surface is entrenched as much as 18 m in the valleys, but is still unentrenched on the fan. The threshold intersection point (Fig. 1.8) has moved progressively downstream; it has been time transgressive. Thus volumes of hillslope sediment reservoirs are large, relative to available stream power. An aggradation event may span 10 ky, during which a diachronous fill terrace tread is formed. This appears to be the case for the fluvial systems underlain largely by sensitive carbonate rocks.

The small Holocene (Q3b) alluvial fans shown on the right side of Figure 3.25 are derived entirely from hillslopes underlain by schist and are entrenched by their trunk stream channels. The re-

sponse time for the schist fluvial systems may have been only 2 ky. Fluvial systems underlain by carbonate rocks have taken much longer to respond to the climatic perturbation than those underlain by schist because (1) larger transport distances are longer in these drainage basins, (2)– thicker hillslope sediment colluvium was available to be stripped, and (3)– moderately larger bedload clasts of marble may have required more time than schist clasts to be transported a given distance down valleys.

Lithologically controlled response times of schist watersheds are applicable to much of the Mojave Desert (Fig. 3.26). The general lack of Pleistocene alluvium in the mountains despite its abundance on piedmonts is indicative of prolonged erosion or equilibrium conditions, or both, prior to the exceptionally strong Pleistocene–Holocene climatic

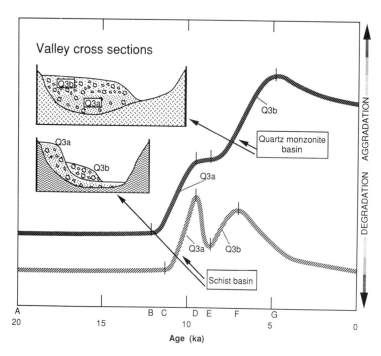

FIGURE 3.26 Time lags of responses to Pleistocene–Holocene climatic change in relatively sensitive (quartz monzonite) and insensitive (schist) drainage basins in hot deserts. Time of presumed climatic change is 12 ka. Top valley cross section shows bouldery Q3b alluvium deposited without interruption on less bouldery Q3a alluvium. Bottom valley cross section shows bouldery Q3b alluvium deposited as an inset terrace after entrenchment of less bouldery Q3a alluvium. Valley cross sections show single fill spanning Q3a and Q3b time for the quartz monzonite basin, and two ages of fill terraces for the schist basin.

perturbation. The horizontal (equilibrium) trend of the aggradation-degradation line (time span A to C) is indicative of equilibrium conditions between 20 and 11 ka. The return of monsoonal thunderstorms resulted in a strong aggradation peak (time D) whose threshold had been passed by 9 ka, thereby allowing most of the Holocene for the development of Q3a soil profiles. The second upswing of the line (time span E to F) represents the second aggradation event after a brief period of stream-channel entrenchment of the Q3a fill. The second event was associated with hotter and drier climates that began at 8.5 to 8 ka. The typical valley floor stratigraphy reveals this Q3b alluvial fill inset into Q3a alluvium. The contrast between the Q3a and Q3b soils is indicative of the time span between the two thresholds and of a slightly higher leaching index between 9 and 8 ka than after 8 ka. The magnitude of the 9- to 8-ka stream-channel degradation between times D and E, and the time between the two threshold peaks D and F, also are functions of the quickness of fluvial

system response to multiple climatic perturbations—the relaxation time of Figure 1.3.

Relaxation times also were affected by amounts of available stream power. Relaxation times tend to be longer in downstream reaches because ephemeral streamflows are less frequent (Figs. 1.6, 1.7). Relaxation times also are longer in relatively more arid basins because less total stream power per unit time is available to entrain and transport bedload supplied for a given rate of erosion of the hillslope sediment reservoir.

Minor departures from the schist plot of Figure 3.26 reflect lithologic controls that cause slightly different geomorphic responses for fluvial systems underlain by gneiss, sandstone, mixed volcanic rocks, and granitic rocks with low mica content. Relative thicknesses and heights of the Q3a and Q3b fill terrace describe such spatial variations in lithology.

The quartz-monzonite line of Figure 3.26 describes the case of extreme sensitivity to the effects of the Pleistocene–Holocene climatic change in

FIGURE 3.27 Flood of Holocene granitic alluvium choking the valleys and spilling onto the piedmont of the western Riverside Mountains. The sheared and altered granitic rocks are sufficiently sensitive to the Pleistocene–Holocene climatic change that all of the Holocene has been needed to transport the large volume of alluvium through the mountain valleys to the piedmont. See Table 2.13 for ages of alluvium.

the Mojave Desert. Some exceptionally sensitive quartz monzonite hillslopes produced such a large surge of debris in response to the Pleistocene–Holocene climatic change that valley floors are choked and Pleistocene alluvial fans are being buried (Fig. 3.27). Aggradation in such basins occurred more rapidly after the 12-ka climatic perturbation (time B compared with time C in Fig. 3.26). This shorter reaction time occurred mainly because of finer materials in the hillslope sediment reservoir. The volume of detritus that had accumulated in the hillslope sediment reservoir was so large that only part of the colluvium had been stripped by 9 ka (time D). Aggradation peaked at 5 ka in a single Holocene threshold (time G) because the time between 9 and 8 ka (D to E) was a time only of reduced rates of aggradation, not a reversal of modes of operation. The system shown in Figure 3.27 probably peaked in the late Holocene. Stratigraphy of the resulting valley-floor alluvium reveals a single depositional unit that lacks buried soils indicative of a depositional hiatus, and Q3b deposits have buried Q3a deposits (see Fig. 3.26 inset). The lithologic controls of geomorphic responses to climatic change in basins underlain by coarse-grained, micaceous, granitic rocks or by thin-bedded limestone thus are characterized by short reaction times and long response times compared with basins underlain by less sensitive rock types such as schist. In Figure 3.26, the relaxation times to the perturbation caused by the onset of monsoonal rains are 2 ky for the schist basin and 7 ky for the quartz monzonite basin.

Chapters 2 and 3 have introduced the subject of geomorphic responses to climatic change by developing concepts of landscape change that are applicable to streams and hills that formerly were semiarid but now are arid or extremely arid. Many of these concepts also apply to landscape changes in the humid realm. The semiarid to subhumid and humid sites of Chapters 4 and 5 provide opportunities for discussion of impacts of climatic change in larger drainage basins in rapidly rising mountains as well as for some interesting modifications of concepts introduced in Chapters 2 and 3.

# Climatic Geomorphology of a Lofty, Semiarid to Subhumid Mountain Range

## 4.1 Introduction

### 4.1.1 CONTEXT, PURPOSE, AND SCOPE

Discussions of geomorphic responses to Pleisto-cene–Holocene climatic change are expanded in Chapters 4 and 5 to include tectonically active large and lofty fluvial systems that produce frequent floods. The arid to semiarid thermic climatic setting (see Table 2.1 for definitions) of Chapter 2 and the extremely arid hyperthermic climatic setting of Chapter 3 were useful for introducing basic definitions; for discussing dating, aggradation, and soils genesis of alluvial geomorphic surfaces; and for outlining important lithologic controls and hillslope process-response models. The next step is to evaluate landscape change in tectonically active watersheds. For completeness we also need to discuss geomorphic responses to climatic changes in a semiarid to subhumid climate (Chapter 4) and in a humid climate (Chapter 5), where rates of weathering and pedogenesis, hillslope erosion, and piedmont deposition may be one to two orders of magnitude more rapid than in tectonically inactive hot deserts. The lofty San Gabriel Mountains in the central Transverse Ranges

of coastal southern California (Fig. 4.1) provide an excellent setting in which to add new dimensions to our understanding of the behavior of hills and streams. The topics covered in this chapter parallel those covered in Chapter 2. However, under each topic a great variety of new subjects is explored.

This chapter is primarily a study of fill terraces. It discusses their topography, sedimentology, and soils, and ages; summarizes the importance of climatic factors in their genesis, and defines a terrace soils chronosequence. It then addresses some key questions: Can the effects of climatic change be separated from those of continuing and large amounts of intermittent uplift along thrust faults? If so, in which parts of fluvial systems might the impacts of climatic change best be evaluated? Do climatic perturbations ever dominate over the effects of rapid concurrent uplift? Can fill-terrace treads be regarded as time lines that pass through tectonically deforming landscapes, or are they time transgressive?

Most watersheds drain southward; they are excellent for studying of the effects of climatic change on subhumid fluvial systems in a tectonically ac-

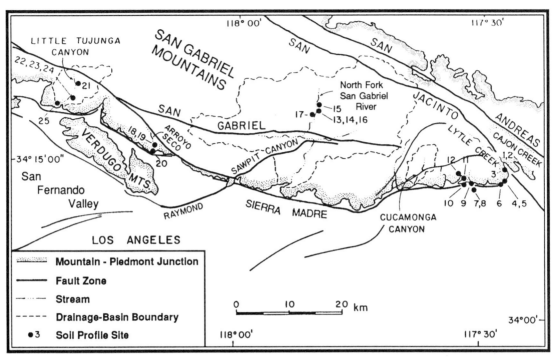

FIGURE 4.1 Locations of the Little Tujunga, North Fork of the San Gabriel River, Cucamonga Canyon, and Cajon Creek terrace study areas, soil description sites, and fault zones of the San Gabriel Mountains, southern California. Numbers 1 through 17 indicate pedon locations; see Tables 4.5 through 4.9 for descriptions.

tive environment. The drainage basins of the North Fork of the San Gabriel River, Little Tujunga Creek, and Cucamonga Canyon (Fig. 4.1) were selected for detailed studies. The superb flight of fill terraces in the North Fork is distant from active range-bounding faults. The other two basins have strath and fill terraces immediately upstream from active range-bounding fault zones. Concurrent climatic change and tectonic base-level fall affected a prominent mid-Holocene terrace (T7) in all three basins. (See Table 1.2 for estimated ages of the Quaternary.)

### 4.1.2 GEOMORPHIC SETTING

The San Gabriel Mountains form the central part of the Transverse Ranges, which extend from the Pacific Ocean into the Mojave Desert (Fig. 4.1).

The range crest increases in altitude toward the east to 3070 m, and most of the crest has altitudes of 1500 to 2000 m. Uplift rates of 1 to 3 m/ky (Bull, 1978; Bull, Menges, & McFadden, 1979; Matti et al., 1982) have exceeded late Quaternary denudation rates (Schumm, 1963). The resulting 100–km-long range forms an impressive orographic barrier that traps precipitation from cyclonic storms as they start to sweep inland from the nearby Pacific Ocean.

Wet years are times of maximum geomorphic work. Rates of fluvial processes are similar to those in weakly seasonal, humid parts of New Zealand (Chapter 5). Both study areas have a high density of landslides and broad, gravelly, braided stream channels that are swept clean of vegetation by frequent floods. The combination of rapid weathering of crushed and sheared granitic and

TABLE 4.1 Summary of geomorphic and geologic characteristics of drainage basins of North Fork of San Gabriel River, Little Tujunga Creek, and Cucamonga Canyon.[a]

| Drainage Basin | Basin Area (km²) | Basin Length (km) | Mean Slope Value[b] | Major Basin Lithologies[c] |
|---|---|---|---|---|
| North Fork of San Gabriel River | 44.3 | 13.6 | 47 | GR 95%, MET 5% |
| Little Tujunga Canyon | 49.0 | 13.4 | 50 | GR 50%, SED 45%, MET 5% |
| Cucamonga Canyon | 34.2 | 10.0 | 79 | MET 60%, GR 40% |

[a] Map analyses by R. H. Peterson and L. D. McFadden.

[b] Mean slope is the average of randomly selected points from detailed topographic maps (Strahler, 1956).

[c] GR—chiefly intermediate to leucocratic plutonic and gneissic rocks; MET—chiefly mesocratic to melocratic metasedimentary and metaplutonic rocks; SED—Cenozoic basin sediments of marine and nonmarine origin.

metamorphic rocks, steep slopes (note the high mean slope values of Table 4.1), and frequent winters with abundant, intense rain cause rapid erosion. More than 400 concrete check dams and debris basins protect urban communities of the greater Los Angeles area from flooding, but flood-hazard control is costly (Cooke, 1984). Data from these sediment traps suggest a present mean denudation rate of >1.5 m/1000 yr for the San Gabriel Mountains.

Dense chaparral brush covers south-facing mountainsides below about 1400 m and is prone to intense and widespread burning during the long summer drought (Fig. 4.20). Conifers and deciduous trees are more common on north-facing slopes which have greater effective precipitation. Alders, willows, cottonwoods, and perennial oaks are common riparian trees.

Highly seasonal precipitation and destructive brush fires contribute to high sediment yields (Wells, 1981; Wells & Brown, 1982). An overall decrease of denudation rate with increasing basin size is shown for the 17 drainage basins represented in Figure 4.2. Variations in slope steepness, erodibility of surficial materials, and frequency of fires in the watersheds also account for much of the variability of basin sediment output during the 40

to 50 years of record. Local denudation rates may exceed 30 mm/yr during the first few years after forest fires. The effect of fire on erosion rates underscores the importance of protective plant cover on hillslope processes. Late Quaternary climatic changes also caused profound changes in types and density of plants that protected hillslopes from rain.

### 4.1.3 CLIMATE

#### 4.1.3.1 Regional Climatology

The interactions of airmasses in this part of North America (Section 2.1.1) cause a semiarid to humid strongly seasonal climate in the San Gabriel Mountains (Fig. 4.3). Winter precipitation increases rapidly with increase in altitude from less than 500 to more than 1000 mm. On the southern flank of the range, the climate ranges from strongly seasonal semiarid and moderately seasonal thermic (Table 2.1) on the piedmont to strongly seasonal subhumid and moderately seasonal mesic in the mountains below about 1400 m. The higher parts of the range are humid and mesic to frigid. A pronounced rainshadow is present on the northern lee side of the mountains, which declines to the arid western Mojave Desert.

A Mediterranean climate is present in which long dry summers are followed by mild wet winters. About 500 to 800 mm of precipitation falls annually at the low altitudes where soil profiles and stream terraces were studied. Adjacent hillslope subsystems generally receive 500 to 1500 mm of winter precipitation annually, much of which occurs as snow. Less than 200 mm of precipitation may fall at a weather station one winter and 1000 mm may fall the next. Orographic uplift of storm clouds triggers intense rainfalls: intensities of 566 and 663 mm have been recorded in 24-hour periods. Until 1956 a weather station in the San Gabriel Mountains held the world's record for 1-minute rainfall intensity.

The latitude at which Pacific Ocean storms originate varies greatly and affects the temperature of precipitation in southern California. Winter storms that originate in the Gulf of Alaska and move down the Pacific coast are cold and commonly bring snow to altitudes as low as 500 to 1000 m in the San Gabriel Mountains. Cumulative precipitation amounts may be large, but snowmelt generally is so gradual that runoff causes minimal hillslope erosion or bedload transport in streams. Many winter storms originate south of the Gulf of Alaska, and some troughs incorporate moisture from subtropical oceans. These warm storms may bring rain to altitudes higher than 2000 m, and intense 30- to 60-mm bursts of rain are typical. Subtropical storms can form west of Mexico in years of high sea-surface temperatures (Douglas, 1976,1981; Pisias, 1978). Intense rainfalls associated with tropical depressions in September and October produced large debris flows in the San Gabriel Mountains during the late 1970s.

### 4.1.3.2 Paleoclimatology

Meager evidence for the times and types of climatic change during the late Quaternary for the central Transverse Ranges can be gleaned from diverse nearby areas. These include coastal marine terraces, the Mojave Desert, the ocean floor of the Santa Barbara Channel of the western Transverse Ranges, and the lofty White Mountains northwest of the Mojave Desert.

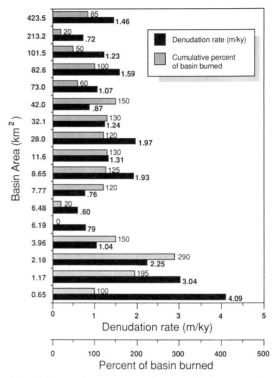

FIGURE 4.2 Comparison of watershed area and denudation rate for 17 drainage basins in the San Gabriel Mountains, based on reservoir sedimentation surveys made between 1920 and 1975. Data from Taylor et al., 1978. Cumulative percent of basin burned exceeds 100 where the sum of areas of repeated burns exceeds the drainage basin area.

The well-dated sea-level highstand that occurred at 129 to 123 ka (Edwards et al., 1987) was 6 m higher than the Holocene highstand and was the only highstand of the past 125 ky associated with sufficiently warm waters to favor growth of solitary corals along the southern California coast (Muhs, 1982,1983,1985; Muhs & Szabo, 1982). Unusually warm sea-surface temperatures at about 125 ka should have favored stronger and more frequent tropical storms in the San Gabriel Mountains.

Sedimentologic, pedogenic, and paleobotanic data from the eastern Mojave Desert provide an

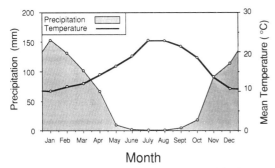

FIGURE 4.3 Monthly variations of precipitation and temperature at an altitude of 600 m in the North Fork of the San Gabriel River, where the climate is strongly seasonal subhumid and moderately seasonal thermic.

internally consistent picture of four types of climate that have characterized the region during the past 20 ky—full glacial, transitional, monsoonal, and interglacial (Section 2.1.2). Plant fossils from packrat middens reveal a glacial-interglacial transitional climate from about 15 to 12 ka. Increase of monsoonal rains between about 12 ka and 7.8 ka separated times of juniper woodland from times of desert scrub since about 7.8 ka in both the eastern (Van Devender, 1973,1977; Van Devender et al., 1987) and western (King, 1976) Mojave Desert. Widespread eolian activity occurred between about 7.5 and 4 ka (Smith, 1967). Then return to slightly wetter conditions allowed vegetation to stabilize many mid-Holocene sand dunes.

Cores of varved marine sediments from a deep basin in the Santa Barbara Channel contain pollen and diatoms that record paleoecologic changes of the past 12 ky. Huesser (1978) concluded that upland coniferous plant communities and wet-cool climates prevailed between about 12 and 7.8 ka and were succeeded by a dominance of oak and aster pollen during a warmer, drier period that culminated about 5.7 ka. Chaparral and coastal sage scrub plant associations became more important at about 2.3 ka. Variations in radiolaria fauna indicate that the warmest Holocene sea-surface temperatures occurred between about 8 and 5.5 ka. Another faunal change at about 5.4 ka is

indicative of a marked cooling of sea-surface temperatures (Pisias, 1978).

Tree rings of bristlecone pines (LaMarche 1973,1974) in the White Mountains provide paleoclimatic information in the arctic-alpine life zone. Dating of dead bristlecone pines above present upper tree line indicates that summer temperatures were about 2°C higher than those of the past few centuries during a mid-Holocene warmer period that started before 7.4 ka and ended at about 4.2 ka. Cooler, wetter climates occurred between 3.5 and 2.5 ka.

The times of past climatic change suggested by the diverse studies are essentially the same in the three areas, which suggests shifts in regional air-mass patterns that would also affect the San Gabriel Mountains. It seems likely that effective precipitation was sufficient to support forests on south- as well as north-facing slopes until about 7.8 ka. The mid-Holocene from about 7.5 ka to 4.2 ka was a time of warmer temperatures than at present, which should have decreased protective plant cover throughout the mountains and probably eliminated trees on hot south-facing slopes at low altitudes. Effective precipitation increased during the late Holocene, but trees are not common on most south-facing slopes below 1300 m.

## 4.2 Terrace Soils Chronosequence

Well-preserved soil profiles on stream terraces in fanhead reaches provide data for describing a chronosequence of aggradation events resulting from climatic perturbations in the San Gabriel Mountains. Soil profiles and cobble weathering stages were described at sites (Fig. 4.1) that have undergone minimal erosion or deposition since formation of the terrace treads (Bull et al., 1979; McFadden et al., 1982). Contrasts in soils genesis also may provide clues about Holocene and Pleistocene climates.

### 4.2.1 SOIL-FORMING FACTORS

Use of a soils chronosequence for defining relative terrace ages assumes that the passage of time accounts for most differences between soil profiles—other soil-forming factors (parent material, climate, topography, and biota) should be similar. Vegetation on young terrace treads prefers well-drained gravel; vegetation density ranges from bushes and annual plants to dense waxy chaparral scrub. Terrace-tread topography consists of planar to gently undulating surfaces that slope 1 to 5 degrees.

Parent materials for soils on stream-terrace treads are functions of hillslope rock types and of fluvial and eolian processes. The downstream half of the Little Tujunga basin is underlain by soft marine mudstones and siltstones and by poorly indurated terrestrial sandstones and conglomerates. Schist underlies the downstream portions of the North Fork and Cucamonga drainage basins. The terrace deposits, in contrast, are mainly leucocratic granitic and gneissic gravel derived from granitic rocks in the headwaters half of each basin. Deposition of granitic-rich bedload results in stream-channel deposits that consist primarily of sandy gravel derived from granodiorite, quartz monzonite, diorite, and granitic gneiss.

Soil climates vary moderately with altitude and distance from the Pacific Ocean. The Little Tujunga, North Fork, and Cucamonga terraces occur at altitudes of 350 to 600 m, 500 to 770 m, and 600 to 850 m, respectively. Mean annual precipitation decreases toward the east, but the 300-m-higher altitude at the Cucamonga Canyon site tends to offset the effects of the regional precipitation gradient. Mean annual precipitation is approximately 500, 700, and 800 mm, respectively, for the Little Tujunga, North Fork, and Cucamonga study areas (McFadden, 1982). A winter precipitation regime tends to maximize water availability for pedogenic processes. McFadden's calculations of leaching indices (Arkley, 1963) for the three sites are 256, 440, and 532 mm. Strong leaching forms 1- to 2-m-thick soil profiles in less than 1 ky and does not permit formation of calcic horizons.

Our working hypothesis is that time, and variations of climate with time, are major reasons for

spatial variations of terrace soil-profile characteristics. Topography and initial parent materials are fairly constant but spatial variations of soil climate, plant communities, and atmospheric dust may cause variations.

### 4.2.2 CRITERIA FOR MAPPING STREAM TERRACES

The soils parameters used for mapping alluvial geomorphic surfaces in the San Gabriel Mountains were different from those used to map the adjacent Mojave Desert. Rock varnish is excellent for distinguishing between terraces of different ages in hot deserts, but cobble weathering stages are more appropriate in subhumid mountains. As in the Mojave Desert, terrace heights provide only general clues as to relative ages and can be misleading. Distinctive topographies of the desert alluvial fans and terraces are a most useful mapping tool, but terrace treads in the San Gabriel Mountains tend to be alike and are obscured by plants. Stratigraphy and sedimentology of terrace deposits of different age generally seem uniform when compared with the extreme contrasts in the lower Colorado River region, but can vary locally (Section 4.3.2). Once again we conclude that soils and weathering data, although different in each study area, are highly useful for defining mappable units of Quaternary alluvium.

#### 4.2.2.1 Summary of Soil-Profile Characteristics

Thick terrace soil profiles of the San Gabriel Mountains range from dark, organic, sandy Holocene soils (entic haploxeroll, typic xerorthent, typic argixeroll) to red, clayey Pleistocene soils (typic haploxeralf, typic palexeralf). (Pedon descriptions for the five soils named above are presented as Tables 4.5 through 4.9.)

Late Holocene soils characteristically lack argillic horizons, but increasingly older soils are characterized by progressively thicker and redder argillic horizons that have more strongly developed grain and ped cutans and blocky to prismatic structure (Table 4.2). The main source of clay in

FIGURE 4.4 Organic clay cutans on cobbles in mollic A horizon of soil-stage 5 on early Holocene T6 tread, pedon 10, Day Canyon alluvial fan (see Fig. 4.1). The scale is numbered in decimeters.

B horizons probably is dust blown from the Mojave Desert by the frequent "Santa Ana" winds, but clay derived from hydrolytic weathering of granitic parent materials is more important than in hot deserts.

Dark cutans that obscure the lithologies of gravel clasts are typical of Holocene soils. Decomposed organic matter accumulates rapidly in loose sandy gravel of the former stream channels, which comprise the initial parent materials. Coatings such as those shown in Figure 4.4 consist of illuvial organic clay (McFadden, 1982) derived from the A horizon and atmospheric dust. Degrees of cutan development on gravel clasts were classified as incipient (lithology recognizable through the coat-ings), moderate (lithology is obscured), heavy (cutan as much as 2 mm thick), and very heavy (cutan >2 mm thick).

The large influx of atmospheric dust that has accumulated in the sandy gravel of Pleistocene soils has been partially weathered to clay, which decreases infiltration capacity. The resulting increase in runoff tends to erode the A horizons. Thin ochric (light-colored) epipedons are characteristic of the Pleistocene soils despite dense chaparral scrub on some terrace treads. Organic matter in these clay-rich soils does not accumulate to depths as great as those in the Holocene loose sandy gravels, but thin A horizons contain 3 to 5 percent organic carbon (Table 4.3).

TABLE 4.2 Summary of field characteristics[a] of soil-stage chronosequence, south side of San Gabriel Mountains of southern California.

| Terrace | A Horizon | | B Horizon | | C Horizon | | | |
|---|---|---|---|---|---|---|---|---|
| | Thickness (cm) | Minimum Hue, Value | Thickness (cm) | Maximum Redness, Chroma | Cobble Weathering Stage[b] | | Maximum Redness | Maximum Depth to Top of C Horizon (cm) |
| | | | | | Leucocratic | Mafic | | |
| Pleistocene | | | | | | | | |
| T1 | 0 | — | 458 | 2.5YR, 8 | 4 | 4 | 10YR, 8/6 | 458[c] |
| T2 | 2–8 | 5YR, 3 | 208–469 | 2.5YR, 8 | 4 | 4 | 10YR, 7/4 | >475 |
| T3[d] | 0–1 | 5YR, 4 | 163 | 5YR, 6 | 3 | 4 | 7.5YR, 6/6 | 163 |
| T4 | 1–7 | 10YR, 2 | 124–246 | 5YR, 8 | 2–3 | 3–4 | 10YR, 7/5 | >254 |
| T5 | 6–80 | 10YR, 3 | 133–380 | 7.5YR, 4 | 2–3 | 4 | 10YR, 7/6 | 460 |
| Holocene | | | | | | | | |
| T6 | 10–35 | 10YR, 1 | 0–224 | 7.5YR, 6 | 2–3 | 4 | 10YR, 7/4 | 240 |
| T7 | 10–40 | 10YR, 2 | 0–71 | 7.5YR, 6 | 1–2 | 2–3 | 10YR, 7/4 | 88 |
| T8 | 28–79 | 10YR, 2 | — | — | 1 | 2 | 10YR, 7/4 | 75 |
| T9 | 0–50 | 10YR, 3 | — | — | 1 | 1 | 10YR, 6/4 | 50 |

[a] Compiled by L. D. McFadden, 1982.

[b] Cobble weathering stages are described in Table 2.4.

[c] Measured from top of buried argillic horizon.

[d] Characteristics of surface profile.

TABLE 4.3 Laboratory data for selected Holocene and Pleistocene soils.[a]

| Pedon[b] | Horizon | Depth (cm) | Percent of | | | Organic Carbon (%) | $Fe_2O_3$ (dithionite) (%) | Clay Composition[c] | | | | |
|---|---|---|---|---|---|---|---|---|---|---|---|---|
| | | | Sand | Silt | Clay[c] | | | Kaolinite | Illite (mica) | Montmorillonite | Vermiculite | Chlorite |
| 10 (T6) Holocene | A | 0–10 | 77.9 | 21.0 | 1.0 | 3.7 | 0.9 | 3 | 3 | 1 | 1 | 0 |
| | A2 | 10–40 | 65.6 | 30.0 | 4.0 | 4.0 | 1.0 | 4 | 2 | 1 | 2 | 1 |
| | A3 | 40–60 | 68.1 | 27.4 | 4.5 | 2.8 | 1.2 | 4 | 1 | 1 | 2 | 0 |
| | AC | 60–89 | 74.1 | 24.7 | 1.2 | 2.1 | 1.0 | 3 | 1 | 1 | 3 | 0 |
| | Cox | 89–119 | 87.9 | 12.1 | trace | 0.2 | 0.9 | 4 | 0 | 2 | 1 | 0 |
| | Cu | 119– | 96.9 | 3.1 | trace | 0.2 | 0.3 | 2 | 1 | 2–3 | 1 | 0 |
| 6 (T2) Pleistocene | A | 0–2 | 60.9 | 36.6 | 2.5 | 4.8 | 2.5 | 3 | 2–3 | 1 | 1 | 1 |
| | Bw | 2–12 | 60.0 | 32.5 | 7.5 | 2.7 | 2.1 | 3 | 2–3 | 1 | 1–2 | 1 |
| | Bt | 12–42 | 57.2 | 26.6 | 16.2 | 0.3 | 3.0 | 3–4 | 2–3 | 1 | 2 | 0 |
| | Bt2 | 42–72 | 43.7 | 25.7 | 30.6 | 0.3 | 4.0 | 4 | 2–3 | 1 | 2 | 0 |
| | Bt3 | 72–162 | 49.9 | 28.7 | 21.4 | 0.1 | 4.2 | 4 | 2 | 1 | 1–2 | 1 |
| | Bt4 | 162–194 | 55.8 | 31.4 | 12.8 | 0.2 | 2.8 | 4 | 1 | 1–2 | 2 | 1 |
| | Bt5 | 194–210 | 63.4 | 25.4 | 11.8 | 0.1 | 2.5 | 3–4 | 0 | 2 | 2 | 0 |
| | | 210– | 64.8 | 27.4 | 7.8 | 0.1 | 1.7 | 3–4 | 1 | 2 | 2 | 0 |

[a] Analyses by L. D. McFadden, 1982.

[b] See Figure 4.1 for locations of pedons 10 and 6.

[c] 0, not detected; 1, trace (<10%); 2, small (20% ± 10%); 3, moderate (40% ± 10%); 4, abundant (60% ± 10%); 5, predominant (>70%).

TABLE 4.4 Terrace (T) chronosequence for San Gabriel Mountains.[a]

| Age | Site | Pedon Location of Fig. 4.1 | Probable Cause of Stream-Channel Incisement[b] |
|---|---|---|---|
| Holocene | | | |
| Active Channel | Gravel bars | | |
| T9 | Sycamore Canyon | 1 | C |
| | North Fork of San Gabriel River south of Bichota Flat | 14 | C |
| | Little Tujunga Canyon | 22 | C and T |
| T8 | East Etiwanda Canyon lowest fault scarp | 9 | T |
| | North Fork San Gabriel River, cut terrace at Tecolote Flat | 13 | C |
| | Sycamore Canyon | 2 | C |
| T7 | Day Canyon fanhead embayment | 11 | C and T |
| | North Fork of San Gabriel River at Bichota Canyon | 15 | C |
| | Duncan Canyon upper piedmont | 4 | C and T |
| | Little Tujunga Canyon | 23 | C and T |
| | Cucamonga Canyon, lower fanhead embayment[c] | | C and T |
| T6 | East Etiwanda Canyon alluvial fan | 7 | T |
| | Arroyo Seco | 18 | |
| | Day Canyon fault scarp | 10 | T |
| | Cucamonga Canyon, upper fanhead embayment[c] | | T |
| Pleistocene | | | |
| T5 | Day Canyon, west side of fanhead trench | 12 | C and T |
| | Arroyo Seco above USFS houses | 19 | C and T |
| | Little Tujunga Canyon at Middle Ranch | 25 | T |
| T4 | Yerba Buena surface | 24 | T |
| | Indian Springs embayment | 21 | T |
| | Cajon Creek and North Fork of San Gabriel River at west end of Bichota Flat (paleosol and underlying deposits is T3) | 16 | |
| T3 | North Fork of San Gabriel River at west end of Bichota Flat (surface soil and deposits is T3) | 16 | C |
| | Mouth of Little Tujunga Canyon | | C and T |

TABLE 4.4 *(Continued)*

| Age | Site | Pedon Location of Fig. 4.1 | Probable Cause of Stream-Channel Incisement[b] |
|---|---|---|---|
| T2 | Arroyo Seco at Gould Mesa | 20 | T and C |
| T1 | Lytle Canyon at Texas Hill | 3 | T and C |
| | North Fork of San Gabriel River at Tecolote Flat | 17 | C and T |

[a] Assuming minimal areal variations of climate and parent material, general age categories of terraces are assigned primarily on the basis of relative ages of soils. See Tables 4.11 and 4.12 for estimated age ranges of terraces.

[b] C, climate change; T, tectonic base-level fall.

[c] Soils examined in dug pits and exposures and soil horizonation noted, but complete pedon descriptions were not made for these terraces.

Iron oxyhydroxides are common soil-coloring agents (Schwertmann & Taylor, 1977; Rowell, 1981; Bloomfield, 1981). Goethite and ferrihydrite produce yellow-brown hues in young soils, and the gradual increase of hematite content with time is responsible for progressively redder hues in older soils. Mid-Pleistocene terraces have maximal B-horizon development; hues generally are 2.5YR to 5YR and thicknesses exceed 4 m (Table 4.2). Laboratory analyses by McFadden (1982,1988; McFadden & Hendricks, 1985) reveal substantial increases with age for most pedogenic constituents of Holocene (pedon 10 on terrace T6) and Pleistocene (pedon 6 on terrace T2) (Table 4.3) soils. Mass summations in grams per square centimeter of pedon area for the T2 (listed first) and T6 soils provide interesting comparisons: dithionite extractable iron oxyhydroxides, 0.6, 11.2; clay, 2.0, 66.2; and organic carbon, 2.4, 0.7.

The pedogenic features that provide relative age control for the alfisol chronosequence in the San Gabriel Mountains are much different from those for the aridisol chronosequence of the Mojave Desert, but are equally interesting. In the next section we use features of dated alfisols, both within the San Gabriel Mountains and with cross-checks from elsewhere in California, to assign general ages to the climatic and tectonic stream terraces of the North Fork of the San Gabriel River, Little Tujunga Creek, and Cucamonga Canyon.

### 4.2.2.2 Stream Terraces and Soils Chronosequence

Sequential pedogenic changes with time are the basis for estimating relative ages of soils. The terrace chronosequence for the San Gabriel Mountains includes five Pleistocene terraces, four Holocene terraces, and the active stream channels (Table 4.4). (See Table 4.2 for a summary of field characteristics and Table 4.3 for a comparison of the laboratory data for Holocene and Pleistocene soils.) Tables 4.5 through 4.9 present pedon descriptions for soils found on five of the nine terraces.

### 4.2.3 AGE ESTIMATES FOR THE TERRACE CHRONOSEQUENCE

Headings in this section use the excellent classification of dating methods of Colman, Pierce, and Birkeland (1987).

### 4.2.3.1 Numerical Ages

Stream terraces are difficult to date. Even radiocarbon dating may not be especially helpful for dating critical stages of terrace formation as is illustrated by Figure 4.27. Charred wood from terrace T9 alluvium at the pedon 1 site (see Fig. 4.1) has a probable radiocarbon age of $0.175 \pm 0.05$ ka. Peat from the base of the T7 fill in the North Fork of the San Gabriel River has a calibrated radiocarbon age of $7.7 \pm 0.3$ ka, and carbonized manzanita seeds from an overlying bed of

TABLE 4.5 Soil-profile description[a] of soil-stage S7 on T9 terrace (pedon 14).

*Location:* Los Angeles County, California; 0.3 km east of Highway 39 and 0.8 km southwest of Bichota Mesa; N75°W of Burro Peak; 34°15'30", 117°51'15"

*Physiographic position:* Streamcut in very recent bouldery terrace, 3–4 m above active channel of North Fork of San Gabriel River; altitude 600 m

*Classification:* Typic xerorthent

| Horizon | Depth (cm) | Description |
|---------|-----------|-------------|
| A1 | 0–11 | Yellow brownish gray to black brown (10YR 4/2 dry, 10YR 2/2 moist) non-gravelly to locally gravelly loamy sand; massive to granular; soft, slightly sticky, nonplastic; common incipient organic clay coatings on pebbles; common roots; clear, smooth boundary |
| C1ox | 11–67 | Gray yellowish orange to brown (10YR 6/4 dry, 10YR 4/4 moist) bouldery sand; massive; soft, nonsticky, nonplastic; few incipient clay coatings on pebbles; clear, smooth boundary |
| C2ox | 67–105 | Gray yellowish-orange to gray yellowish-brown (10YR 6/3 dry, 10YR 4/3 moist) sandy gravel; massive; loose; nonsticky, nonplastic; abrupt wavy to irregular boundary |
| C3 | 105– | Light yellow-brownish-gray to yellow-brownish gray (10YR 7/2 dry, 10YR 4/2 moist) sand; massive; loose; nonsticky, nonplastic |

Cobble weathering stages

All lithologies, 1. Some mafic plutonic, schist, very rarely, incipient, 2

[a]Described by L. D. McFadden.

silty sand grew at $7.4 \pm 0.2$ ka. These radiocarbon analyses are summarized in Table 4.10. Radiocarbon dating was important for assessing ages of the T3B and T6 terraces in Cajon Creek (Section 4.4.2; Weldon, 1986).

### 4.2.3.2 Calibrated Ages

Ages for stream terraces can be calibrated where specific soil properties can be dated in type areas and then used in areas that lack independent age controls. Rock-varnish chemistry and cobble weathering-rind thickness dating methods (Sections 2.4.1.2.1, 2.4.1.2.2) are novel and interesting examples of calibrated age techniques. The San Gabriel Mountains study allows discussion of equally fascinating new ways of dating alluvial geomorphic surfaces–the passage of sound waves through weathered boulders and use of uniform rates of horizontal fault displacement of fluvial landforms.

*4.2.3.2.1 Boulder Acoustic Wave Speeds* Richard Crook measured weathering-induced variations in acoustic velocities of granitic boulders and developed a sensitive tool for dating stream-terrace deposits (Crook et al., 1978; Crook & Kamb, 1980). Acoustic velocities are measured with a micro-seismic timer by mounting a sensing transducer on one side of a boulder and introducing shock waves into the boulder at four to six points with hammer blows to a steel ball held against the boulder surface. The timer measures first arrival time of P waves. Travel times of four to six measurements are used to regress a travel-time profile. Under ideal conditions, boulder acoustic wave speeds are reproducible with a standard deviation of about 0.05 km/s. Mean acoustic velocities for groups of 20 boulders are reproducible to within 1 percent. Thus precision and reproducibility are minor sources of error and scatter compared with the uncertainties associated with sampling of

**TABLE 4.6** Soil-profile description[a] of soil-stage S6 on T8 terrace (pedon 13).

*Location:* Los Angeles County, California; 0.3 km east of Highway 39 and 0.8 km southwest of Bichota Mesa; N75°W of Burro Peak; 34°15′30″, 117°51′15″

*Physiographic position:* Steep streamcut in bouldery deposits of cut terrace, about 15 m above active channel of North Fork of San Gabriel River; altitude 612 m

*Classification:* Entic haploxeroll

| Horizon | Depth (cm) | Description |
|---|---|---|
| A1 | 0–57 | Yellowish-brown to very dark brown (10YR 4/3 dry, 10YR 2/3 moist) granular sandy loam; moderate subangular blocky (moist) to crumb; soft, slightly sticky, nonplastic; incipient to moderate organic clay pebble coatings; many roots; locally bioturbated; abrupt wavy boundary |
| AC | 57–79 | Pale brown to dark brown (10YR 5/3 dry, 10YR 3/3 to 2/3 moist) gravelly to very gravelly sand; weak subangular blocky to massive; loose, slightly sticky, nonplastic; incipient to moderate organic clay pebble coatings; common distinct dark mottles; common roots; abrupt wavy to locally irregular boundary |
| C1ox | 79–128 | Yellowish-brown to dark yellowish brown (10YR 5/4 dry, 10YR 4/4 moist) bouldery coarse sand; weak subangular blocky to massive; nonsticky, nonplastic; incipient coatings on upper surface of pebbles; abrupt, wavy boundary |
| C2 | 128– | Very pale brown to pale brown (10YR 7/3 dry, 10YR 6/3 moist) sandy gravel; massive; loose, nonsticky, nonplastic; occasional pockets of iron-oxide-stained sand (7.5YR 5/8-8 dry, 7.5YR moist) |

Cobble weathering stages

Leucocratic plutonic lithologies, 1; occasional mafic metamorphics, 2 to 1

[a]Described by L. D. McFadden and W. B. Bull.


**TABLE 4.7** Soil-profile description[a] of soil-stage S5 on T7 terrace (pedon 15).

*Location:* Los Angeles County, California, 0.2 km directly east of Highway 39, 0.3 km N10°E of Bichota Mesa; N15°E of Burro Peak; 34°16′, 117°51′

*Physiographic position:* Steep escarpment/hillslope formed on terrace gravels, about 40 m above present active channel of North Fork of San Gabriel River; altitude 722 m

*Classification:* Typic argixeroll

| Horizon | Depth (cm) | Description |
|---|---|---|
| A1 | 0–17 | Yellowish-brown to brown (10YR 5/3 dry, 10YR 3/4 moist) sandy loam; massive to granular; soft, nonsticky, nonplastic; clear, wavy boundary |
| Bt | 17–88 | Gray-brown to strong brown (7.5YR 6/4 dry, 7.5YR 4/6 moist) very gravelly sandy loam; massive to weak subangular blocky; hard, slightly sticky, slightly plastic; moderate clay coatings on larger pebbles; clear, wavy boundary; common very incipient grain argillans, few small pores |
| C1ox | 88–144+ | Gray yellowish-orange to yellow-brown (10YR 7/4 dry, 10YR 4/6 moist) very gravelly sand; massive; loose, nonsticky, nonplastic; few incipient clay coatings on pebbles |

Cobble weathering stages

B horizon, leucocratic plutonic rocks, 1; mafic rocks, 2 to 1; C horizon, coarse grained diorite, mica schist, 2 to 3, rarely 4; leucocratic lithologies, 1–2

[a]Described by L. D. McFadden and C. M. Menges.

TABLE 4.8 Soil-profile description[a] of soil-stage S4 (surface) and S3 (buried) for T4 terrace tread and buried T3 terrace tread (pedon 16).

*Location:* Los Angeles County, California; 1 km north of Tecolote Flat on Highway 39, ~20 m above west side of highway and 60 m above present active channel of North Fork of San Gabriel River; 34°15′45″, 117°51′15″

*Physiographic position:* Steep, partly artificially modified hillslope capped with terrace gravels; altitude 655 m

*Classification:* Surface profile, typic haploxeralf; buried profile, typic haploxeralf

| Horizon | Depth (cm) | Description |
|---|---|---|
| O1 | .5–0 | Organic matter, abrupt planar boundary |
| B1 | 0–20 | Reddish-brown to dark reddish-brown (5YR 5/4 dry, 5 YR 3/4 moist) sandy clay loam; moderate to coarse subangular blocky; slightly hard, slightly sticky, slightly plastic; common roots; abrupt, wavy boundary |
| B21t | 20–70 | Reddish-yellow to yellowish-red (5YR 6/6 dry, 5YR 4/6 moist) clay loam; strong coarse subangular blocky; hard, sticky, plastic; many moderately thick ped argillans and grain argillans; clear, wavy (?) boundary |
| B22t | 70–88 | Pink to yellowish-red (5YR 7/4–7.5YR 5/4 dry, 5YR 4/8 moist) gravelly clay loam; moderate to coarse subangular blocky; hard slightly sticky, plastic; many moderately thick ped argillans and grain cutans; clear, wavy boundary |
| B23t | 88–127 | Reddish-yellow to yellowish-red (5YR 7/6 dry, 5YR 5/8 moist) cobbly to nongravelly sandy clay loam; moderate angular blocky; hard, sticky, plastic; abrupt, irregular boundary |
| B24T | 127–163 | Reddish-yellow to strong brown (7/5YR 8/6 dry, 7.5YR 5/6 moist) gravelly sandy loam; moderate subangular blocky; slightly hard, slightly sticky, slightly plastic; few thin ped cutans coatings; abrupt, wavy boundary |
| Cox | 163–237 | Reddish-yellow to strong brown (7.5YR 6/6 dry, 7.5YR 5/6 moist) bouldery sand; massive; soft, slightly sticky, nonplastic; abrupt, wavy boundary |
| IIB21tb | 237–280 | Light brown to strong brown (5YR 6/4 dry, 5YR 4–5/6 moist) clay loam; moderate to coarse angular blocky; slightly hard to hard, sticky, plastic; few thin grain and ped cutans; clear, wavy boundary |
| IIB22tb | 280–300 | Pink to yellowish-red or strong brown (7.5YR 7/4 dry, 5YR to 5YR 5/6 moist) gravelly sandy loam; moderate subangular blocky; slightly hard, slightly sticky, slightly plastic; very few, very thin ped and grain cutans; abrupt, irregular boundary |
| IIB3b | 300–320 | Pink to strong brown (7.5YR 7/4 dry, 7.5YR 5/6 moist) bouldery sand; massive to weak subangular blocky; slightly hard, slightly sticky, nonplastic; clear, wavy boundary |
| IIC1oxb | 320–366 | Very pale brown to reddish-yellow (10YR 8/4 dry, 7.5YR 6/6 moist) silty sand; massive; soft, slightly sticky, nonplastic; abrupt boundary |
| IIC2oxb | 366– | Yellow to yellowish-brown (10YR 8/6 dry, 10YR 5/6 moist) sand; massive; loose, nonsticky, nonplastic |

Cobble weathering stages

Surface profile, all mafic lithologies, 4; leucocratic plutonic and gneiss lithologies, upper B horizon 3 to 4, lower B and Cox, 2 to 3. Buried profile, all mafics, 4; leucocratic rocks, 3 to 4

[a] Described by L. D. McFadden and W. B. Bull.

TABLE 4.9 Soil-profile description[a] of soil-stage S1 on T1 terrace (pedon 17).

*Location:* Los Angeles County, California; 0.1 km west of Highway 36, 0.3 km N10°E of Tecolote Flat; 34°15'45", 117°51'40"

*Physiographic position:* Very steep hillslope formed on terrace gravels about 100 m above North Fork of San Gabriel River; altitude 667 m

*Classification:* Typic palexeralf

| Horizon | Depth (cm) | Description |
| --- | --- | --- |
| 01 | 2–0 | Organic matter |
| A1 | 0–6 | Dark reddish-brown to black (5YR 3/3 dry, 5YR 1/1 moist) silt loam; massive; slightly soft, nonsticky, nonplastic; many roots; abrupt, smooth boundary |
| B21t | 6–66 | Reddish-brown to red (5YR 5/6 dry, 2.5YR 4/6 moist) clay loam; moderate to coarse subangular blocky; very hard, very sticky, very plastic; many moderately thick to thick grain and ped cutans; clear, smooth boundary |
| B22t | 66–160 | Reddish-brown to red (5YR 5/6 dry, 2.5YR 4/6 moist) very gravelly clay loam; massive; hard, sticky, plastic; many moderately thick grain cutans; clear, smooth boundary |
| B23t | 160–240 | Gray-orange to reddish-brown (5YR 6/6 dry, 5YR 4/6 moist) very gravelly sandy clay loam; massive to coarse subangular blocky; hard, sticky, plastic; common thin grain cutans; gradual, smooth boundary |
| B24t | 240–315 | Strong orange brown to light brown (7.5YR 7/6 dry, 7.5YR 5/6 moist) very gravelly sand loam; massive; hard, sticky, plastic; common thin grain cutans; distinctly mottled due to saprolitic weathering of granitic cobbles; gradual, smooth boundary |
| B25t | 315–475 | Light-orange to strong orange-brown (7.5YR 8/6 dry, 7.5YR 6/6 moist) very gravelly sandy loam; massive, hard, sticky, slightly plastic; few thin grain cutans, distinctly mottled due to saprolitic weathering of granitic clasts; gradual, smooth boundary |
| B3ox | 475– | Gray-orange to gray-brown (7.5YR 7/4 dry, 7.5YR 5/4 moist) very gravelly loam sand; massive; slightly hard, slightly sticky, slightly plastic; few incipient coatings on clasts; distinct strong mottling due to saprolitic alteration of granitic clasts |

Cobble weathering stages

Upper B horizon, leucocratic plutonics and mafic lithologies, 4; lower B horizon, 4 except for occasional quartz-rich metamorphic rocks, 2 to 3

[a] Described by L. D. McFadden and W. B. Bull.

boulders from alluvium. The technique is moderately time consuming–about 20 to 30 boulders at each of four to six sites in a typical day.

It is important to limit variations of acoustic velocities that would result from sampling different boulder lithologies (Fig. 4.5). Crook and Kamb measured acoustic velocities of boulders of Mount Lowe granodiorite, which is a widespread and easily recognized lithology in the San Gabriel Mountains. Acoustic velocities ranged from 3 km/s for boulders with stage 1 (Table 2.6) weathering characteristics to 1 km/s for stage 3 boulders. Similar velocities were measured by Gillespie (1982) and Crook (1986) in a rigorous study of complexes

TABLE 4.10 Summary of radiocarbon analyses from sites along the south side of the San Gabriel Mountains.

| Sample | Latitude | Longitude | Material | Conventional Age (ka)[a] | $\delta^{13}C(\permil)$ | Calendric Age (ka)[b] |
|---|---|---|---|---|---|---|
| USGS 774A | 34°15.33' N | 117°51.67' | Manzanita seeds from basal deposits of terrace 7 aggradation event | 6.45 ± 0.070 | −24.59 | 7.327 ± 130[c] |
| USGS 774B | 34°15.33' N | 117°51.67' | Peat from basal deposits of terrace 7 aggradation event | 6.80 ± 0.080 | −27 | 7.594 ± 180[c] |
| USGS 775 | 34°12.33' N | 177°25.33' | Charred root from terrace 9 alluvium | 0.175 ± 0.050 | −27 | 0.188 ± 45[d,e] |

[a] Half-life of 5568 years is used; 1 sigma counting errors are shown. Age before 1950 A.D. has been corrected for carbon isotope fractionation using the $\delta^{13}C$ value normalized to −25‰.

[b] Calibrated $^{14}C$ age to account for variations in the specific activity of $^{14}C$ in atmospheric $CO_2$, using the program of Stuiver and Reimer (1986).

[c] Calendric age = 1949 + B.C. age.

[d] Calendric age = 1950 − A.D. age.

[e] Dendrochronological calibration of the $^{14}C$ age resulted in multiple calendric ages of 150, 183, 187, 205, and 277 years. 58% probability that the age is between 1722 and 1811 A.D.

FIGURE 4.5 Differences in the weathering of large boulders in Holocene alluvium of Cucamonga Canyon. The stage 1 boulder of granitic gneiss reflects sunlight from abrasion-polished surfaces. The dark stage 3 diorite boulder is so weathered that it has lost most of its stream-worn shape after being exhumed.

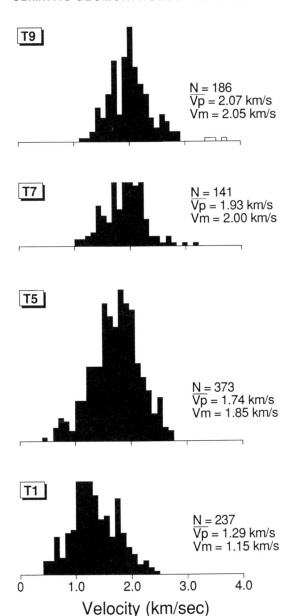

FIGURE 4.6 Bar graphs of acoustic wave speeds of boulders of Mount Lowe granodiorite in four ages of terrace and fan deposits between Big Tujunga and San Gabriel canyons. Vp, mean P-wave velocity; Vm, modal P-wave velocity. See Figure 4.11 for the ages of the four terraces. From Crook and Kamb, 1980.

of glacial moraines along the east side of the Sierra Nevada of California. Acoustic velocities of granitic rocks on moraine crests ranged from about 4 to 0.3 km/s.

A range of acoustic wave speeds for boulders in active channels occurs because of variations of weathering and fracturing of boulders derived from many hillslope sources. Different boulders were supplied to the streams at the times of deposition of the T9 and T7 terrace gravels. T7 terrace deposits represent the initial stripping of surficial colluvium from thick hillslope sediment reservoirs that had been subjected to prolonged weathering. T9 terrace deposits in part represent fresher materials derived from deeper levels of the hillslope sediment reservoir. Old terrace deposits are another source of weathered boulders.

Boulder acoustic wave speeds form a useful basis for discriminating between terrace deposits of different ages. Clasts of Mount Lowe granodiorite were sampled from eight drainage basins between Big Tujunga and San Gabriel Canyons. Twofold to fivefold ranges occur in distributions of acoustic wave speeds that tend to become more polymodal with increasing relative age (Fig. 4.6). Both modes and mean velocities of acoustic wave speeds decrease significantly for older relative ages.

When calibrated, boulder acoustic wave speeds have the potential for dating deposits with ages of 1 to 1000 ka. The semi-logarithmic graph for the Mount Lowe granodiorite (Fig. 4.7) suggests that it may be difficult to differentiate between deposits with similar ages that are older than 100 ka.

Age control is available for four terraces for which boulder acoustic wave speeds were measured (Fig. 4.7). Several radiocarbon ages have been obtained for the T9 terrace deposits. The two radiocarbon ages from the base of the 50-m T7 fill in the North Fork suggest an age of about 5 ka for the surficial deposits, whose boulders provided Crook a mean velocity of 1.89 km/s. An age estimate for T5 terrace deposits in Eaton Canyon is based on a radiocarbon age and deposition rates. Three sets of samples for remnant magnetism analyses were collected from the North Fork T1 terrace deposits. Two sets collected from within the soil profile have a normal polarity; the one set from below the soil profile has a reversed polarity. Crook and Kamb concluded that the T1 aggradation event occurred before the Brunhes-Matuyama magnetic polarity reversal at 790 ka (Section 2.4.1.3.1) and that the soil profile formed since then. Boulders within the reversed-polarity part of the terrace deposits had a mean velocity of only 1.07 km/s.

One way of expanding the versatility of this dating technique would be to measure acoustic wave speeds for several lithologies with distinctly different weathering rates. For a given chronosequence, a rapid weathering lithology could be used to date young terraces and a slow weathering lithology could be used to date old terraces.

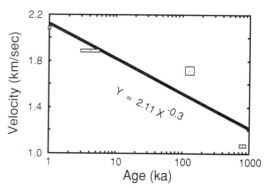

FIGURE 4.7 Variation of acoustic wave speeds of boulders of Mount Lowe granodiorite from four dated terraces of the San Gabriel Mountains. Dimensions of rectangles indicate estimated uncertainties. From Figure 7 of Crook, 1986.

*4.2.3.2.2 Constant Fault-Slip Rates* Ray Weldon radiocarbon-dated 13 samples from swamp deposits, stream channels, and deposits and treads of fill terraces (Weldon & Sieh, 1985). He also measured amounts of horizontal fault displacement of landforms by the San Andreas fault at the Cajon Creek site at the east end of the San Gabriel Mountains (Fig. 4.1). Movement along this segment of the San Andreas fault does not occur as creep. It is seismogenic with short recurrence intervals. Six earthquakes during the past 1 ky suggest an earthquake recurrence interval of only 0.15 to 0.2 ka. His dating shows that the rate of right-lateral slip along the San Andreas fault has been a remarkably constant $24.5 \pm 3.5$ m/ky for the past 14.4 ky. Weldon and Humphreys (1986) concluded that the slip rate may have been constant for 4 Ma since the opening of the Gulf of California.

Calibration by dating of landforms and deposits that have been offset by the fault provides a means of estimating ages of other landforms that have been offset by the same segment of the fault but for which age control is not available. The key assumption is that the rates of fault slip are uniform relative to the age of the landforms being dated.

Because of the short earthquake recurrence interval of the San Andreas fault, even a 1-ka landform can be considered to have been offset at roughly a constant rate. The uncertainties involved in the radiocarbon dating, in its calibration (Porter, 1981; Stuiver, 1982; Stuiver & Reimer, 1986), and in measuring the fault offsets indicate that ages estimated with this technique are accurate to within $\pm 15$ percent.

Calibration of earth deformation provides an exceptionally useful way of dating terraces. By measuring the amounts of offset of basal deposits and of terrace treads, Weldon was able to estimate ages of initiation and termination of the T6–T7 aggradation event at Cajon Creek, and thus provided key temporal information about geomorphic responses caused by Pleistocene–Holocene climatic change (Section 4.4.2.).

Using the assumption of uniform rates of horizontal fault displacement, Weldon obtained a calibrated age for the 85-m-thick T4 aggradation event in Cajon creek. The riser crest has been displaced 20 m vertically and 1300 to 1400 m horizontally.

$$\frac{1350 \text{ m}}{24.5 \text{ m/ky}} = 55 \text{ ka} \pm (0.15 \times 55 \text{ ky} = 8 \text{ ka}) \quad (4.1)$$

Constant earth deformation can also be used in areas of vertical tectonic activity to date terraces of fluvial (Rockwell et al., 1983,1985; Keller et al., 1985) and marine (Bull, 1984,1985; Bull & Cooper, 1986) origin, where uniform rates of uplift seem probable and heights are known for a terrace flight.

### 4.2.3.3 Correlated Ages

Two types of correlated ages were used; both rely on dating done elsewhere. The first is the Brunhes-Matuyama magnetic polarity reversal that was used as a calibrated age for boulder acoustic wave speeds for T1 terrace deposits in the North Fork of the San Gabriel River. The second is discussed in Section 4.2.3.5, and is a correlation of alfisol soil-profile characteristics in the San Gabriel Mountains with similar soils elsewhere in California.

### 4.2.3.4 Relative Ages

Weathering and pedogenic characteristics are most useful for defining relative ages of individual terraces and for correlating flights of stream terraces. Useful characteristics include darkness and organic content of A horizons; redness, thickness, and clay and iron oxyhydroxide contents of B horizons; and weathering stages of cobbles in the B and C horizons.

The use of several characteristics is preferable because a single criterion can be misleading. Thickness and darkness of A horizons vary considerably with variations in the accumulation/oxidation ratio of organic matter and infiltration/runoff ratios. B-horizon characteristics may be misleading if terrace treads occur below hillslopes or older terraces with argillic horizons. Red clay from old pedogenic sources may be transported short distances and translocated into young B soil horizons. The resulting B horizons appear too red, clayey, and thick compared with other relative-age indicators such as stage of cobble weathering.

The semiarid to subhumid, strongly seasonal climate of the San Gabriel Mountains provides an opportunity to expand initial discussions of relative soil ages (Section 2.4.1.4) with brief discussions of cobble weathering stages (Table 2.6) and iron oxyhydroxides.

*4.2.3.4.1 Cobble Weathering Stages*   Cobble weathering indices require minimal time to collect and are useful in defining terrace chronosequences in climatic settings that range from extremely arid to extremely humid. Cobble weathering in hillslope sediment reservoirs ranges from incipient to advanced. It is desirable to examine at least 20 cobbles of a specific lithology in alluvial gravels derived from many hillslope sources. Unfractured small cobbles at the bottom of the Figure 4.8 photograph are composed of leucocratic lithologies such as granitic gneiss. More mafic rocks, such as the fractured diorite, weather much more rapidly than granitic gneiss. The resulting changes in volume are caused primarily by alteration and expansion of biotite and hydrolytic weathering of calcium plagioclase, which cause fracturing and ultimately disintegration of the rock into grus. Similar marked differences in weathering of large boulders in Holocene alluvium are illustrated in Figure 4.5.

*4.2.3.4.2 Soil-Profile Development*   The systematic changes in composition and content of iron oxyhydroxides in alluvial soils of the Transverse Ranges are important components of soils chemistry that are strongly influenced by the passage of time (McFadden & Hendricks, 1985). Iron-bearing minerals are converted into *stable pedogenic iron oxyhydroxides* (or for the sake of brevity, *Fe oxides*); the principal Fe oxides are ferrihydrite, goethite, and hematite. Soils in strongly seasonal thermic climates become progressively redder, because warm climates favor formation of hematite (Schwertmann et al., 1982). Mesic humid climates (Chapter 5) favor development of yellow-brown soils colored by goethite.

Two wet chemical extraction procedures were used (McKeague & Day, 1966; McKeague et al., 1971; Schwertmann, 1973). Oxalate extraction ($Fe_o$ for brevity) is used to determine those pedogenic iron oxyhydroxides that consist mainly of ferri-

FIGURE 4.8 Variation in weathering of cobbles in the Cox horizon of soil stage 5 on early Holocene T6 tread, pedon 10, Day Canyon alluvial fan. Cobble weathering stages are defined in Table 2.4. Stages 1 and 2 leucocratic granitic and gneissic rocks are unweathered or have only a few weathering fractures. The stage 3 coarse-grained diorite has abundant fractures and emits a punky sound when struck with a hammer. The scale is numbered in decimeters.

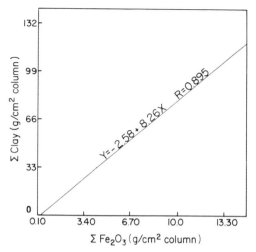

FIGURE 4.9 Relation between total mass of pedogenic clay and dithionite-extractable iron oxides in the soils of the San Gabriel Mountains. From Figure 5 of McFadden and Hendricks, 1985.

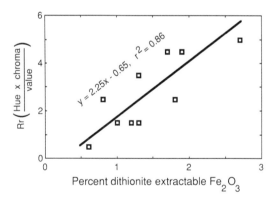

FIGURE 4.10 Relation between soil reddening (Rr is a redness index) and percent total dithionite-extractable iron oxides ($Fe_d$) for the late Pleistocene soil stage 4 on the T4 tread in the North Fork of the San Gabriel River. From Figure 6 of McFadden and Hendricks, 1985.

hydrite and organic Fe complexes. Dithionite extraction ($Fe_d$ for brevity) is used to determine the total amount of pedogenic iron oxyhydroxides after removal of magnetite.

Weakly crystalline ferrihydrite is most abundant in latest Pleistocene and Holocene soils, and goethite and hematite become progressively more important in older soils. $Fe_o/Fe_d$ ratios describe irreversible transformation of ferrihydrite to hematite with increase in soil age. $Fe_o/Fe_d$ ratios generally are lower in Bt than in Cox horizons.

Rates of increase of $Fe_d$ contents of the soils initially are rapid but decrease exponentially with increase in time. McFadden and Hendricks found that $Fe_o$ accumulation can be described by

$$\log Fe_d = -2.15 + 0.55 \log t \qquad (4.2)$$

where t is time. This equation describes the rate of increase of total pedogenic iron oxides in soils of the chronosequence. Concurrent formation and translocation of Fe oxides and pedogenic clay results in strong correlations between total masses of $Fe_d$ and clay (Fig. 4.9) and between $Fe_d$ and soil redness (Fig. 4.10).

### 4.2.3.5 Summary of Ages and Comparisons with Other Chronosequences

The chronosequence for seven stages of terrace soil-profile development (Fig. 4.11) spans a million years and is based on 25 pedons (Fig. 4.1) described by Bull, Menges, and McFadden (1979), and analyzed in the laboratory by McFadden (McFadden, 1982; McFadden & Tinsley, 1982). The oldest soil is at the pedon 3 site in Lytle Canyon. Total mass summations of iron oxyhydroxides and clay are much larger in the pedon 3 soil than for any other T1 soil (McFadden, 1982), which indicates an age substantially greater than 1 Ma.

The terrace chronosequence consists of nine terraces (Fig. 4.11). Some terrace stratigraphic ages are based on the sum of estimated ages of surface and buried soil profiles, such as for the compound fill of terraces T3 and T4 in the North

Fork whose relict and buried soils are described in Table 4.8.

The terrace chronosequence is not the same as the soils chronosequence. Weathering and soils parameters may lack the sensitivity needed to separate terrace treads that differ little in age. This leads to a tendency to group suites of soils for terraces (Fig. 4.11) or glacial moraines (Burke & Birkeland, 1979) with similar ages, but which can be clearly separated topographically.

Local faulting may create terraces with similar ages that are topographically distinct but pedogenically similar. Such terraces are common in reaches of increased stream power where stream slopes have been steepened by tectonic deformation and subsequent fluvial adjustments. The effect of a tectonic perturbation may be to (1) hasten crossing of the threshold of critical power in streams that are aggrading slowly in response to climatic change, (2) initiate downcutting by streams in static equilibrium, or (3) accelerate stream-channel downcutting in degrading streams. A soils chronosequence would place the several terraces with similar soil profiles in one class, but the important tectonically controlled differences in terrace heights would be acknowledged in the terrace chronosequence. Examples include Cucamonga Canyon, where both terrace T6 and terrace T7 have the general characteristics of S5 soils, and Little Tujunga Creek, where terraces T4 and T5 have the general characteristics of S4 soils.

Stream-terrace treads of a given aggradation event should not be regarded as being exactly synchronous, even in tectonically stable mountain ranges (Fig. 3.25). Fill-terrace tread ages may vary between adjacent canyons because of different response times to the same climatic perturbation. Local tectonic deformation tends to make a given terrace even more diachronous. Times of terrace-tread formation also may vary in the downstream direction (Section 4.4). Where reaction and re-

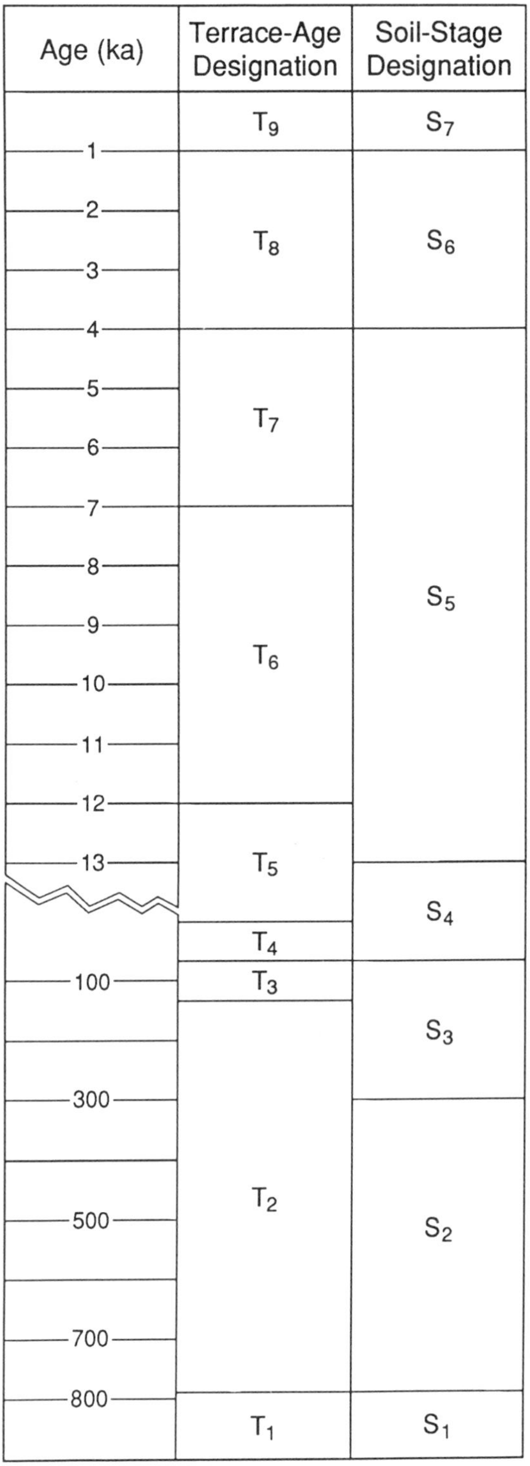

| Age (ka) | Terrace-Age Designation | Soil-Stage Designation |
|---|---|---|
| 1 | $T_9$ | $S_7$ |
| 2 3 | $T_8$ | $S_6$ |
| 4 5 6 7 | $T_7$ | |
| 8 9 10 11 12 | $T_6$ | $S_5$ |
| 13 | $T_5$ | |
| | $T_4$ | $S_4$ |
| 100 | $T_3$ | $S_3$ |
| 300 500 700 800 | $T_2$ | $S_2$ |
| | $T_1$ | $S_1$ |

FIGURE 4.11 General ages of soil stages and stream terraces of the San Gabriel Mountains. Modified from Figure 2 of McFadden et al., 1982.

sponse times are short, alluvial-fan and fill-terrace tread ages can be synchronous (Section 5.2.2.)

The stream terrace chronosequence described in Figure 4.11 and Table 4.4 is compared with four other chronosequences of alfisols in subhumid mesic to semiarid thermic parts of California (Tables 4.11 and 4.12). Changes in alfisols with time in a humid climatic setting of coastal California are discussed by Chadwick and colleagues (1990) and by Merritts and colleagues (in press). The closest is in the northeastern San Gabriel Mountains where Weldon and Sieh (1985) dated alluvial surfaces along the San Andreas fault by radiocarbon ages of sag-pond organic materials and fault offset rates. Age ranges for alluvial geomorphic surfaces appear to be similar for the northeast and south sides of the San Gabriel Mountains, especially when slightly slower rates of pedogenesis are taken into account for the northeast sites, which are in a cooler (higher) and drier (rain shadow) part of the range.

The data from Rockwell and colleagues (1985)

in the western Transverse Ranges are based on nine radiocarbon ages and uniform rates of uplift on flexural-slip faults. The times of terrace-tread formation appear to be similar in the western and central Transverse Ranges.

The Wheeler Ridge-San Emigdio soils chronosequence of Keller and colleagues (1985) is in the northern Transverse Ranges adjacent to the southern San Joaquin Valley. Numerical age estimates for the tectonic and climatic stream terraces were obtained by $^{14}C$ and $^{230}Th/^{234}U$ isotopic analyses. Calibrated age estimates were made that assumed uniform rates of uplift for terraces that have been raised to different positions in the landscape.

The alluvial geomorphic surfaces along the semiarid east side of the San Joaquin Valley about 500 km north of the San Gabriel Mountains have been studied in detail by many workers. This soils chronosequence has excellent age control provided by $^{14}C$, K/Ar, and paleomagnetic methods (Harden, 1982). The soils are similar to southern California

Table 4.11 Comparison of estimated age ranges of stream-terrace soils described in chapter 4 with other soil chronosequences formed in strongly seasonal semiarid to subhumid, moderately seasonal thermic climate of the Transverse Ranges of Southern California.[a]

| San Gabriel Mountains, South Side | | | San Gabriel Mountains, Northeast Side | | | Central Ventura Basin, Western Transverse Ranges | | |
|---|---|---|---|---|---|---|---|---|
| Terrace | Soil Profile[b] | Age Range, ka | Terrace | Soil Profile | Age Range, ka | Terrace | Soil Profile | Age Range, ka |
| T9 | A-Cox-Cu | <1 | | | | Qt1,2,3 | A-Cox-Cu | <0.5 |
| T8 | A-AC-Cox-Cu | 1–4 | | | | | A-AC-Cox | |
| T7 | A-Bt-Cox | 4–7 | T4 | A-AC-Cox | 4–11 | Qt4 | A-AC-Cox | 8–12 |
| T6 | A-Bt-Cox | 12–7 | T3 | A-Bt-Cox | 5–14 | | | |
| T5 | A-Bt-BC-Cox | 20–12 | T1 | A-Bt-Cox | 7–20 | Qt5 | A-Bt-Cox | 15–30 |
| T4 | A-Bt-BC-Cox | 50–60 | Qoa-d | A-Bt-BC-Cox | 55±8 | Qt6 | A-Bt-Cox | 38–100 |
| T3 | A-Bt-BC-Cox | 120–130 | | | | Qt7 | A-Bt-Cox | 160–200 |
| T2 | A-Bt-BC-Cox | 700–790 | | | | | | |
| T1 | A-Bt-BC-Cox | >790 | | | | | | |

[a]Northeast side chronosequence is from Weldon and Sieh, 1985. Ventura Basin chronosequence is from Keller, 1979, and Rockwell et al., 1983, 1985.

[b]Abbreviations for soil-profile characteristics are defined in Table 2.3.

TABLE 4.12 Comparison of estimated age ranges of stream-terrace soils described in chapter 4 with other soil chronosequences formed in strongly seasonal arid to semiarid, strongly seasonal thermic climate of San Joaquin Valley of central California.[a]

| San Gabriel Mountains | | | San Joaquin Valley of Central California | | | | | |
|---|---|---|---|---|---|---|---|---|
| South Side | | | Wheeler Ridge | | | East Side | | |
| Terrace | Soil Profile[b] | Age Range, ka | Terrace | Soil Profile | Age Range, ka | Terrace | Soil Profile | Age Range, ka |
| T9 | A-Cox-Cu | <1 | | | | Post-Modesto III | A-Cu A-Cox-Cu | <1 |
| T8 | A-AC-Cox-Cu | 1–4 | | | | Post-Modesto II | A-Cox-Cu | 3 |
| T7 | A-Bt-Cox | 4–7 | Q1 | A-Ck-Cu | 4–10 | Post-Modesto I | A-AC-Cox | 4–9 |
| T6 | A-Bt-Cox | 12–7 | | | | Modesto | A-AC-Cox A-Bt-Cox | 9–70 |
| T5 | A-Bt-BC-Cox | 20–12 | | | | Upper River-bank | A-Bt-Cox | 130–260 |
| | | | Q2 | A-Bt-Bk-Cox | 16–28 | | | |
| T4 | A-Bt-BC-Cox | 50–60 | Q3 | A-Bt-Bk-Cox | 28–59 | | | |
| T3 | A-Bt-BC-Cox | 120–130 | Q4 | A-Bt-Bk-Cox | 93–146 | | | |
| T2 | A-Bt-BC-Cox | 700–790 | Q5 | K-Cox | 218–540 | Lower River-bank | A-Bt-Cox | 450 |
| T1 | A-Bt-BC-Cox | >790 | | | | Turlock Lake | A-Bt-Cox | >600 |

[a]Wheeler Ridge chronosequence is from Keller et al., 1985. East side chronosequence is from Marchand and Allwardt, 1980.

[b]Abbreviations for soil-profile characteristics are defined in Table 2.3.

alfisols; A-Bt-Cox soil profiles become progressively redder with age.

The possibility that climatically induced aggradation events are forced by variations in the earth's orbital parameters was developed in Sections 2.1.2.2 and 2.4.2. Although tectonic perturbations complicate the causes of terrace formation in the Transverse Ranges, pulses of aggradation may be linked to the astronomical clock. The age control present in Table 4.11 and 4.12 for four chronosequences in southern and central California does not conflict with the idea of regional aggradation events at roughly 120, 55, and 10 ka.

## 4.3 Stream Terraces

### 4.3.1 CLIMATIC AND TECTONIC STREAM TERRACES

Times of attainment of the base level of erosion are characterized by formation of flood plains and lateral erosion that bevels strath (bedrock) or cut (alluvium) surfaces. Subsequent climatic perturbations may terminate the equilibrium conditions by causing either aggradation of valley-floor alluvial fills or stream-channel incision and formation of terraces. Many stream-terrace treads (but not all; see Section 4.4) can be considered time lines in fluvial landscapes. They provide opportunities to separate effects of climatic perturbations from effects of pulsatory tectonic movements.

Fill terraces were differentiated from strath terraces (Section 1.3.1.1) on the basis of thickness of alluvium as exposed in streambanks and road cuts and as measured by shallow seismic refraction studies. Beveled bedrock surfaces of paired strath terraces were defined on the basis of lateral continuity along and across a valley. Unpaired straths have formed locally where streams temporarily impinged on hillslopes and cut laterally into bedrock. Such unpaired terraces are of little use to the climatic or tectonic geomorphologist; they generally represent an unwelcome complication.

The three sites studied here are in different tectonic settings. The terraces of the North Fork of the San Gabriel River are 15 to 20 km north of the range-bounding faults. At the beginning of the study, the San Gabriel and nearby faults (Fig. 4.12) were thought to have been inactive since the early Quaternary; therefore the North Fork terraces appeared to have the potential of being the result solely of climatic change. As the study progressed it became readily apparent that uplift along the range-bounding faults played a role in the North Fork and that internal faults also were active during the middle Pleistocene.

The terraces of Little Tujunga Creek are upstream from the surface ruptures of the 1971 San Fernando Valley earthquake, and the Cucamonga Canyon terraces are cut by Holocene multiple-rupture-event fault scarps along a rapidly rising mountain front. Matti et al. (1982) described the soils, stratigraphy, and 18 surface-rupture events since 13 ka along the Cucamonga fault zone.

Field studies included surveying longitudinal profiles of strath and fill levels of former valley floors, comparing them with the present valley-floor profiles, and studying the terrace deposits and soil profiles to better understand the times and mechanisms of terrace formation. The fill terraces were the result of climatic perturbations that were dominant even where concurrent uplift tended to promote degradation.

### 4.3.2 NORTH FORK OF THE SAN GABRIEL RIVER

The North Fork of the San Gabriel River (Fig. 4.1) has an impressive flight of terraces (Fig. 4.12) with 40- to 60-m-thick fills that were deposited at about 800, 120, 55, and 6 ka. T2, T5, and T6 terraces are not present. These episodes of major valley-floor aggradation attest to marked variations in the capacity of the river to transport bedload. Well-preserved terrace remnants occur between Bichota Canyon and the West Fork of the San Gabriel River despite erosion of a broad valley floor by the present stream, which flows on a strath mantled with large boulders. The flight of terraces is preserved because the stream has migrated toward the southeast. The North Fork site is the only drainage basin analyzed where terraces older than 50 ka can be traced for 4 km along the valley (Figs. 4.13 and 4.14).

A history of the North Fork terraces starts with beveling of the strath surface below the T1 fill (Fig. 4.15) in the early Pleistocene. The equilibrium conditions represented by the strath were followed by at least one episode of valley-floor aggradation, as is indicated by 40 m of alluvial fill deposited on the T1 strath at approximately 800 ka. Then erosion apparently predominated until terrace T3 time at about 130 ka, when 60 m of T3 valley fill was deposited. Renewed channel downcutting below the T3 terrace tread permitted development of the T3 haploxeralf soil during the next 60 to 70 ky. This period of soil formation ended with another aggradation event, when T4 gravels more than filled the incised channel and deposited several meters of gravel on top of the strongly developed T3 soil profile. Renewed channel downcutting allowed the T4 haploxeralf soil to form. Both soils are described in Table 4.8. The time needed to form both soils is substantial—estimated at about 50 to 60 ky—but is much less than the age represented by the soil profile on the T1 fill, which has a 5-m-thick Bt horizon (Table 4.9). After terrace T4 time, erosion persisted until the middle Holocene as suggested by the lack of

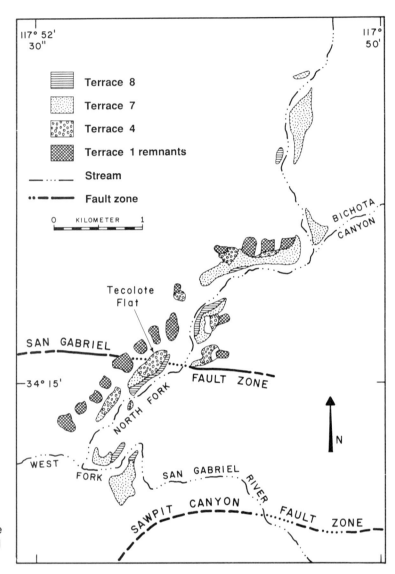

FIGURE 4.12 Terraces of the North Fork of the San Gabriel River.

T5 and T6 terraces; both of these terraces are found in the more tectonically active parts of the San Gabriel Mountains. Terrace T7 records 50 m of valley-floor aggradation in the middle Holocene. T8 is a fill-cut terrace that records temporary equilibrium conditions during degradation of the T7 fill. Terrace T9 consists of discontinuous patches of young gravels with incipient soil profiles; it also appears to be a fill-cut terrace.

Gravel is only a small component of the total sediment load being transported by the present streams of the San Gabriel Mountains. More than 450 debris-catchment structures protect urban communities from mudflows and floodwaters.

FIGURE 4.13 Terraces of the North Fork of the San Gabriel River at Tecolote Flat. Terrace T8 is cut into the T7 fill at the left side of the photo. A cabin is shown for scale left of center. Figure 4.17 shows the area to the right of center.

Studies of the volume, density, and particle-size distribution of fluvial deposits periodically excavated from the basins indicate that only about 25 percent of the sediment transported by San Gabriel Mountains streams consists of gravel; the remaining 75 percent consists of sand, silt, and clay. Thus the 40 to 60 m of aggradation gravels are clear evidence for major depletions of hillslope sediment reservoirs in the North Fork fluvial system.

The concept of the threshold of critical power is readily applicable to these stream terraces. Channel downcutting occurred when stream power exceeded critical power—either when (1) sediment load decreased, (2) discharges increased, (3) gradient increased, or (4) a combination of these occurred. Fill-terrace deposits result from selective deposition of gravel during times when stream power is insufficient to transport all the bedload (Fig. 4.16). Valley-floor aggradation occurred when more gravel was supplied from hillslopes. Strath and cut surfaces were created during times of equilibrium when stream power equaled critical power. Uplift has increased relief

FIGURE 4.14 View of the terraces of the North Fork of the San Gabriel River, looking upstream toward the Tecolote Flat terraces (right of center) shown in Figure 4.13.

of the mountain range, steepened streamflow gradients, and caused intermittent tectonically induced channel downcutting. Frequent flood discharges on steep gradients provide an excess of stream power, so that large streams downcut rapidly to reestablish type 1 dynamic equilibrium conditions (Section 1.3.1.2) before the next pulse of uplift or aggradation event.

The active channel and the T8 terrace deposits are representative of streambed materials during times of equilibrium. The present stream can transport boulders larger than 2 m. The channel gravels are extremely poorly sorted and contain 13 percent sand: median particle size is 110 mm. The sandy

gravel of terrace T8 is similar to the stream-channel sediment because both deposits are thin lag gravels in which boulders from the hillslope sediment reservoir and alluvium were concentrated as the stream cut through the T7 and T3 alluvial fills. The median particle size for the T8 sample is 32 mm; other T8 localities have 4-m boulders.

Sedimentology of the deposits of the three fill terraces provides interesting clues regarding bed-load production and transport. Histograms of the T1 and T7 fills are remarkably similar despite their early Pleistocene and mid-Holocene ages. Both deposits consist of tightly packed, well-sorted, water-laid gravels that contain only 3 percent sand.

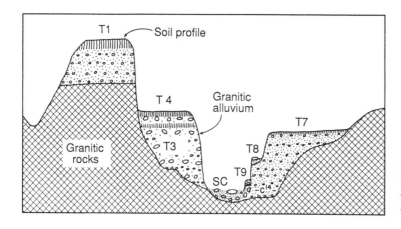

FIGURE 4.15 Generalized valley cross section showing fill terraces of the North Fork of the San Gabriel River.

The dominant size class of both samples is 32 to 64 mm and the median particle sizes are 30 mm for the T1 fill and 28 mm for the T7 fill. These similarities in the outputs of the hillslope subsystem are suggestive of similar modes of operation of the fluvial system at 6 ka and at >800 ka. The T3 deposits are much different. The modal size class is 64 to 120 mm, but a third of the beds were deposited as debris flows.

The T7 fill dominates the view of the fill-terrace deposits at Tecolote Flat (Fig. 4.17); it extends from the stream channel to an undissected planar terrace tread. Seismic refraction studies reveal an undulating planar bedrock surface–a strath–only

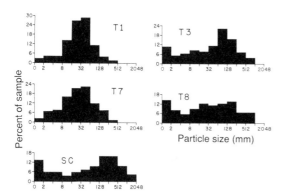

FIGURE 4.16 Histograms of terrace gravels, North Fork of the San Gabriel River.

2 to 3 m beneath the foreground channel gravels. Peat and manzanita seeds for radiocarbon dating (Table 4.10) came from basal beds. Most of the overlying gravels are well sorted and moderately fine grained, but abundance of boulders increases toward the top of the fill. Increase in boulder abundance is suggestive of (1) more selective gravel entrainment, possibly due to more winnowing as rates of aggradation decreased, (2) an increase in boulders supplied from hillslope sediment reservoirs as colluvium was partially stripped and underlying bedrock was exposed to erosion, (3) a change in source of boulders such as increased landslide activity, or (4) a combination of these. The remnant of T3 fill at the right side of Figure 4.17 has a 7.5YR hue in contrast to the 10YR hue of the T7 fill. The alluvial cap of the T8 alluvium is only a 2- to 4-m-thick bouldery bed on a surface cut into the T7 fill.

Longitudinal profiles of the present and past valleys of the North Fork are roughly parallel (Fig. 4.18). Present valley-floor gradient decreases from 0.085 to 0.073 at the Bichota Canyon junction. The decrease in gradient probably reflects a substantial increase in the discharge component of stream power (Eqs. 1.3, 1.4) downstream from the junction for a stream that is at the base level of erosion.

Details of the longitudinal profiles are interesting. The profile of the terrace T7 tread shows a

FIGURE 4.17 Quaternary stratigraphy of the fill terraces at Tecolote Flat. The gravels of the T7 fill surround and cap a former hill of T3 fill. Radiocarbon sample sites near the base of the T7 fill date initiation of the T7 aggradation event. See Figure 4.13 for overall view of the terrace flight.

decrease in gradient similar to that of the present stream channel, but the tread represents a threshold when the stream switched from aggradation to degradation in the mid-Holocene. This threshold also may have been a brief period of equilibrium if the stream gradually approached and stayed at the same altitude. The profile of the T4 tread diverges downstream with respect to the T7 profile. The T4 tread is just above the T7 tread downstream from Bichota Canyon. T4, T7, and T8 terraces are preserved as large planar remnants. Remnants of terrace T1 are numerous but have been eroded to accordant ridge and hill tops; few planar remnants remain. The T1 valley floor downstream from the San Gabriel fault zone was tectonically rotated upstream prior to terrace T3 time.

### 4.3.2.1 Responses of Geomorphic Processes to Climatic Change

Four 40- to 60-m-thick terrace fills in the North Fork bespeak major climatic perturbations that forced the mode of stream operation far to the aggradational side of the threshold of critical power. These terraces are not of tectonic origin because aggradation associated with base-level rise, such as fault movements or landslide dams in the down-

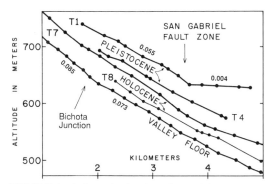

FIGURE 4.18 Longitudinal profiles of the present and past valleys of the North Fork of the San Gabriel River. Numbers are gradients.

stream reaches, would be revealed by accumulations of fine-grained sediments. Instead, the valleys of the San Gabriel River were aggraded with stream gravel. T7 fill and correlative deposits occur throughout the central and western Transverse Ranges (Table 4.10). The regional extent of the aggradation event implies a regional perturbation such as the latest Pleistocene–Holocene climatic change.

Regional paleoclimatic data (Section 4.1.3.2), together with the general format of the process-response model of Figure 2.38, can be used to outline a possible scenario of geomorphic responses to climatic change that resulted in the T7 aggradation event. The latest Pleistocene and early Holocene were times of stable hillslopes below treeline. Aggradation did not begin in the North Fork until 6.4 ka; before then the stream was cutting a strath (type 1 dynamic equilibrium). The angular, fine-grained gravel of the T7 fill suggests that splitting of granitic rocks by freezing and thawing was during the time of the latest full-glacial climate. Apparently neither periglacial nor glacial processes were important in the highest parts of this watershed, because sediment yields were not increased sufficiently to cause valley-floor aggradation. The excellent sorting of these water-laid gravels suggests fluvial erosion from hillslope sediment reservoirs instead of mass

movements of landslide debris into stream channels. Hillslopes were subject to stripping with the onset of climatic change that increased erosiveness of precipitation-induced runoff events or decreased resistance to erosion through changes in the protective vegetation cover. Both types of change probably occurred in the San Gabriel Mountains.

Changes in the type of air-mass circulation may have affected geomorphic processes that caused stripping of part of the hillslope colluvium and deposition of T7 alluvium. Mean annual precipitation need not have changed. Accelerated hillslope erosion might be favored by (1) a decrease in frequency of cold storms with resulting thinner winter snowpacks, (2) an increase in frequency of warm storms that would cause intense rainfalls or rapid melting of snow, or both, and (3) changes in seasonal distribution of rain that would strengthen and lengthen the annual summer drought.

Decreases in vegetative protection for the hillslope sediment reservoir might have occurred in two ways–decrease in plant density or changes in plant species. Plant cover may have become thinner because of decreases in precipitation or increases in temperature, or both. The studies of eolian deposits, pollen, plant macrofossils, and radiolaria (Section 4.1.3.2) all indicate a warmer and drier climate between about 8 and 4 ka.

Mid-Holocene erosion was too brief to completely empty thick hillslope sediment reservoirs. Although bedrock outcrops are common on hillslopes (Figs. 4.13 and 4.14) most slopes presently are mantled with colluvium that supports a dense growth of chaparral. Less bedrock may have been exposed during the latest Pleistocene. Partially stripped and revegetated hillslopes with triangular talus remnants (Fig. 4.19) suggest a recent episode of accelerated erosion that did not completely deplete hillslope sediment reservoirs because of reestablishment of denser protective plant cover during the late Holocene.

Climatic change commonly alters species compositions of plant communities. Drier and warmer mid-Holocene climates should have resulted in shifts of plant communities to higher altitudes or

FIGURE 4.19 Partially stripped and revegetated colluvium on triangular talus facets in the headwaters of Big Tujunga Canyon.

to wetter microenvironments. At present, coniferous forests predominate at the higher altitudes, particularly on the north-facing slopes. Deciduous trees are common in moist habitats at lower altitudes. Maple, sycamore, and alder trees presently are restricted chiefly to riparian settings or to cool-moist hollows on north-facing slopes. Early Holocene deciduous and coniferous forests on the south-facing hillslopes were replaced with dense chaparral scrub that now is the dominant plant community on hot, dry slopes.

Change to waxy and highly flammable plants such as chamise should have greatly increased frequency of fires. Burned hillslopes favor revegetation by chaparral; this self-enhancing feedback mechanism favors progressive expansion of the chaparral plant community (Hanes, 1971,1977). Brush fires every few decades during the mid-Holocene would have temporarily removed vegetation and greatly accelerated erosion of hillslope sediment reservoirs. The view of a burned hillslope in Figure 4.20 is representative of the increased potential for erosion after a fire when colluvium is exposed to intense rains. Sediment yields during subsequent winter storms can be increased 10- to 30-fold for 3 years after a fire (Fig. 4.2). Increased frequency of removal of hillslope vegetation by fires greatly increases critical power and thus tends to promote valley-floor aggradation.

Erosion is particularly spectacular when major storms bring downpours to burned slopes. Fires

FIGURE 4.20 Devastation of vegetation by fire that occurred near the end of a long summer drought.

burned steep hillslopes near the range-bounding faults in 1968, and Scott (1971) summarized the effects of two 100-year rainfall events that occurred in January and February of 1969 (Fig. 4.21). Erosion was relatively small near the ridgecrest of a typical hillslope, large on the midslope, and exceptionally large on the footslope.

A most interesting aspect of Figure 4.21 is that even erosion during a single season continues to reflect tectonically induced degradation that is concentrated along the stream channel and the adjacent footslope. The convex hillslope profile reflects long-term uplift along faults 1 km downvalley, which keeps the small stream strongly to the degradational side of the threshold of critical power. Nonuniform denudation continues to favor hill-

slope convexity—the waxing slope of Penck (1953, p. 159)—and increase landslide susceptibility. An equally interesting conclusion is that the impressive sediment yield that resulted from the combination of fire and two exceptionally large rainfall events was not sufficient to induce valley-floor aggradation in this small watershed–the stream subsystem was too far removed from the threshold of critical power.

Major valley-floor aggradation is much more likely in reaches that are distant from active faults or in reaches of large streams that are capable of rapid tectonically induced downcutting and attainment of equilibrium conditions. The North Fork is a good example. Streams with large, frequent floods commonly have sufficient stream power to attain equilibrium conditions, and thus are more likely than small headwaters tributaries to undergo either aggradation or accelerated degradation as a result of a strong climate-change perturbation.

A summary of interactions between hillslope and stream subsystems of the North Fork uses a modified version of the general model of Figure 2.38. For mid-Holocene aggradation to occur, it is assumed that:

1. Climatic change at 8 to 7 ka resulted in less snowfall and more intense rainfall
2. The climate became sufficiently drier or warmer for plant communities on the south-facing hillslopes to change from woodland to highly flammable chaparral.
3. Lightning-ignited fires removed flammable vegetation and exposed steep hillslopes to many brief episodes of rapid erosion.

Major sediment-yield increases overwhelmed the capacity of the North Fork to transport bedload; 60 m of aggradation ensued. Self-enhancing feedback mechanisms reduced the amount of colluvial detritus in the hillslope sediment reservoir, but progressive increases in area of exposed bedrock increased the flashiness of runoff, which eroded soils adjacent to outcrops. Thus progressive increases in stream power and decreases in critical power occurred. The threshold of critical power eventually was recrossed; degradation of T7 valley fill began and continued until a new base level of erosion was attained. A partial return to a cooler or moister climate during the late Holocene increased plant density on partially stripped hillslopes, which has tended to maintain the degradational mode of operation.

The T1, T3, and T4 aggradation events reflect impacts of earlier climatic change. The remarkably similar particle-size distributions of the well-sorted T1 and T7 gravels are suggestive of similar erosion, transport, and deposition. Climatic change and interactions of geomorphic processes may not have been the same, but sedimentologic outputs were remarkably similar. The poorly sorted water-

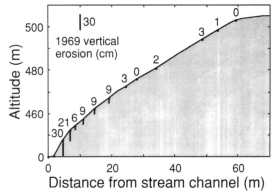

FIGURE 4.21 Erosion of a burned hillslope by two major storms during the winter of 1969. Modified from Figure 2 of Scott, 1971.

laid gravels and abundant debris-flow beds of the T3 deposits indicate a different type of climatic change and fluvial system response than for the T1 and T7 aggradation events.

Debris flows are common now and result from three processes. They occur when exceptionally intense, prolonged rain falls on burned-over silty colluvium that was largely saturated by prior rains. Mass movements of saturated footslope materials on tectonically steepened hillsides continue down stream channels as debris flows (Scott, 1971). A second mechanism occurs in the northeast San Gabriel Mountains (Sharp & Nobles, 1953; Johnson, 1970). Highly sheared Pelona schist along the San Andreas fault zone is the source material for mass-movement-generated debris flows during times when rapid melting of above-normal snowpacks saturates hillslope materials. The third process is liquefaction of seismogenic landslides that occur during wet winters. Debris flows in the T3 fill could have been caused by all three processes, but intense rains and earthquakes are the most likely causes.

The climatic terraces of the North Fork of the San Gabriel River also reflect the influence of tectonic perturbations. Tectonic influences include (1) middle Pleistocene rotation of the downstream part of the T1 terrace, (2) long-term tectonically induced valley downcutting, and (3) overall increase in valley-floor gradient from 0.055 (early Pleistocene T1) to 0.075 (middle Holocene T7). Continuing uplift of the San Gabriel Mountains along the range-bounding faults has affected even distant reaches such as the North Fork. For terrace flights that reveal stronger tectonic influences we proceed to a discussion of the Little Tujunga and Cucamonga Canyon sites.

### 4.3.3 CLIMATIC TERRACES IN TECTONICALLY ACTIVE REACHES

#### 4.3.3.1 Little Tujunga Creek

Little Tujunga Creek enters the San Fernando Valley near the west end of the San Gabriel Mountains in a tectonically active setting. Multiple mountainous escarpments are present and each topographic front is thrust faulted. Geomorphic tectonic analyses indicate that both the range-bounding fronts and one internal front have been the sites of recurrent tectonic movement during the late Quaternary.

The Little Tujunga Creek landscape contrasts markedly with the North Fork landscape, which is a simple valley without abrupt changes in cross-sectional topography. Little Tujunga Creek consists of alternating reaches of narrows and embayments. Narrow valleys occur where the stream cuts through rising mountain fronts. Broad topographic embayments occur between mountain fronts, and a piedmont extends downstream from the range-bounding front. Stream terraces are broad expanses in embayment reaches but tend to be intermittent in narrows reaches.

Tectonically induced channel downcutting along Little Tujunga Creek reestablished equilibrium longitudinal profiles repeatedly during the late Quaternary. Even tributary valleys of embayment reaches have straths and strath terraces. The prominent T4 strath terrace extends up tributary valleys, but maximum heights above streambeds are along the trunk channel. Strath-terrace height decreases upstream because the tectonically induced stream-channel incision that converted a strath to a strath

terrace is a consequence of tectonic base-level fall in downstream reaches. Effects of tectonically induced increases in relief typically decrease upstream in an exponential fashion.

The weakly developed (A-Bw-Cox) terrace T7 soil profile clearly indicates that the climatic perturbation responsible for a valley-floor aggradation event occurred during the Holocene. The T7 tread represents a threshold crossing, but the T7 gravels were deposited on a complex strath surface that may be younger than 60 ka.

Tributary streams tend to reflect climatically induced changes in hillslope subsystems better than trunk stream channels. A climatic perturbation may result in a valley fill that thickens rapidly downstream in tributary reaches and then is constant or decreases in thickness downstream. For example, the North Fork has much better defined fill terraces than does the main valley of the San Gabriel River.

The net effect of the contrasting tectonic and climatic modes of formation of the T4 and T7 terraces is that the relative altitudes of the treads of the two terraces vary in different parts of the drainage net. The treads differ in height by as much as 33 m along the trunk stream channel but become progressively closer up tributary valleys. The T4 tread in Figure 4.22 is only a few meters higher than the much younger T7 tread, but the relative ages of the two terraces are readily discerned on the basis of their distinctive soil profiles. The view in Figure 4.22 shows extensive deep dissection of the embayment since deposition of the T7 fill. The large volume of materials eroded since about 4 to 6 ka is on a scale that rivals the erosion of the 5- to 6.8-ka fill-terrace deposits of the North Fork of the San Gabriel River. Accelerated late Holocene erosion along Little Tujunga Creek was caused by climatic change that increased stream power and decreased resisting power; both occurred at the same time that tectonically induced increases in stream power were occurring. The combination forced the mode of operation strongly to the degradational side of the critical-power threshold.

FIGURE 4.22 View looking downvalley (southeast toward trunk channel of Little Tujunga Creek) showing dissection of terraces and underlying Saugus Formation in an embayment reach. A remnant of the T4 terrace is left of center; a remnant of the T7 terrace to the right of center has a dark mollic A soil horizon.

### 4.3.3.2 Cucamonga Canyon

Cucamonga Canyon crosses multiple thrust-faulted mountain fronts and, relative to Little Tujunga Creek, has more rugged narrows reaches and restricted embayments reaches. Closely spaced fault zones, rapid uplift rates, and resistant lithologies influence geomorphic responses to climatic change as reflected in stream terraces. Prominent T6 and T7 terraces are present; remnants of Pleistocene terraces are represented by a few high benches capped with red alluvium.

The terraces sweep across a well-defined fan-head embayment that extends 4 km upvalley from the range front (Fig. 4.23). Beveled bedrock is as much as 70 m above the active channel for the T6 terrace and 50 m for the T7 terrace. Relative to the North Fork of the San Gabriel River (Fig. 4.12), strath terraces are much more common in Cucamonga Canyon (Fig. 4.24). Both terrace treads are cut by multiple-rupture-event fault scarps of the Cucamonga fault zone (Eckis, 1928; Morton & Miller, 1975; Matti et al., 1982). The data of

FIGURE 4.23 View of T6 and T7 terraces in the fanhead embayment of Cucamonga Canyon. Seismic surveys revealed 1.8 to 2.6 m of gravel above the strath being cut by the active channel in the foreground. The triangular facets are on an erosional mountain front which was cut by the stream of Cucamonga Canyon.

Matti and colleagues suggest a mean uplift rate of 2.8 m/ky across the fault zone during the past 13 ky.

An irregular strath complex beneath 5 to 30 m of terrace gravels (Fig. 4.25) records repeated attainment of equilibrium conditions during the Pleistocene. Straths are fairly smooth when formed, as indicated by the uniform thickness of 1.8 to 2.6 m of gravel above the strath beneath the active channel (Fig. 4.23). Most of the relief of the bedrock–alluvium contact is the result of multiple

times of strath formation and of stream-channel incision into straths. Part of the relief may be due to faulting. A minor part of 3 to 25 m variations in thickness of T7 alluvium above the strath complex upstream from front 23C is due to projections of strath altitudes into the line of section. Individual straths are continuous locally, such as between distances 2.7 to 3.7 km on Figure 4.25. The much higher strath in the adjacent downstream reach is older.

The tread of the T7 terrace diverges and then

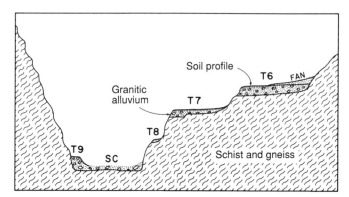

FIGURE 4.24 Generalized valley cross section showing stream terraces of Cucamonga Canyon.

converges with the longitudinal profile of the present valley floor (Fig. 4.25). T7 terrace height between fronts 23A and 23B increases from 14 to 58 m. The height of the single strath increases 10–fold in this reach. This divergence may be the result of tectonically induced downcutting in response to (1) continuing uplift in a broad zone of earth deformation at front 23B, (2) rotation of the fault block between fronts 23A and 23B, and (3) uplift at fronts 23C and 23D. The T7 terrace tread converges 23 m with the channel between fronts 23B and 23C. Shallow seismic surveys revealed that thickness of stream-channel alluvium in-

creases abruptly from 2 to 10 m at the Cucamonga fault zone.

The T6 tread occurs at heights of as much as 100 m above the active channel of Cucamonga Canyon. Both the tread and strath diverge with respect to the present channel profile between fronts 23A and 23B and converge between fronts 23B and 23C, where they are truncated by a fault scarp with 12 m of vertical displacement.

Most of the strath complex and bedrock gorge of Cucamonga Canyon were eroded during the late Pleistocene only to be covered by fluvial gravels during the latest Pleistocene–Holocene aggrada-

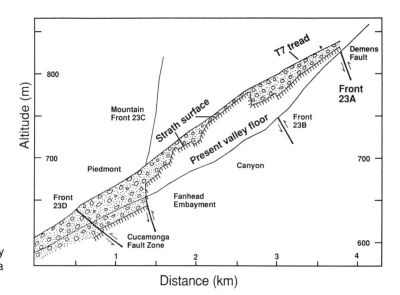

FIGURE 4.25 Longitudinal profiles of the present valley and T7 tread in Cucamonga Canyon.

tion event. Evidence for such extreme fluctuations in stream-channel altitude includes paleosols and comparisons of vertical displacements along active faults and tectonically induced downcutting.

Canyon downcutting into bedrock was caused by local base-level falls that resulted from uplift along faults. The strath complex beneath the T6 aggradation gravels is as much as 70 m above the active channel. At front 23C the strath surface is displaced about 25 m (Fig. 4.25), the T7 tread about 4 m and the T6 tread about 12 m. The discrepancy between 12 m of vertical displacement during the past 12 ky and 70 m of tectonically induced downcutting is best explained by having most of the canyon downcutting occur during the Pleistocene.

Additional evidence for a Pleistocene bedrock gorge is found in fanhead trench exposures of alluvium downstream from front 23C. Red, clayey sand with common stage 3 weathered cobbles has a truncated argillic horizon that is buried by 25 m of Holocene gray gravel with many stage 1 and a few stage 2 weathered cobbles. The Pleistocene red sand shows that at least part of the strath complex formed before burial by the Holocene aggradation event.

The late Quaternary history of Cucamonga Canyon is a story of thresholds and rapid rates of geomorphic processes that have been affected by strong climatic and tectonic perturbations. The data presently available suggest the following scenario. Gravel-capped straths formed repeatedly during the Pleistocene only to be incised and converted to strath terraces as a result of tectonically induced downcutting caused by pulses of uplift on the Cucamonga and other fault zones. Each strath terrace represented a pause in long-term downcutting that also created a deep inner gorge.

The Pleistocene–Holocene climatic change greatly increased hillslope sediment yields, which initiated an aggradation event. Instead of spreading out in broad embayments, such as along Little Tujunga Creek, aggradation was confined and approximately 100 m of gravel was deposited in Cucamonga Canyon. Alluvium filled the bedrock gorge and completely buried the strath complex.

The crossing of the threshold of critical power that formed the T6 tread may have been timed by movements on the Cucamonga fault zone if uplift increased stream power sufficiently to change the mode of stream operation from aggradation to degradation. Active faulting continued and about 8 m of vertical displacement of the T6 tread occurred before the T7 tread was formed.

The T7 terrace formed during a period of renewed aggradation, which also may have been terminated in part by uplift. Subsequently, 4 m of vertical displacement has occurred at front 23C, and 6 m at front 23D.

The stream then reexcavated the bedrock gorge and is now deepening it. Channel downcutting rates were greater than 10 m/ky. Rapid acceleration of excavation of the last 40 m of valley fill during the late Holocene was the result of a combination of thrust faulting at fronts 23C and 23D, which caused tectonically induced downcutting that steepened stream-channel slopes, and climate-change-induced self-enhancing feedback mechanisms on the hillslopes (Section 4.3.2.1). The first process increased stream power and the second decreased critical power; both tended to increase the ratio of stream power to critical power. Acceleration of degradation may also be a reflection of a large increase of the reaction time between fault displacement and the start of tectonically induced downcutting that occurs because a pulse of aggradation that is dominant favors the opposite mode of stream operation. The stream had to wait until aggradation was finished before it could catch up with its long term tectonically induced downcutting.

Similar responses to climatic change and tectonic perturbations occurred in nearby Day Canyon, but not in Deer Canyon. Day Canyon is 10 km east of Cucamonga Canyon; it has a fanhead embayment in which faulted T7, T6, and T4 terraces rise above a deep fanhead trench. The late Quaternary history of Day Canyon appears to be similar to that postulated for Cucamonga Canyon, perhaps because the characteristics of the fluvial system and the location of the Cucamonga fault zone are similar.

Aggradation has continued to the present in Deer Canyon (5 km to the east), whereas in Cucamonga and Day canyons aggradation switched to degradation in the middle Holocene. A large, unentrenched alluvial fan extends into the canyon mouth as a broad embayment and T9 terraces continue up the canyon. The Deer Canyon fluvial system has a much longer response time to complete the aggradation event induced by the Pleistocene–Holocene climatic change because the Cucamonga fault zone is not located at the mountain-piedmont junction. The fault has been just as active as at Day and Cucamonga canyons 5 km to the east and west, but it crosses the aggrading reach of the Deer Canyon alluvial fan. Vertical fault displacements tend to decrease aggradation rates for a short distance upstream from the fault and to increase them downstream. Thus the effects of the climatic change have been modulated by uplift along a highly active fault zone in much different ways in the Deer and Cucamonga Canyon fluvial systems. Tectonic base-level fall in Cucamonga Canyon has shortened the time needed to cross the threshold of critical power, but alluvial-fan deposition is a base-level rise that has favored continued aggradation in Deer Canyon.

### 4.3.4 SUMMARY OF RESPONSES OF STREAMS TO CLIMATIC CHANGE

Abundance, ages, heights, and types of terraces vary from canyon to canyon in the San Gabriel Mountains because (1) the stream terraces are the result of both tectonic and climatic perturbations and (2) reaction and relaxation times to climatic perturbations vary in this complex, lofty mountain range. Similar suites and ages of terraces would be present in the drainage basins if the aggradation events were entirely the result of regional climatic change and system response times were similar. Similar topographic and lithologic characteristics of the drainage basins also would be required. Response times vary between drainage basins, mainly because of the ways in which geomorphic processes affecting yields of water and sediment from hillslopes vary with altitude. Furthermore, each fluvial system has been affected by tectonic base-level fall, but the times of faulting have not been synchronous and locations of faulting within the fluvial systems vary greatly.

Terraces in the San Gabriel Mountains are primarily the result of climatic or tectonic perturbations. Longitudinal profiles of strath terraces that diverge from the valley-floor profile to an active fault are considered to be of tectonic origin. Fill terraces that converge with the valley-floor profile and are undeformed where they pass through fault zones are considered to be of climatic origin. Where both tectonic and climatic perturbations are likely, distance from an active fault zone (tectonic control) or distance from headwaters hillslopes (climatic control) may determine which is the dominant independent variable. Fill-cut terraces such as T8 are complex-response terraces.

The most likely causes of the stream-channel entrenchment that converted straths or valley fills into terraces are noted in Table 4.4 for the terrace chronosequence. Treads of fill terraces cut by prominent fault scarps are considered to have an age determined primarily by uplift (T6 terrace in Cucamonga Canyon). Strath terraces terminated by range-bounding faults are also considered to be of tectonic origin (T9 terrace in Little Tujunga Creek). Fill terraces in headwaters reaches and upstream from tectonically inactive mountain fronts are most likely of climatic-change origin; examples include the T4 and T7 terraces of the North Fork of the San Gabriel River. Where fill terraces occur upstream from active fault zones it can be difficult to discern which type of perturbation was responsible for entrenchment of the terrace tread (T7 terrace at Little Tujunga and Cucamonga Canyons). Climatic change caused the aggradation, but initiation of entrenchment followed by increasing rates of canyon downcutting probably was determined by a combination of tectonic and climatic factors. T1 and T2 terraces owe their great heights above the active channels to long-term tectonically induced downcutting, even though their deposits and treads had climatic origins.

Five episodes of 10- to 50-m-thick valley floor backfilling appear to be widely spaced in time. These are recorded by the T1 (deposited before the 790 ka Brunhes-Matuyama magnetic-polarity reversal) T2, T3, T4, and T7 fill terraces. Other, weaker aggradation events may have occurred but only the higher, larger volume fill events are likely to be preserved in the active erosional environment of the San Gabriel Mountains.

Although the Transverse Ranges are adjacent to the Mojave Desert, we should not expect a direct transfer of the Figure 2.38 process-response model intended for hot deserts to the semiarid to subhumid, mesic San Gabriel Mountains. Lofty mountain ranges are geomorphically much more complex than low desert hills. The North Fork ranges in altitude from 500 to 2600 m. The rugged 39 km$^2$ basin of Cucamonga Canyon rises from 634 m at its mouth to 2878 m. Drainage basin reliefs of >2000 m suggest that geomorphic processes may be spatially complex. Area is uniformly distributed with increasing altitude in most of the Cucamonga Canyon basin (Fig. 4.26B). Vigorous erosion of materials of moderate resistance is sufficiently advanced, despite continuing rapid uplift, that steep and sharp-crested slopes prevail above the narrow valley floors throughout the drainage basin. Such a similarity in overall slope morphology reflects the importance of tectonically induced downcutting, but does not reflect a spatial similarity of operating geomorphic processes.

The relative importance of geomorphic processes changes with altitude (Fig. 4.26A). Only highly generalized comparisons can be made between present and past geomorphic processes at low and high altitudes. The following discussion of only three processes–periglacial processes, chemical weathering, and the relative importance of snow as compared to rain–merely underscores the need to consider altitudinally controlled variations in geomorphic processes. For these general purposes let us assume that a 25- to 18-ka full-

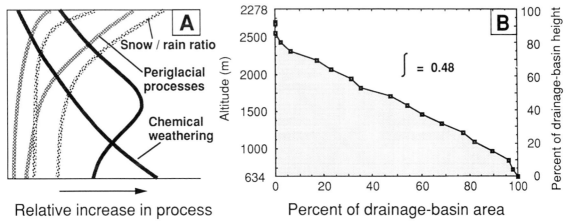

Relative increase in process        Percent of drainage-basin area

FIGURE 4.26 Changes of geomorphic processes and of basin area with increase in altitude in the San Gabriel Mountains, using the Cucamonga Canyon drainage basin as an example. A. Relative importance of several geomorphic processes with increase in altitude. Thicker patterned lines indicate full-glacial conditions and thinner lines indicate late Holocene interglacial conditions. B. Areal distribution of altitude in the 34 km$^2$ drainage basin. Mean slope is 0.79.

glacial climate was colder and slightly dryer than the present climate, and that treeline was depressed from 2800 to 1800 m.

Snowfall presently is insignificant in the foothills and is dominant at high altitudes; precipitation at intermediate altitudes fluctuates between rain and snow during most winters. Snowline was substantially lower during full-glacial times and most of the range received frequent snowfalls. Freeze–thaw and periglacial processes are important now at the highest altitudes, and most likely were dominant hillslope processes above 1800 m during full-glacial times.

Chemical weathering shows interesting altitude-controlled variations depending on whether weathering is limited by temperature or soil-leaching moisture. Moisture availability increases and temperature decreases with increasing altitude; the net result is an altitudinally controlled peak of optimal chemical weathering (Fig. 4.26A). At present, chemical weathering rates at low altitudes reflect optimal temperatures but are limited by less than optimal moisture conditions. Chemical weathering at high altitudes is limited by progressively lower temperatures. Cooler full-glacial conditions tended to greatly increase effective precipitation for soil leaching, but cold temperatures made chemical weathering at high altitudes less important than at present. The peak of optimal chemical weathering in the general illustrative model of Figure 4.26A was at altitudes lower than 600 m during times of full-glacial climate and at altitudes of 1300 to 1400 m during the late Holocene climate.

The large range of drainage basin altitudes within the San Gabriel Mountains provides many combinations of altitude area-controlled climatic geomorphic settings. Each altitude climatic zone had a different potential or actual time of threshold crossing in response to Pleistocene–Holocene climatic change. Downstream reaches integrate the effects of all source reaches. Thus one should not expect times of valley-floor aggradation or incision to be exactly synchronous in the different fluvial systems.

Even partial consideration of altitudinal control of geomorphic processes in the 300- to 3000-m altitude range of the San Gabriel Mountains suggests some initial general hypotheses regarding the timing and magnitudes of aggradation events.

1. Small low-altitude watersheds may have responded to the Pleistocene–Holocene climatic change in a manner similar to that of Mojave Desert mountains, which have only one-third the relief of the San Gabriel Mountains. If so, maximum hillslope sediment yields occurred during the latest Pleistocene and early Holocene.

2. Small high-altitude watersheds probably had maximum hillslope sediment yields during full-glacial times, because protective forests returned to most slopes above full-glacial treeline during the Holocene. Steep valley-floor gradients near the range crest would have offset the tendency for aggradation caused by increased sediment yields associated with periglacial processes, but streams may have been close to threshold conditions.

3. Large streams that drain hillslopes from the range crest to the foothills probably had a complex response that was a combination of hypotheses 1 and 2. The net response would be a function of the relative proportions of basin subareas characterized by different climatic-geomorphic process. For example, only 35 percent of the Cucamonga basin occurs above 1800 m, so the effect of climatic change along equilibrium reaches near the mouth of the basin would be controlled mainly by the discharge of sediment and water from slopes between 1000 and 1800 m, which comprise 60 percent of the watershed (Fig. 4.26B). Thus one should expect to find Holocene instead of full-glacial valley fills in the North Fork of the San Gabriel River and in similar streams.

High-altitude hillslope processes may not be the chief factor that determines the mode of operation in much lower trunk stream reaches, but they may affect reaction times, aggradation rates, and downcutting rates after a fill-terrace tread is formed. A combination of high-altitude sediment yield during full-glacial times followed by Holocene sediment-yield increases at lower altitudes could cause valley-floor aggradation to occur earlier than in fluvial systems that were entirely below full-glacial treeline. The full-glacial pulse of increased bed-load transport rate would precondition the stream subsystem—move it closer to threshold conditions—so that reaction times to subsequent climatic perturbations would be shorter. Timing of aggradation events could overlap in some drainage basins, which would result in an extended period of alluviation of valley floors.

Changes in hillslope microclimates during the past 25 ky probably varied with slope aspect. Watersheds that drain colder north-facing hillslopes, such as Cajon Creek, might have had geomorphic responses to climatic change that favored aggradation pulses at 20 to 10 ka. Basins that drain warmer south-facing hillslopes, such as the North Fork, might have had aggradation pulses at 7 to 4 ka. Such contrasts would be greatest for slopes underlain by granitic rocks, which are common and especially sensitive to variations in sediment yield resulting from climatic change (Section 3.3). Thus the age ranges suggested by the soils chronosequence reflect not only lack of precision and accuracy of dating methods and field measurements but also real variations of aggradation times.

Whether T4 fill buries T3 fill in a given drainage basin is dependent on (1) the relative magnitudes of the two aggradation events for a given fluvial system and (2) the amount of tectonically induced erosional lowering of the valley floor between the two aggradation events. T4 deposits bury the T3 terrace tread in the North Fork of the San Gabriel River, Cajon Pass (Weldon, 1986), and in San Gorgonio Canyon to the east of Cajon Pass (Harden et al., 1986). The two fills occur as distinct terraces along Little Tujunga Creek because channel downcutting induced by large amounts of late Quaternary uplift along the range-bounding fault was sufficient to prevent T4 aggradation from burying the T3 terrace tread.

T3 aggradation may have occurred during the major interglacial at about 120 ka. Soils on T3 terraces in the San Gabriel Mountains have roughly the same strength of development as soils on alluvium deposited on nearby 120 ka marine terraces described by Lajoie and colleagues (1979, 1982) and by Keller (1979). The T3 terrace soils are in a stronger leaching environment, but this factor is largely offset at the marine-terrace sites, which receive more sodium from atmospheric sources; both factors tend to increase rates of soil-profile development. T4 and buried T3 soils appear to be similar at each of the three localities where T4 gravels bury the T3 soil (McFadden & Weldon, 1987). Thus the T3 aggradation event appears to be approximately twice the age of the 55–ka T4 event. The presence of abundant debris-flow beds in the T3 fill deposits of the North Fork also supports the possibility of an aggradation event during a time of more frequent tropical storms, such as the period of warmer sea-surface temperatures during the 120-ka interglacial.

The effects of climatic change on the operation of the fluvial system of the North Fork are representative of most of the San Gabriel Mountains. Geomorphic responses to climatic change are more obvious in the North Fork because climatic perturbations clearly dominated over the effects of distant tectonic perturbations. Tectonic effects were stronger relative to the influences of climatic change in those reaches of Little Tujunga and Cucamonga Canyons that are upstream from active faults. However, the climatically controlled increase in sediment yield associated with the T7 fill was the dominant perturbation during the mid-Holocene, even near the active range-bounding faults.

We conclude that a major Holocene aggradation event occurred in many of the canyons of the San Gabriel Mountains, but important unanswered questions remain. How much variation is there between canyons for the times of initiation of valley-fill deposition? Within a given valley how much variation is there in the times of terrace-tread formation? Although the next section may not provide all the answers, it attempts to address some most interesting details of geomorphic responses to climatic change.

## 4.4 Time-Transgressive Threshold-Intersection Points

Stream terraces sweep through mountain canyons and across piedmonts in many climatic settings. Terrace treads commonly parallel each other and the longitudinal profile of the active channel. Such smooth profiles of former valley floors may indeed strike the casual observer as being time lines passing through the erosional landscapes that surround them. The use of stream-terrace treads as planar reference surfaces that are assumed to have formed at a specific time is important to tectonic geomorphologists who wish to decipher late Quaternary histories of faulting and folding. But are the treads or straths of climatic or tectonic terraces sufficiently synchronous to be regarded as time lines? We address this question in order to clarify the widely used assumption of terrace-tread synchroneity used in analyses of fluvial landscapes, and to learn more about responses of streams to climatic changes. First we consider spatial variations in rates of formation of strath and fill-cut (equilibrium) terraces and of fill (threshold) terraces. Then we consider the implications of an important study of Cajon Creek made by Ray Weldon.

### 4.4.1 SYNCHRONOUS AND DIACHRONOUS TERRACES

#### 4.4.1.1 Strath and Cut Terraces

Determining if stream-terrace treads are synchronous or diachronous is easier for degradation (strath and fill-cut) terraces than for aggradation (fill) terraces. Let's consider three phases of strath formation:

1. Genesis of fluvial beveled bedrock surfaces
2. Attainment of a continuous strath surface
3. Incision of the active channel into the strath to create a terrace

Straths occur in reaches that have attained static or type 1 dynamic equilibrium. Many straths gradually extend upstream into reaches that are still degrading because of less stream power than in downstream reaches. The degrading reach is slowly converted into an equilibrium reach. Such straths are time transgressive in regard to their time of formation. Straths are common in ephemeral streams of arid regions but probably require more time to form than along the more powerful streams of humid regions.

Modern straths that extend for many kilometers beneath stream-channel gravels of powerful rivers are clear evidence for straths being a synchronous fluvial landform. Most straths are synchronous landforms regardless of whether or not strath genesis was diachronous. Strath formation may end at the beginning of an aggradation event, which buries the beveled bedrock surface (Fig. 4.15; Sections 5.2.2, 5.2.4), or because of stream-channel incisement into the surface (Sections 5.2.3, 5.4.3).

A synchronous strath becomes a synchronous strath terrace only if the entire reach is incised at the same time. Although it is difficult to date times of initiation of entrenchment of strath and fill-cut terraces, it seems reasonable that powerful streams of humid regions can quickly initiate channel degradation over an entire reach of a strath. Relative increases in stream power caused by increases of stream discharge throughout a drainage basin should affect the entire strath reach at the same time. Cataclysmic flows can be important (Bull, 1988). The strongly seasonal winter precipitation regime of the semiarid to subhumid San Gabriel Mountains may produce a frequency of flood discharges similar to that of rivers in weakly seasonal humid mountains.

Strath entrenchment is not synchronous where incisement that terminates type 1 dynamic equilibrium is the result of faulting that increases stream power by local increase of gradient. Strath entrenchment will migrate upstream from a zone of tectonic base-level fall as headcuts and rapids. Rates of nickpoint migration are slow for ephemeral streams flowing on resistant rock and rapid for large perennial streams flowing on weak materials.

These same arguments pertain to fill-cut terraces. These complex-response terraces are more likely to have a synchronous time of incision because alluvium has less resistance to erosion than most rock.

Synchroneity is relative to terrace age for both static and dynamic equilibrium terraces. I would regard a 200-year time span for initiation of strath incision to be diachronous for a 1-ka terrace (20%) and synchronous for a 100-ka terrace (0.2%).

### 4.4.1.2 Fill Terraces

It is difficult to show that fill terraces are synchronous. Some are not. Aggrading valley floors may extend for many kilometers, but unlike the case with strath genesis, the level of streamflow is rising rather than remaining at about the same level. Rates of aggradation typically increase and then decrease as the threshold of critical power is approached. Threshold-intersection points (Fig. 1.9) at the upstream and downstream ends of an aggrading reach shift with time as the level of valley fill rises; this leads to spatial variations in times of initiation of aggradation. Initiation of stream-channel incisement generally is a momentary transition between two disequilibrium modes of streamflow—aggradation switching to degradation—rather than the end of a protracted time span (Section 5.4.4.1), as in the case of strath-terrace formation.

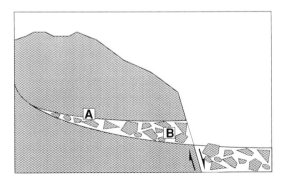

FIGURE 4.27 Hypothetical example of a fill terrace truncated by a range-bounding normal fault. A and B are sites of radiocarbon-dated wood that was deposited with the alluvium.

We need to examine fill terraces in detail because they seem less likely to represent time lines in fluvial landscapes than strath or cut terraces. The discussion starts with general characteristics and dating problems and then considers response times of aggrading fluvial systems in arid and humid regions. Rates of aggradation generally seem to be nonuniform, and may vary with variations of input processes. Two geometries of valley fills and different combinations of stream power and resisting power also affect the time-transgressive nature of threshold-intersection points associated with fill terraces.

Figure 4.27 shows many characteristics of fill terraces. All valley fills and fill terraces pinch out in the upstream direction, some in the trunk channel and others in the headwaters rills. Even where aggradation of headwaters footslopes is occurring it is impossible for aggradation to extend to ridgecrest divides. The aggradation surface therefore diverges downstream from the longitudinal profile of the preceding valley floor. Typical decreases in gradient of 5 to 30 percent represent a substantial local decrease in stream power that tends to promote further aggradation. In Figure 4.27 longitudinal profiles divergence continues to an active range-bounding fault. Maximum divergence of the terrace tread and present stream longitudinal profiles identifies the location of the tectonic perturbation responsible for terrace-tread entrenchment. The effects of such perturbations decrease exponentially upstream.

Dating of thick fill terraces, such as the one depicted in Figure 4.27, is fraught with pitfalls. Radiocarbon ages are not ideal even in those rare instances where wood has been deposited in terrace gravels. Dated wood at locality B only provides an age for the lower part of the valley fill; it does not tell us when aggradation began or ended. Two ages at locality B would provide added confidence regarding age of the fill and a mean rate of gravel deposition between the two samples. But we would still know little about when aggradation began or ended, unless we were willing to make the decidedly hazardous assumption that rates of deposition were constant. Interpolation of depositional rates is a much safer procedure than extrapolation of rates. To date the time of terrace-tread formation one might seek datable material from the uppermost bed of the valley fill, or attempt to date the exposed terrace tread. However, an age of the tread at locality A might not represent the tread age elsewhere. Aggradation could continue at A after stream-channel incisement of the tread has occurred at B. As in the case of faulted strath terraces, the time needed for the effects of the tectonic perturbation to move from reach B to reach A would be a function of available stream power to do erosional work, resistance of alluvium to erosion, and distance between the reaches.

Some workers prefer to estimate terrace-tread ages by dating surficial materials that change with the passage of time. For example, cobble weathering-rind thicknesses and soil-profile characteristics can be calibrated by radiocarbon (Section 5.2.3.1) and thermoluminescence dating methods. Dating accuracy is dependent on the inherent inaccuracies of both the geomorphic and the isotopic or radiogenic methods.

Resolution of dating methodologies, relative to actual time spans of terrace-tread formation, largely determines our perception of treads as synchronous or diachronous. Age estimates of 5 ka and 10 ka for a fill terrace tread in two reaches would suggest a twofold age difference in the time of terrace formation, and we would conclude that a diachronous terrace tread was present. If the actual age range of an older terrace tread was 100 ka and 105 ka, we would have no way of precisely determining these older ages with confidence and would conclude that the tread was most likely part of a synchronous terrace. Calibrated ages of terrace treads commonly lack the precision needed to identify diachronous terraces. Thus synchronous terraces are those that appear to have a single time of tread formation within the constraints of the dating methods used.

Climatic-change-induced aggradation is different in humid and arid fluvial systems and in de-

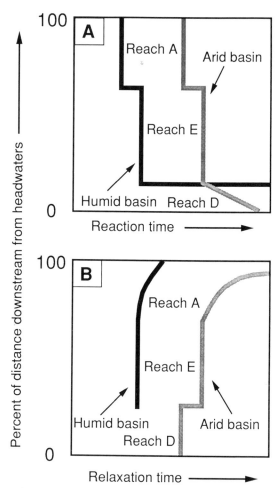

FIGURE 4.28 Response times (sum of reaction and relaxation times) for a climatic-change perturbation in degrading (D), equilibrium (E), and aggrading (A) reaches of humid and arid fluvial systems. A. Reaction times for initial aggradation to occur after a climatic perturbation. B. Relaxation times (time needed to complete aggradation).

grading, equilibrium, and aggrading reaches. Figure 4.28 considers the response times for these six types of reaches. Streamflow is the connecting link between different reaches in both arid and humid fluvial systems. Changes in water and sediment discharges pass downstream quickly as streamflows but reaction and relaxation times (Fig. 1.3) vary with modes of stream subsystem operation. In both the A and B parts of Figure 4.28, it is assumed that the climatic perturbation decreases stream power or increases resisting power, or both, sufficiently to induce aggradation in equilibrium reaches.

In a humid basin, water and sediment are conveyed through all three reaches with different consequences (Fig. 4.28A). The aggrading reach reacts first; it already has a stream power/resisting power ratio of less than 1, so the reaction causes an increase in deposition rate as the reach moves farther away from threshold conditions. The base-level rise caused by aggradation has a minimal effect on the adjacent upstream reach (Leopold & Bull, 1979). Relatively larger annual stream power (compared with that of ephemeral streams) also causes the reaction to occur rapidly in the equilibrium reach. Aggradation begins after the capacity for equilibrium adjustments between variables has been exceeded. The decrease in stream power/resisting power ratio moves the degrading reach closer to threshold conditions and thus causes a decrease in downcutting rate.

The aggradational and equilibrium reaches of the arid basin have responses similar to those of the humid basin, but reaction times are longer because, with relatively smaller annual stream power (smaller and less frequent flows), more time is needed to convey the increased sediment load through the system. At a still later time aggradation is initiated in the degradational reach, perhaps triggered by the aggradational base-level rise in the former equilibrium reach. The reaction time is time transgressive in the degrading reach because the initial stream power/critical power ratio was progressively larger upstream.

Relaxation time is the time span between initiation of aggradation and subsequent return to the mode of operation or rate of geomorphic process that prevailed before the perturbation (Fig. 4.28B). The greater capacity for work in humid systems results in shorter relaxation times in these systems than in arid systems. The former equilibrium reach

has the shortest relaxation time, and the upstream portion of the aggradational reach is similar. Some delay in relaxation time is noted at the downstream end of the aggradational reach as the climate-change-induced increase in bedload is temporarily stored en route, only to be entrained again by subsequent streamflows. Such slugs of bedload tend to move downstream in a kinematic manner (Langbein & Leopold, 1968).

The arid basin also has an interesting pattern of relaxation times that is a function of how far removed each reach was initially from threshold or equilibrium conditions. The former degradational reach was last to initiate, and first to terminate, the aggradational episode. It was never far to the aggradational side of the critical-power threshold. Aggradation continues for a substantially longer time in the former equilibrium reach of the arid system than in the humid system example. A stronger delay in relaxation time also is noted for the aggradational reach. Bedload is temporarily stored en route, only to be entrained again by subsequent infrequent ephemeral streamflows.

Consideration of reaction and relaxation times provides insight about synchroneity of fill terraces in arid and humid fluvial systems. Initiation of aggradation after a climatic perturbation seems to occur with progressively longer reaction times in a sequence that goes upstream through aggradational, equilibrium, and degradational reaches. Relaxation-time comparisons indicate that the time needed to complete aggradation of valley fills may increase in the downstream direction where streamflow consecutively passes through reaches that initially were characterized as degrading, equilibrium, and aggrading. Both reaction and relaxation times are longer in arid than in humid systems. Figure 4.28 suggests that fill-terrace aggradation begins in downstream reaches and is terminated first in upstream reaches; thus the time span represented by alluvium of a given aggradation event should increase downstream.

The characteristics of fill-terrace deposits in New Zealand, western North America, Iceland, and the Sinai Peninsula of Egypt suggest that aggradation rates tend to decrease as the threshold of critical power is approached (Fig. 4.29). Sedimentology is easier to evaluate than chronology, which consists merely of informed estimates in most cases. Case A describes situations where nonalluvial processes supply large amounts of bouldery detritus to streams. Examples include when a landslide moves onto a valley floor or when rapid retreat of a glacier exposes highly unstable moraines to meltwater streams. The resulting valley fill tends to be poorly sorted and commonly contains large boulders. Aggradation begins abruptly and is rapid initially, but decreases progressively as the source of added bedload is eliminated. Abundance of the boulders increases toward the top of the fill, apparently because of selective deposition of the coarsest bedload. Bedding and sorting of fluvial gravels generally improve toward the top if large boulders are not present. Both features indicate decreases in aggradation rate. About 10 m of aggradation may require less than a year (for example, the 1988–1989 aggradation event on the Waiho River, which drains the Franz Josef glacier in New Zealand).

Case B describes the fairly common situation of accelerated fluvial erosion of detritus stored in hillslope sediment reservoirs. Climatically induced stripping of hillslope colluvium may begin slowly but tends to accelerate as a result of self-enhancing feedback mechanisms (Fig. 2.38). Valley-floor aggradation progressively increases, and the steepest part of graph B occurs at the time of maximum departure from preaggradation threshold or equilibrium conditions. Then aggradation rates decrease until the time of fill-terrace tread formation. Water-laid deposits tend to become more bouldery near the top of the fill because of an increase in abundance of boulders supplied from partially denuded hillslopes as deeper levels of colluvium and associated bedrock are exposed to erosion. Abundance of debris-flow beds, if present, also decreases toward the top (Fig. 4.17). Expansion of gullies followed by revegetation also produces a case B pattern of aggradation rates.

The inferred style of aggradation described in

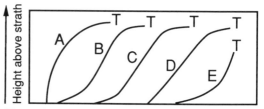

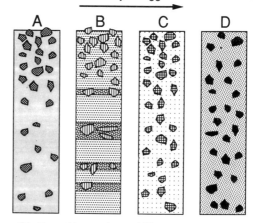

Stratigraphic sections of fill-terrace deposits

FIGURE 4.29 Styles of sedimentology and general changes in rates of deposition of alluvial fills in valleys affected by (A) massive increases in nonalluvial sources of bedload such as landslides and rapid retreat of valley glaciers, (B) stripping of hillslope sediment reservoirs as a result of Pleistocene–Holocene climatic change, (C) change from fluvial to periglacial hillslope processes, (D) stream capture that causes decreases in discharge of water or increases in discharge of sediment, or both, and (E) encroachment of sand dunes onto bare bedrock slopes of an arid watershed. T is the threshold of critical power.

case C is caused by climatically induced changes from fluvial hillslope erosion to periglacial processes (Section 5.2.1). Sediment yields tend to be larger during times of periglacial than during times of fluvial processes. The onset of a colder climate, and the resulting decrease in vegetation cover, initiate gradually accelerating aggradation rates. Aggradation rates may remain fairly constant un-

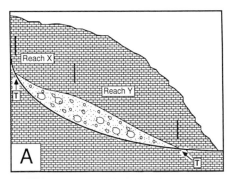

Fill incision caused by less resisting power

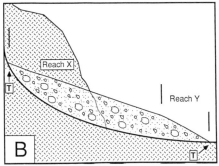

Fill incision caused by less stream power

FIGURE 4.30 Two types of fill terraces associated with different causes of fill incision. Reach X was degrading before the aggradation event and reach Y was at equilibrium before the event. T is threshold-intersection point. A. Situation favoring valley-fill incision resulting from decrease in resisting power. B. Situation favoring valley-fill incision resulting from increase in stream power.

der the new regime if supplies of potential detritus for periglacial processes are not limited, but they may decrease if supplies decrease or when the climate becomes warmer and fluvial erosion becomes more important. Such gravel deposits tend to be massive with more distinct bedding near the base and top of the section.

Cases D and E are not common. The hypothetical restraints on case D specify a stream-capture-induced decrease in the ratio of stream power/resisting power. This abrupt perturbation is a change in regime that may cause uniform rates of aggradation until terminated by internal changes in the

system such as changes in gradient caused by aggradation. Presumably the resulting deposits have uniform characteristics until near the end of the episode of aggradation. Case E represents the introduction of large amounts of gravel-free sand into a bare, rocky basin of an arid region. Runoff strips part of the dune sand from bedrock slopes, but small annual stream power is unable to cope with such a major addition of sediment load. Aggradation is rapid; the terminating threshold may occur abruptly with cessation of eolian sand influx.

The variety of aggradational settings portrayed in Figure 4.29 suggests that aggradation of alluvial fills in mountain valleys may accelerate, decelerate, or be uniform. Fills that have slow terminal rates of aggradation are the most likely to have roughly synchronous times of tread formation.

The overall geometry of a valley fill formed during aggradation episodes commonly is a long, narrow lens that tapers and pinches out at both ends (Fig. 4.30). Even alluvial fans have lens-shaped geometries (Bull, 1977a). Aggradation terminates at stable base levels such as adjacent equilibrium reaches and waterfalls (Fig. 3.21). Threshold-intersection points mark transitions between reaches with different modes of operation. Migration of threshold-intersection points as a stream incises into a valley fill is clear evidence for diachronous terrace-tread formation.

The geometry of a lens of alluvium deposited on a prior concave longitudinal stream profile consists of an upstream reach that is gentler and a downstream reach that is steeper than before aggradation (Figs. 1.9 and 4.30). The location of the transition between reaches X and Y partly determines the mechanics of terrace-tread incision. The transition may be near the upstream (Fig. 4.30A) or downstream (Fig. 4.30B) end of the lens of alluvium. For the purposes of this discussion of a tectonically stable stream, reach X is assumed to be degrading into bedrock before and after the aggradation event, and reach Y is assumed to return to the same base level of erosion as before the event.

Variations of stream-channel incision rates for the four reaches of figure 4.30 vary greatly between different field settings, but four possible scenarios are outlined in the hypothetical graphs of Figure 4.31. The lower left corner of each plot represents the time of terrace-tread formation at the initiation of channel incision into the valley fill, and the upper right represents some time after the stream has eroded through the valley fill.

Both Y reaches achieve equilibrium again after incision to the base of the fill—the horizontal portions of plots A and C. Aggradation-induced increases in gradient contribute to the tendency for initial terrace-tread entrenchment. Plot A starts with a slow downcutting rate that accelerates before decreasing to zero with the return of equilibrium conditions. The maximum rate coincides with maximum departure from equilibrium or threshold conditions. In plot A, downcutting rates are slow initially as sediment load and resisting power gradually decrease and cause the stream to switch modes of operation. (See Figure 5.18 for examples of this style of stream-channel downcutting.)

Reach X (plot B) has a similar pattern of downcutting, except that it continues to downcut into bedrock at a reduced rate after the stream has cut through the valley fill.

Reach Y of Figure 4.30B is the steepest of the four reaches, which makes it susceptible to stream-channel entrenchment. Incision begins abruptly as a headcut, which efficiently confines streamflow in a narrow channel and establishes a self-enhancing feedback mechanism that maintains the degradational mode as the headcut moves upvalley to reach X. Reach Y returns to equilibrium (plot C of Fig. 4.31) and after a similar headcut entrenchment, reach X (plot D of Fig. 4.31) returns to its former degradational mode.

The scenarios of Figures 4.30 and 4.31 indicate that treads of fill terraces tend to be diachronous for streams with limited stream power. Threshold-intersection points may migrate downvalley as upstream reaches become entrenched primarily as a result of increase in discharge of water or de-

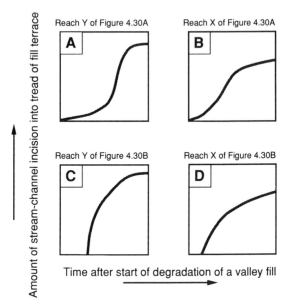

FIGURE 4.31 Variations in channel-incision rates into alluvial fills of reaches that tend to be at equilibrium (A and C), or in a degrading mode (B and D), before and after a hypothetical aggradation event. See text for discussion.

crease in bedload, or both. An even more likely type of threshold-intersection point migration occurs where development of headcuts establishes strong self-enhancing feedback mechanisms and the entrenched reach extends upstream. The examples presented in this section have provided conceptual background. Now we proceed to a study of late Quaternary aggradation in one fluvial system—Cajon Creek.

### 4.4.2 DIACHRONOUS FILL-TERRACE DEVELOPMENT IN CAJON CREEK

Cajon Creek is unusual. It drains 186 km$^2$ of Cajon Pass and the adjacent eastern end of the San Gabriel Mountains. The low pass formed by right-lateral movement on the San Andreas fault (Fig. 4.32) during the past 4 my separates the lofty San Bernardino and San Gabriel mountains (Weldon & Humphreys, 1986; Meisling & Weldon, 1989).

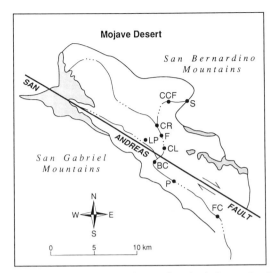

FIGURE 4.32 Map of Cajon Creek drainage basin showing the San Andreas fault and locations of places referred to in the text. BC, Blue Cut; CCF, Crowder Creek Forks; CL, Cleghorn Canyon; CR, Crowder Canyon; F, Flat Creek; FC, Freeway Crossing; LP, Lone Pine Canyon; P, Pitman Canyon; S, Summit. Altitudes above 1500 m are shaded.

Highlands are mainly along the southwest side of the drainage basin. At about 500 ka, stream capture extended Cajon Creek into the Mojave Desert rainshadow of the San Gabriel Mountains. The northern headwaters now receives mean annual precipitation of only 250 mm, but the southwest side and basin mouth receive 500 to 800 mm (Ahlborn, 1982). This unusual distribution of precipitation favors an unusually high rate of increase of stream power in the downstream direction. Schist southwest of the fault furnishes most of the gravel deposited in the late Quaternary valley fills. Loose Cenozoic basin-fill deposits northeast of the fault supply large amounts of sandy sediment to the stream.

Concepts introduced in the preceding section can be tested in the Cajon Creek fluvial system because of the excellent study by Ray Weldon

(1986,1989). The following figures are from his dissertation. For coherence, I use the terrace and soils chronology of Figure 4.11 and have inserted several small graphs. Weldon's analysis of fill-terrace aggradation and degradation addresses the thought-provoking question, How time transgressive are the fill terraces of Cajon Creek? The following discussion focuses on aggradation and incision of the T6 terrace of latest Pleistocene and early Holocene age.

The ages of the late Quaternary terraces are constrained by 18 radiocarbon ages, and are crosschecked by measurements of the amounts of offset of alluvial geomorphic surfaces by the uniformly slipping San Andreas fault (Section 4.2.3.2.2). Inset T6 deposits of the inner gorge of Cajon Creek have a large volume relative to the minor T7 fill (Fig. 4.33). The T6 tread is about 60 m below the T4 tread and 37 m above the active channel. Numerous fill-cut terraces formed after the threshold associated with T6–tread incisement.

Longitudinal profiles of Cajon Creek and the T4 and T6 terrace deposits reveal lenses that taper upstream and downstream (Fig. 4.34). Long-term tectonically induced channel downcutting of about 1 m/ky has exposed the base of the composite T3–T4 terrace fill. The volume of the T3–T4 fill may be an order of magnitude larger than that of the T6 fill. The base of the T6 fill is exposed just above Cajon Creek in the headwaters third of the basin and along tributary streams just before they join the trunk channel in the middle third of the basin. Thus the streambed profile approximates the base of the T6 fill. The T6 fill is much thinner and the amplitude of its lens shape is not nearly as large as that of the composite T3–T4 terrace fill. The smooth profile suggests that formation of the T6 terrace tread was synchronous, but the summary of Weldon's analysis that follows reveals the time-transgressive nature of T6 aggradation and degradation.

A convenient method of comparing the spatial changes of a terrace tread relative to the longitudinal profile of a present valley floor is simply to

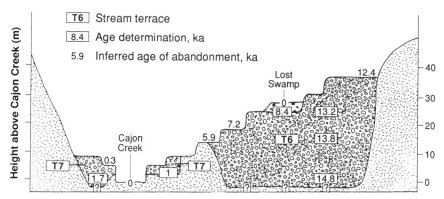

FIGURE 4.33 Cross section through the inner gorge of Cajon Creek showing radiocarbon-dated fill and cut terraces and inferred times of terrace-tread abandonment. From Figure 3-2 of Weldon, 1986.

note the heights of the terrace tread above the stream-channel datum with increase in distance downstream. This technique is especially useful for terrace studies where topographic maps are not available.

T6 tread heights are small in headwaters and basin-mouth reaches (Fig. 4.35), but 35-m heights in middle reaches record a substantial departure from the equilibrium conditions that are presumed to have prevailed for Cajon Creek before the T6 aggradation episode. The smooth plot of T6 tread height indicates that vertical movements along the San Andreas fault were minor.

Times of aggradation of T6 fill are based on the age control summarized in Figure 4.36, whose three curving lines show Weldon's interpretation of spatial variations in the times of initiation of aggradation, aggradation midpoint, and terrace-tread formation. Synchroneity would be shown by horizontal lines; diachronous behavior of the fluvial system increases with line slope. Samples for six radiocarbon ages were collected from 12 km of the 22-km-long lens of T6 fill.

Three radiocarbon ages at the base of the T6 fill indicate that between 14.8 and 11.6 ka the loci of initiation of aggradation migrated about 11 km

FIGURE 4.34 Longitudinal profiles of the T3-T4 and T6 terraces and the valley of Cajon Creek. Both terrace fills are of thin to negligible thickness upstream and downstream from the reaches of maximum thickness in the central part of the drainage basin. From Figure 3-5 of Weldon, 1986.

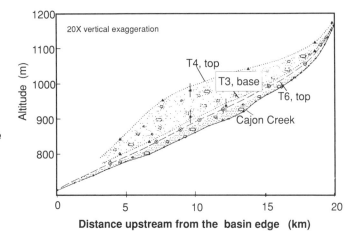

Distance upstream from the basin edge (km)

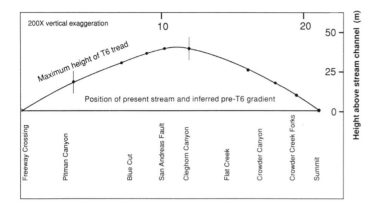

FIGURE 4.35  Height (vertical exaggeration 200X) of the T6 terrace tread above Cajon Creek, with the locations of place names along Cajon Creek that are referred to in the text. Terrace profile and thickness do not change at the San Andreas fault. From Figure 3-6 of Weldon, 1986.

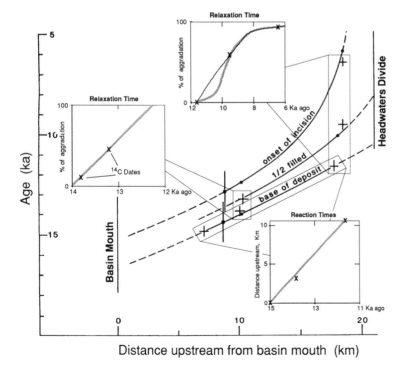

FIGURE 4.36  Variations in the ages of the T6 deposits with distance upstream from the mouth of Cajon Creek. +, C-14 age; ●, age extrapolated from pair of C-14 age; †, age based on offset across the San Andreas fault divided by known slip rate. From Figure 3-7 of Weldon, 1986.

upstream—a rate of 3.5 km/ky. Thus the diachronous reaction time of the Cajon Creek fluvial system to the climatic perturbation responsible for the aggradation event was progressively longer with increasing distance upstream. The reaction time has the same style as for arid reach D in Figure 4.28A. The Figure 4.28A model is that the longer times needed for upstream reaches to begin aggradation reflect progressively larger stream power/ resisting power ratios in the upstream direction prior to the climatic perturbation.

The best age control regarding the times and rates of aggradation is provided by three radiocarbon ages at the 18–km reach of Figure 4.36. Only minor interpolation is needed to estimate the midpoint time, and minor extrapolation for the tread-formation time. The threefold longer time needed for deposition of the upper half of the fill clearly defines decrease of aggradation rates. The inset graph shows two possible aggradation rate curves to compare with those of Figure 4.29.

Two closely spaced ages of terrace deposits at the 10-m reach of Figure 4.36 require minor extrapolation, to estimate the base of the deposit, and major extrapolation, to estimate the age of the terrace tread. Weldon used a uniform mean rate of deposition between the two samples (see inset graph) as the basis for both extrapolations. If the

3:1 ratio of the 18 km site is also used at the 10-km site, the terrace-tread extrapolated age would be 10.9 instead of 12.4 ka. Adoption of a model of decreasing aggradation rates for the 18- and 10-km reaches decreases the difference in relaxation times between the two reaches. The "onset of incision" line would roughly parallel the "base of deposit" line, so the threshold-intersection point is still clearly time transgressive. The San Andreas fault is only 1 km downstream from the dated 10-km reach. Fault-offset age estimates of the start and end of T6 aggradation fit Weldon's model quite well but have large age uncertainties.

Weldon's extrapolations of the curves of reaction (base) and relaxation (tread) times to the mouth and headwaters reaches suggest that T6 aggradation began about 17 ka at the mouth and then migrated upstream, reaching the headwaters at about 6 ka. Aggradation had ended by 10 ka for the 15 km upstream from the mouth. The next three figures outline Weldon's hypothesis for upstream migration of the lens of T6 valley fill.

The T6 aggradation event migrated upstream even though much of the sediment was derived from tributaries in the central and upstream parts of the basin. Age control (Fig. 4.36), geometry of the lens of alluvium (Fig. 4.35), and degradational history (Fig. 4.33) provide assumptions and con-

FIGURE 4.37 Relative levels of Cajon Creek at 1-ka intervals during the past 17 ky. Data points are calculated from Figures 4.33, 4.34, and 4.35, as described in the text. Local stream gradients changed 15 to 35 percent as the aggradation migrated upstream. Downstream reaches of terrace tread were being incised before aggradation had started in upstream reaches. From Figure 3-10 of Weldon, 1986.

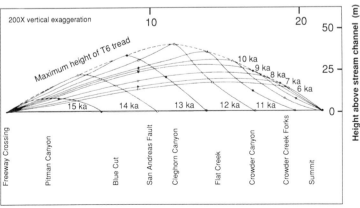

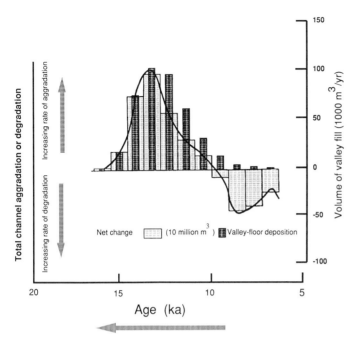

FIGURE 4.38 Aggradation and degradation along Cajon Creek during the past 17 ky. Aggradation peaked at 13.5 ka when channel aggradation was about 50 percent more than the total sediment load of the present stream. The solid line shows net changes in amounts of valley fill for each 1-ka interval. From Figure 3-11 of Weldon, 1986.

straints that were used to draw longitudinal valley profiles at 1–ky intervals. Horizontal lines drawn through Figure 4.36 indicate where aggradation had just begun and where it had just terminated; the "1/2 filled" line provides a third point. Downcutting histories provide fourth and fifth points for some plots.

The lenticular shape of the T6 valley fill (Fig. 4.35) requires that the profiles of Figure 4.37 pinch out upstream and downstream. Aggradation decreased valley gradients by 25 to 35 percent in the reaches upstream from the loci of maximum aggradation at the 10 times; gradients increased by 15 to 25 percent in adjacent downstream reaches. Erosion of the valley fill began at 13 ka near the basin mouth, but aggradation had yet to begin in the headwaters of Cajon Creek.

One cross-check for Weldon's model is to compare soil profiles on 12- to 6-ka T6 treads. Unfortunately, soil-forming factors other than time are important at the two reaches. The downstream (12-ka) reach has a subhumid climate and schist gravel parent material. The upstream (6-ka) reach

has a semiarid climate and sandy parent materials with substantial atmospheric dust derived from the adjacent Mojave Desert. Inputs of atmospheric silt are a dominant factor in genesis of latest Quaternary soils (Harrison et al., 1990).

The areas under the 10 profiles of Figure 4.37 can be used to estimate the changing volumes of T6 valley fill when the third dimension of width of the deposit is taken into account. Most of the fill was stored along the valley floors of Cajon Creek and its two largest tributaries, Lone Pine and Crowder Canyons. Additions of new fill and net amounts of fill for each 1-ky increment (Fig. 4.38) are equal until erosion of fill began in downstream reaches at about 13 ka (calendric age of approximately 15.2 ka using the corrections of radiocarbon ages described by Bard et al. [1990]). This was also the time of maximum rates of valley-floor aggradation.

The T6 aggradation event spanned 11 ky, but the total volume of valley-floor gravels increased only during the 6 ky between 17 and 11 ka and peaked sharply at 14 to 13 ka. The Holocene has been characterized by net degradation except for a brief pulse of T7 aggradation. Deposition of alluvium induced by the Pleistocene–Holocene climatic change seems to have required much less time than subsequent channel incision and removal of the fill.

Weldon also calculated variations of total sediment load being conveyed at two times by Cajon Creek. Cajon Creek is ungauged but may be compared with similar nearby watersheds. Long-term sediment yields in well-studied watersheds of similar size suggest a present mean annual sediment yield from Cajon Creek of about 75,000 m³. At the time of peak aggradation at 13.5 ka, just the bedload component of total load that was being selectively deposited as valley fill was about 80,000 m³ per year. Holocene degradation of valley fill averaged about 30,000 m³ per year, which is equal to four-tenths of the total load derived from the hillslope subsystem.

The total amount of sediment load carried by Cajon Creek during the T6 aggradation event greatly exceeded the part being deposited as fill terrace gravels. Taylor's (1981) analysis of the sediment yield of Transverse Ranges watersheds shows that basins underlain by schist or granitic rocks produce about 70 percent sand, silt, and clay; this percentage increases to more than 95 percent for basins underlain by marine and continental sedimentary rocks. If, during the late Quaternary, Cajon Creek deposited all sediment coarser than sand (and if the coarse fraction comprised 10% of the total load), the maximum annual rate of sediment load conveyance was roughly 800,000 m³. This 10-fold increase in sediment load (compared with that of the present) is a rough measure of the magnitude of fluvial-system response to the late Quaternary climatic perturbation. Another measure of the size of the perturbation is the maximum rates of aggradation, which were 20 times the long-term tectonically induced channel downcutting rate of about 1 m/ky. Once again it is clear that the effects of climatic perturbations can overwhelm the effects of concurrent tectonic deformation.

Weldon's summary of diachronous T6 aggra-

dation and stream-channel incisement (Fig. 4.39) shows that deposition started in the reach farthest from the source of most bedload. This reach seems analogous to reach Y of Figure 4.30B. Substantial aggradation had occurred by 13 ka, and even by 10 ka depths of channel incision into the T6 tread were small. Although a pulse of aggradation continued upstream into the headwaters reach like a kinematic wave (Langbein & Leopold, 1968), most of the Holocene was characterized by channel incision and removal of the T6 valley fill.

I would like to add a few comments to Weldon's interesting case history of Cajon Creek. The T6 fill event started in the basin-mouth reach, because that reach was closest to the threshold of critical power. Using the Figure 4.28A model, only the basin-mouth reach would be affected until the strength of the perturbation increased sufficiently to cause aggradation in upstream reaches that presumably were on the degradational side of the threshold of critical power.

Two mechanisms might cause a degrading reach to cross the threshold of critical power. Increases in the amount or size of bedload would increase resisting power, and decreases in slope would tend to decrease stream power. Either change would decrease the ratio of stream power/resisting power, and aggradation would begin when the ratio became less than 1.0. Maximum aggradation rates would occur during times of smallest ratios.

The lens shape of fill-terrace deposits (Figs. 1.9, 4.30, 4.35) tends to create two self-enhancing feedback mechanisms. Decrease in gradient on the upstream half of the lens promotes formation of a gentle aggradational slope as the threshold-intersection point migrates upstream. Increase of gradient in the downstream half of the lens, and the deposition of bedload in the upstream half, both tend to increase the stream power/resisting power ratio and promote stream-channel incisement into the recently deposited fill. The relative strengths of these two self-enhancing feedback mechanisms determine if the downstream and upstream threshold-intersection points migrate upstream at the same rate, or whether aggradation or

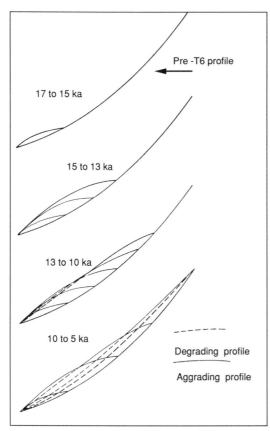

FIGURE 4.39 Successive profiles of T6 aggradation and degradation along Cajon Creek. Upstream reaches aggraded after downstream reaches had switched to the degradational mode of operation. From Figure 3-13 of Weldon, 1986.

degradation becomes dominant because one feedback mechanism is stronger than the other.

Initiation of aggradation at about 17 ka implies a relation of the sediment-yield increase responsible for the aggradation event to termination of full-glacial climates in North America from 25 to 15 ka that peaked at 18 ka (calendric ages of

approximately 28.0, 18.2, and 21.5 ka using the corrections of radiocarbon ages described by Bard et al. [1990]). Cold, dry, northeast-facing hillslopes of the Cajon Creek basin may have been sites of active periglacial processes. Major sediment-yield increases, and maximum aggradation rates, occurred during the transitional climates of the latest Pleistocene. Accumulation of hillslope detritus for the aggradation event may have occurred during full-glacial times and been stored until times of greater stream power associated with a wetter climate. Sediment-yield increases from the small high-altitude area (Fig. 4.32) apparently were not strong enough to change modes of streamflow operation until the increase of bedload arrived at the basin-mouth reach.

Several fascinating problems remain. We do not know why the perturbations that caused the T6 aggradation event in Cajon Creek were much stronger than the T7 perturbations, or why the T6 event appears to be missing in the North Fork of the San Gabriel River, where the T7 event was dominant. Nor do we know if the unusual distribution of altitude and precipitation within the Cajon Creek basin played a role in fluvial-system response to late Quaternary climatic change. Weldon's study focuses our attention on an important problem in fluvial geomorphology, and should lead to other studies that will determine how representative Cajon Creek is compared with other basins in arid and humid regions.

The climatic geomorphology of the San Gabriel Mountains is indeed complex because many processes change with altitude and tectonic setting. In the next chapter we examine the responses of a colder and more humid fluvial system of another lofty mountain range to the effects of late Quaternary climatic changes, at a site where both horizontal and vertical earth deformation are important controls on stream behavior.

# Climate and Landscape Change in a Humid Fluvial System

## 5.1 Introduction

Our survey of the impacts of Pleistocene–Holocene climatic change on hills and streams in different parts of the world concludes with this chapter, which focuses on geomorphic responses in an unglaciated humid watershed and its piedmont stream terraces during the past 40 ky. Both the watershed and piedmont reaches of the Charwell River will be analyzed. Hillslopes in this humid mesic watershed had maximum sediment yields during times of harsh full-glacial climates; hillslope sediment reservoirs became more stable during the milder climates of the Holocene. (See Table 1.2 for definitions of temporal terms.) This chapter sets the stage for Chapter 6, which compares the effects of changing climates in basins that range from extremely arid to humid.

Climatic change in mesic to frigid regions may affect the relative importance of distinctly different types of geomorphic processes as periglacial (and even glacial) processes become important modes of hillslope degradation and control fluvial deposition (Frye, 1961; Whitehouse, 1979). Initiation of periglacial hillslope processes can profoundly affect rivers by causing major increases in bedload

transport rate. Alternation between modes of periglacial and fluvial erosion in alpine valleys is such an extreme result of climatic change that it may be regarded as a replacement of geomorphic processes. Erosional histories in such geomorphic systems are polygenetic. An example of an arid *polygenetic system* is the Sahara Desert, where eolian processes have replaced former fluvial processes as the dominant geomorphic agent (Haynes, 1982). Much of the Charwell River watershed may have polygenetic characteristics because of the watershed's sensitivity to late Quaternary climatic changes.

The Charwell River study area provides an unusual opportunity to evaluate multiple degradation and aggradation events through study of flights of stream terraces. Important goals of this chapter are to:

1. Describe the types of valley fill that accumulated during climate-change-induced aggradation events because these deposits record the outputs of the fluvial system. Changes in water and sediment discharge had profound effects on the stream subsystem, which has a flight of four major

aggradation terraces. Each aggradation event ended with incisement of an aggradation surface, which formed a tread of a climatic terrace. This initiated a degradation event characterized by multiple complex-response terraces.

2. Assess the likely impacts of late Pleistocene and Holocene climatic changes on the hillslopes. Stratigraphic information from the piedmont reach will be used to develop process-response models for watershed hillslopes during changes to full-glacial and interglacial climates.

3. Evaluate the effects of concurrent climatic and tectonic controls on the latest Pleistocene–Holocene degradation event. Dating of degradation terrace treads reveals variations of downcutting rates that describe relative changes of stream power and resisting power.

4. Elucidate the attainment of equilibrium conditions or crossing of the threshold of critical power in a landscape subject to rapid change because of steep slopes, soft materials, and powerful streams.

Geology, climate, and geomorphic processes are summarized first in the context of Quaternary evolution of the Charwell landscape.

### 5.1.1 CHARWELL RIVER FLUVIAL SYSTEM

The Charwell River drains 40 km² of the rugged Seaward Kaikoura Range in northeastern South Island, New Zealand (Fig. 5.1). The basin mouth at the highly active Hope fault is only 16 km from the Pacific Ocean, but the stream flows a circuitous 50 km to the ocean down the Charwell and Conway rivers. The watershed rises from an altitude of 450 m at its mouth to as high as 1605 m. The rugged mountains of the watershed reach contrast with the low relief of dissected terraces in the 6-km-wide piedmont reach (Figs. 5.2–5.5).

The drainage basins of the rugged Seaward Kaikoura Range are underlain by folded and faulted, massive to medium bedded graywacke (and minor argillite) of the Pahau terrane (Lensen, 1962; Bradshaw et al., 1980; Bishop et al., 1985; Silbering

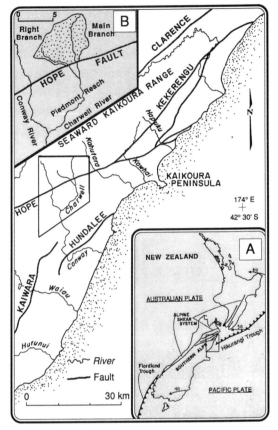

FIGURE 5.1 Larger rivers and fault zones near Kaikoura Peninsula, northeastern South Island, New Zealand. The Hikurangi Trough (insert map A) occurs where oceanic lithosphere abuts continental lithosphere. Insets show (A) locations of study area in New Zealand and (B) the drainage basin and piedmont reaches of the Charwell River (map B). Vectors of relative plate motion are from Walcott, 1984.

et al., 1988) of the Mesozoic Torlesse supergroup (Campbell & Coombs, 1966). Bedding thickness generally is thin to medium with a few thick beds of massive sandstone. The rocks underlying the Charwell River watershed have approximately uniform lithologic and structural properties, except in the crush zone of the range-bounding Hope fault.

The highly fractured, crushed, and sheared na-

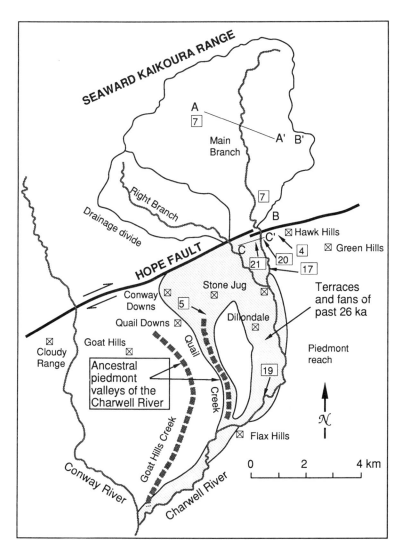

FIGURE 5.2 Base map showing watershed and piedmont reaches of the Charwell River and the locations of places discussed in the text. Numbers and lines of topographic sections refer to figures in this chapter. Topographic cross-valley profiles for the lines of sections are shown in Figures 5.7 and 5.21. X marks locations of named farms.

ture of these silty sandstones promotes effectiveness of hillslope freeze-thaw processes and failure by mass movements. There seem to be unlimited sources of bedload for streams; they extend far below thin hillslope soils. Frequent intense rainfalls and mean drainage-basin slopes greater than 0.55 also contribute to sediment yields that are among the highest known (O'Loughlin & Pearce, 1982). Sediment yields probably were even larger during full-glacial times of regional aggradation

of valley floors. Each aggradation event had a duration, intensity, and sedimentology that reflected the behavior of the watershed reach in response to changes in precipitation and temperature that were strong enough to cross thresholds of fluvial-system operation (Bull, 1979; Bull & Knuepfer, 1987). Base-level falls resulting from uplift along range-bounding Hope fault provided sites for accumulation of fan deposits (Bull, 1977a,b).

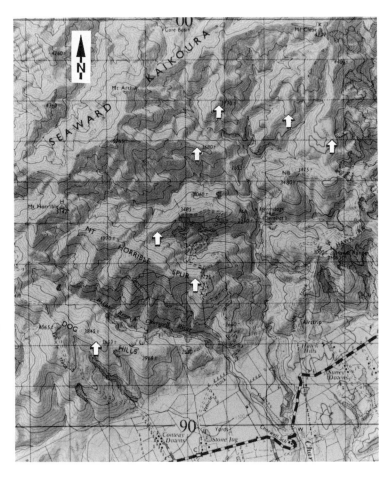

FIGURE 5.3 Topographic map of the Charwell River upstream from the Hope fault. Arrows point to remnants of the inferred 305 ka marine terrace, which is part of the flight of terrace remnants used to estimate the uplift rate for this part of the Seaward Kaikoura Range. 100-foot contour interval; 1000 yard grid.

The muddy sandstones probably are highly susceptible to periglacial weathering and mass-movement processes during times of cold climate. Closely spaced fractures allow water to infiltrate, and freezing splits sandstone blocks. Mechanical and chemical weathering produces a mixture of sand, silt, and clay that favors development of subsurface layers of seasonally or perennially frozen ground during times of colder climate when saturated conditions conducive for solifluction and other periglacial processes were likely.

Remnants of periglacial deposits are common in the mountains and in hills that rise above the piedmont stream terraces, but topographic setting and lithology do not favor formation or preservation of periglacial features such as cryoplanation terraces, nivation hollows, and frost-wedge polygons (Washburn, 1980; Péwé, 1983). Poorly sorted sheets of massive to thick-bedded silty gravel (*diamicts*) occur as colluvium on footslopes and as deposits that largely fill hillslope gullies that presumably formed as a result of fluvial erosion during preceding interglacial climates. These silty hillslope materials appear to have originated as periglacial mudslides, as described elsewhere by Chandler (1972) and Hutchinson (1974). Sheets of younger light brownish-gray diamicts underlain by older light yellowish-brown diamicts suggest

FIGURE 5.4 Aerial view of the unglaciated Charwell River drainage basin. The river widens abruptly where it leaves a confining gorge and crosses the range-bounding Hope fault (foreground). The steep slopes, narrow canyons, and fairly high drainage density all indicate rapid erosion of fractured greywacke in a tectonically active humid watershed. Most of the beech forest has been cleared for sheep pasture; the highest remnants are near the natural treeline at 1200 m.

distinct episodes of intense solifluction activity in the Seaward Kaikoura Range. Similar deposits have been described elsewhere in New Zealand and in Australia and are considered to be solifluction deposits (Stevens, 1957; Cotton & Te Punga, 1955; Wasson, 1979; Soons, 1980).

Other origins, such as debris flows, should be considered for the diamicts. Debris flows are interbedded with water-laid deposits, but only in the alluvial-fan deposits of small drainage basins that drain the 1-km-wide crush zone of the Hope fault. Only water-laid deposits occur in the valley fills deposited by the Charwell River during late Pleistocene aggradation events. Thus the typical hill-

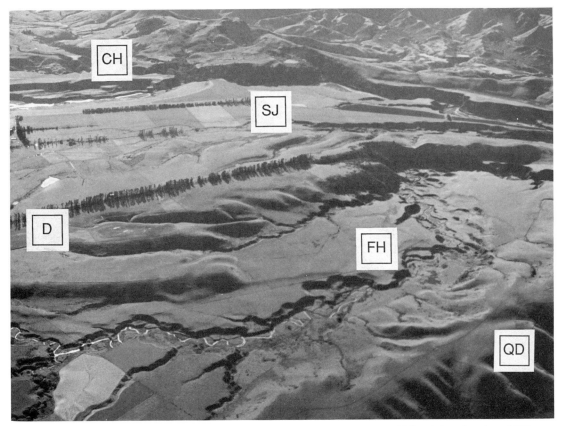

FIGURE 5.5 Aerial view of the piedmont reach of the Charwell River near Dillondale. The river (CH) presently flows at the base of the hills in the background in a valley that is entrenched into the smooth latest Pleistocene Stone Jug (SJ) aggradation surface. The broad valley in the foreground with the underfit stream of Quail Creek was occupied by the Charwell River during the Flax Hills aggradation event (FH). Right lateral movements along the Hope fault caused the stream to switch to its present position when aggradation filled the piedmont valley adjacent to the mountain front. The Dillondale (D) and Quail Downs (QD) aggradation surfaces were formed during earlier periods of full-glacial climate.

slope colluvial stratigraphy suggests a former dominance of periglacial mass-movement processes that moved much of the detritus in the hillslope sediment reservoir on crestslopes and midslopes to the footslopes. Water flows in adjacent stream channels transported large amounts of bedload to piedmont depositional sites.

Watershed soils thinner than 1.5 m supported a dense forest of beech trees before they were largely burned by Polynesians who discovered New Zea-

land about 1200 years ago (McGlone, 1983). Beech forests have not been reestablished in many of these burned areas. European colonization converted piedmont and mountain forests into pastures. Treeline is at about 1200 m, above which subalpine grasses and shrubs grow among craggy outcrops.

Alternating aggradation and degradation occurred repeatedly during the late Quaternary in a 6-km-wide structural trough downstream from the

Hope fault. Alluvial fans that grade downstream into fill terraces are the dominant landform in the piedmont reach. Each episode of aggradation presumably resulted from sediment-yield increases during times of full-glacial climate. Intervening episodes of stream-channel downcutting to a new base level of erosion are recorded by flights of degradation terraces that mark pauses in downcutting.

## 5.1.2 CLIMATE

The climate of New Zealand is controlled primarily by its position in southern temperate latitudes and by the maritime influence of vast oceans around fairly small islands. Virtually all of New Zealand is within 100 km of the ocean. Prevailing westerly winds bring a procession of anticyclones and depressions, commonly at 8- to 10-day intervals. Anticyclones rarely persist for more than 2 weeks. The yearly pattern of climate at Christchurch (Fig. 5.6, 140 km south of the Charwell River, is representative of the seasonal variations of temperature and precipitation at lower altitudes.

Orographic influences of the Southern Alps greatly affect the moisture-laden airstream (Griffiths & McSaveney, 1983; Whitehouse, 1985). Strong hot winds that descend the eastern slopes of the Southern Alps are common in the spring and summer. The initial phases of the passage of a typical storm over the South Island cause a strong northwesterly airflow that brings heavy rain to the

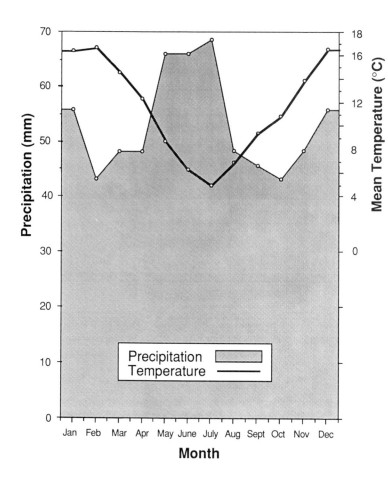

FIGURE 5.6 Monthly variations of precipitation and temperature at an altitude of 36 m in Christchurch, South Island, New Zealand. The average climate is weakly seasonal subhumid and moderately seasonal mesic. Data from U.S. Department of Commerce, 1972,1975.

northwest side of the Southern Alps, which has a mean annual precipitation of 5000 to 15,000 mm. Maximum mean annual precipitation measurements have been made that exceed 11,000 mm (Whitehouse, 1985). The southeast side gets strong winds but little precipitation. Passage of the cyclonic disturbance results in a shift to a southerly airflow that causes clouds to bank up against the eastern slopes of the Southern Alps, bringing widespread precipitation; little precipitation falls on western slopes. The analysis by Griffiths and McSaveney (1983) of the distribution of mean annual precipitation shows that rainfall distribution is a function of altitude and distance from the crest of the Southern Alps.

The Charwell River drainage basin climate is weakly seasonal humid—mean annual precipitation of 1200 to 2000 mm—and moderately seasonal mesic to frigid (see Table 2.1 for definitions of climatic terms). Mean January temperatures range from less than 12°C to 14°C, and mean July temperatures range from much less than 1°C to about +3°C. Piedmont climate at the Charwell River has a mean annual precipitation of 1000 to 1400 mm, a mean annual temperature of 9.5 to 10.5°C, and mean January and July temperatures of 14.5 to 15.5°C, and 3.5 to 4.5°C, respectively (New Zealand Meteorological Service, 1985).

### 5.1.3 PALEOCLIMATOLOGY

#### 5.1.3.1 Latest Full-Glacial Climate

The Charwell River study provides an opportunity to describe and compare the effects of climatic change during the latest Pleistocene intense stadial glaciation (peaking at about 20 ± 2 ka) and during the preceding stadial glaciation (peaking at about 35 ka). Nearly all prior New Zealand studies concern the latest Pleistocene full-glacial climates, which appear to have been especially harsh. Studies of snowlines (Porter, 1975b; Soons, 1979; McGlone, 1980) show that snowline depression was 800 to 830 m in the Southern Alps and on the volcanos of the North Island. Estimates of depression of mean annual temperatures are 4.5

to 5.0°C (substantially more if precipitation was less than at present). Treeline in the Seaward Kaikoura Range probably was depressed to altitudes below 300 m. During the latest Pleistocene "climatic conditions were bleak" (Moar, 1980). Climates were sufficiently colder or drier to eliminate forests from most of the South Island and from the southern part of the North Island, and the rain-shadow effect of the Southern Alps may have been increased.

Paleoclimatic evaluations of New Zealand plant macrofossils and microfossils are summarized by McGlone (1988) for the glacial maximum, 22 to 14 ka; the late glacial, 14 to 10 ka; and the Holocene; 10 to 0 ka. Tree pollen is generally less than 20 percent in full-glacial samples; shrubs and grasses are the dominant taxa. Beech tree (*Nothofagus fusca* type; *N. menziesii*) pollen is present in virtually all samples, but this wind-dispersed pollen generally is greatly overrepresented. Pollen from all other trees must have been rare in a landscape dominated by grasslands and open herbfields. Widespread forests were eliminated but communities of small trees, shrubs, and herbs managed to persist in sheltered spots, (McGlone, 1988; Soons & Burrows, 1978).

Regeneration of trees during Pleistocene–Holocene transitional climates may have been adversely affected by hard frosts associated with unseasonable outbreaks of polar air and lower-than-present mean annual precipitation, combined with summer drought, fire, increased windiness, and competition from grasses (Hume et al., 1975; Burrows, 1977; McGlone et al., 1978; Petit et al., 1981; McGlone & Topping, 1983; McGlone & Bathgate, 1983; Wardle, 1985; Jane, 1986; McGlone, 1988). Extensive full-glacial icefields began to retreat by 14 ka but minor glacial readvances continued until 9 ka (Burrows & Russell, 1975; Chinn, 1975,1981; Burrows et al., 1976; Burrows, 1979; Birkeland, 1982,1984b). Rapid retreat of large valley glaciers in the Southern Alps suggests that an important climatic-change threshold was crossed at about 14 ka (calendric age of approximately 16.6 ka using the corrections of radiocarbon ages described by Bard et al. [1990]). In the South Island, full-glacial plant communities of sparse grassland herbfields became grassland and partial shrubland after 12 ka as harsh climates began to ameliorate. Strong westerly airmass circulation continued to 9.5 ka (McGlone, 1988).

Synchronous changes in pollen spectra in the South and North Islands at about 10 ka are best ascribed to an abrupt regional climatic change characterized by greater biomass. Beech forests returned to the northern South Island by 9.5 ka. Dense beech and podocarp-hardwood forests developed in less than 400 years with the onset of warmer Holocene climates at about 10 ka.

Factors affecting reestablishment of hillslope forests include:

1. Characteristics of transitional soil conditions and plant associations between 14 and 10 ky ago.

2. Slow seed-dispersal rates for trees such as beech (*Nothofagus*) and rapid dispersal rates for podocarps.

3. A delay caused by the time needed to regenerate hillslope colluvium that had been largely stripped by rapid degradational processes during full-glacial times.

At sites of potentially high soil leaching, plant growth would be improved by exposure of an unweathered substrate (minimal leaching of plant nutrients).

A lack of mid-Holocene neoglacial deposits suggests warmer climates than at present between about 8 ka and 5.5 ka. Radiocarbon dating of tree fossils from sites in the eastern South Island and southern North Island are suggestive of a climate that was 1°C warmer and possibly wetter than at present (Moar, 1966; Stevens, 1974). Pollen studies by McGlone and others indicate increasing temperatures and rainfall during the first half of the Holocene, and for the east coast of the South Island an intensification of summer drought and northwesterly winds during the late Holocene. McGlone's assessment of the general nature of late Quaternary climatic change along the east coast of the South Island is outlined in Table 5.1.

It is difficult to assess the relative importance of temperature and precipitation changes during

TABLE 5.1 General changes in plant communities and inferred climates during late Quaternary along east coast of South Island, New Zealand.

| Age (ka) | Paleovegetation | Paleoclimate |
|---|---|---|
| Glacial maximum (25–15) | Grassland herbfield and bare ground dominate | Cool, dry, variable and windy |
| Late Glacial I (15–12) | Shrubland replaces or partially supplants grassland | Still cool, but rainfall increasing |
| Late Glacial II (12–10) | Tall shrubland-low forest develops | Substantial warming but also increased rainfall |
| Early Holocene (9.5–7.5) | Podocarp-hardwood forest abruptly develops | Sudden warming; rainfall 30% below today's value but more frequent than previously |
| Mid to late Holocene (7.5–2.5) | *Nothofagus* (beech) forest spreads | Gradual increase in rainfall toward values of present |
| Late Holocene (2.5–present) | Natural fire deforests central Otago; *Nothofagus* spreads | Northwesterly wind system intensifies; summer drought becomes feature in some areas |

*Source:* M. S. McGlone, Botany Division, New Zealand Department of Scientific and Industrial Research, 1988, and written communication, January 24, 1986.

full-glacial and interglacial climates. Moar, and McGlone and Topping, pointed out that there may be no modern analog for a complex climatic regime that resulted in widespread plant communities of subalpine grasses and bushes. Decreases in temperature were important, but precipitation may have increased northwest of the Southern Alps and decreased to the southeast, where locally saline soil conditions appear to have formed during times of drier full-glacial climates. Soons (1979) underscored the difficulty of trying to separate the relative affects of changes in precipitation, temperature, and storminess in terms of their effects on glaciers and plants.

Given the present lack of local paleobotanical data, only regional trends of latest Pleistocene climate change can be defined for the Charwell area. Two types of change seem reasonable. Temperature had to decrease because treeline altitude is controlled mainly by temperature and windiness, not by precipitation. With colder temperatures, an increase in precipitation should have resulted in glaciation. The lack of glacial landforms and deposits in the Charwell basin is best attributed to concurrent substantial decrease in latest Pleistocene precipitation. Thus the latest full-glacial climate appears to have been cold and dry compared with the preceding full-glacial climate or the present.

Spatial extremes of South Island climatic variations are illustrated by the present large valley glaciers that descend along the northwestern flank of the Southern Alps to within 300 m of the sea and the apparent lack of extensive Pleistocene glaciation in the Seaward Kaikoura Range, even at altitudes above 2000 m. Cold-dry full-glacial climates, which may in part have been due to an increase in rainshadow effects, inhibited expansion of glaciers in the Seaward Kaikoura Range (Suggate, 1965; Soons, 1979; Moar, 1980; O'Loughlin & Pearce, 1982).

The nature of transitions from full-glacial to interglacial climates is clear in the stratigraphic and geomorphic record of the Charwell River. Valley-floor aggradation ended as streams crossed the threshold of critical power. Return to a degradational mode of operation left the tread of an extensive fill terrace, which thus marks termination of the climatically induced aggradation event.

### 5.1.3.2 Prior Full-Glacial Climate

Conditions during periods of full-glacial climate older than the cold, dry, latest full glacial were assessed by pollen analyses of samples of carbonaceous silts from the piedmont reach of the Charwell River. The samples were collected by James Crampton (1985) and analyzed by D. C. Mildenhall of the New Zealand Geological Survey. Sample-site altitudes are about 350 m. One sample was collected below a peat layer at the base of aggradation gravels that throughout the region have a range of radiocarbon ages of 36 to 48 ka (Table 5.3). About 42 percent of the pollen was herbaceous, with 36 percent *Cyperaceae* and 36 percent *Nothofagus fusca* group. The sparse forest pollen and presence of subalpine plant pollen imply a cold temperate (perhaps glacial) climate for a sedge swamp surrounded by a beech forest. An older sample contains 69 percent *Cyperaceae,* 18 percent *Gramineae,* and only 4 percent *Nothofagus fusca* group, with 92 percent herbaceous pollen. This suite is interpreted as representing a sedge swamp with a distant beech forest during cold glacial times. Most taxa define an alpine to subalpine plant community. These paleobotanical analyses probably are representative of glacial or near-glacial conditions in the piedmont reach of the Charwell River for much of the time span between about 80 and 30 ka.

### 5.1.4 TECTONIC SETTING

The 220-km-long Hope fault (Fig. 5.1) is the most active splay of the Alpine shear system in the Marlborough region (Knuepfer, 1984). The fault bounds the south side of the Seaward Kaikoura Range and thus separates the watershed and piedmont reaches of the Charwell River. It strikes at about 73° across the South Island from the Alpine fault on the northwest side of the Southern Alps to the southwestern end of the Hikurangi Trench. It is part of a diffuse transpressional transform boundary between the Australian and Pacific plates that links the Tonga–Kermadec–Hikurangi and Puysegur trenches. Oblique plate convergence is large (about 47 m/ky at 264° ± 2° for the Hope fault, Walcott, 1979) with a compressional component of as much as 22 m/ky that tends to raise the Southern Alps at 5 to 8 m/ky and thicken their crustal root (Walcott, 1984; Allis, 1981, Bull & Cooper, 1986).

Estimates of long-term fault slip rates were obtained using radiocarbon-calibrated ages of graywacke cobble weathering rinds (Whitehouse et al., 1986; Knuepfer, 1988) collected from offset stream channels, stream-terrace treads, and alluvial fans. Knuepfer (1984) estimated a mean horizontal slip rate for an offset channel on an 11 ka Charwell River terrace as 32 ± 17 m/ky. Van Dissen (1989) measured the tectonic displacements for an offset stream terrace of Sawyers Creek, 6 km to the east, whose tread has a cobble weathering rind age of 4.6 ± 0.9 ka. The riser crest has been offset 150 ± 20 m and the tread separated 11 m vertically. Van Dissen calculated mean slip rates to be 33 ± 13 m/ky horizontally and 2.4 ± 0.6 m/ky vertically. I found that apexes of four offset alluvial fans west of the main branch of the Charwell River have weathering rind ages of 3.8 ± 0.5 ka to 12.9 ± 2 ka, and mean rates of horizontal displacement of 30.5, 31.1, 32.9, and 33.3 m/ky. The two lower slip-rate values are at sites where minor movements have also occurred along a fault splay within the crush zone of the fault, thereby causing a slight reduction in the slip-rate estimate for the range-bounding fault. McCone (written communication, 1989, Univ. Canterbury) measured displacements for a stream channel that had been offset 155 m near Robson Creek, 9 km to the west, whose tread has a cobble weathering rind age of 5.77 ka. His estimate of long-term slip rate is 29.9 m/ky. Thus mean long-term horizontal slip rate of 33 ( + 2–4) m/ky seems reasonable for the Conway segment of the Hope fault, which includes the Charwell River. The slip rate is similar to that of the San Andreas fault in California (Section 4.2.3.2.2).

A tectonic setting of rapid horizontal displacement is largely responsible for the Charwell River being a most unusual and valuable study area. The drainage basin of the river has been progressively shifted to the northeast relative to the piedmont reach. Former piedmont valleys of the Charwell River (Figs. 5.2, 5.5) are preserved because of the rapid right-lateral tectonic translocation. Each former valley resulted from a major shift (avulsion) of the river to the east. The river is confined in a 50- to 65-m deep valley during times of maximum degradation such as the present (Fig. 5.21). A tendency for avulsion of the stream channel is induced by continued lateral fault movements between the drainage-basin and piedmont reaches, but cannot occur during times of channel entrenchment.

Optimal conditions for the piedmont reach to adjust its position relative to the tectonically translocated drainage basin occurred during times of maximum aggradation when streamflows issuing from mountain canyons spread out on unentrenched alluvial fans. At these times the river was free to seek a new course farther to the east on the piedmont reach, to correct for the amount of tectonic translocation that had occurred since the preceding avulsion. Then degradation created a new piedmont valley about 2 km east of the prior valley. Streams that do not cross an active strike-slip fault are much different. Aggradation and degradation episodes generally are stacked on top of each other in a bewildering complex of largely eroded stratigraphic and geomorphic records.

The tectonic setting at Charwell River tends to preserve complete records of both aggradation and degradation in tectonically translocated valleys through which the river no longer flows. Quail Creek (Figs. 5.2, 5.5) is an example in which both the Flax Hills (48 to 31 ka) aggradation event and the subsequent set of degradation terraces are preserved without burial by the Stone Jug (26 to 14 ka) aggradation event, which occurred mainly to the east of Dillondale ridge. Flax Hills deposition occurred on both sides of Dillondale ridge. On the east side it is largely buried by Stone Jug deposits or is eroded by Stone Jug degradation terraces. Both the aggradation surface and degradation terraces of Flax Hills age are preserved on the west side of the ridge.

The Charwell River flows due south from the Seaward Kaikoura Range and then turns to the southwest, cutting across and exposing the stratigraphic record contained in its ancestral valleys. These valleys have been tectonically translocated as much as 3 to 4 km. This tectonic setting is ideal for geomorphic, stratigraphic, pedogenic, paleoclimatic, neotectonic, and paleobotanical studies of the past 100 ka, because separate chapters in the history of the river have been set aside and largely preserved in separate former river valleys.

An important factor in piedmont terrace preservation in the reach adjacent to the Hope fault is the rapid rate of tectonic translocation of the piedmont reach relative to the stream issuing from the mountains. Flights of terraces on the *leading edge*— the true left side (looking downstream) of a stream— are destroyed when moved opposite a canyon mouth. Terraces on the *trailing edge*—the true right side—

tend to be temporarily preserved as they are moved away from the stream that formed them. Stream terraces are preserved longer (1) where distances between canyon mouths are large, (2) where rates of tectonic offset are slow, and (3) at distances of more than 5 km downstream from the range-bounding fault.

Long-term (>20 ky) uplift rates also may be inferred from studies of marine terraces along the Kaikoura coast; they also occur as remnants along ridgecrests within the Seaward Kaikoura Range (Figs. 5.3, 5.7). Bull (1985) and Bull and Cooper (1986) used altitudinal spacings of remnants of marine terrace platforms to correlate flights of New Zealand terraces with dated global marine terraces at New Guinea (Chappell & Shackleton, 1986). Coastal uplift rates opposite the Charwell River appear to be uniform at a given site, but vary between sites from 0.3 to 1.1 m/ky during the past 300 ka (Bull, 1984, Fig. 11).

Remnants of marine terraces also occur as notched spur ridges in the Seaward Kaikoura Range northwest of the Hope fault. An example is the set of eight prominent notches in different parts of the watershed at an altitude of $1130 \pm 15$ m on the topographic map of Figure 5.3. Remnants of this bench are thought to represent a shore platform

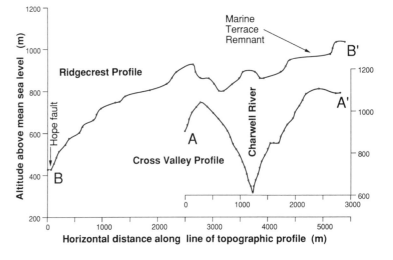

FIGURE 5.7 Topographic profiles along ridgecrest and across valleys of the Charwell River drainage basin. See Figure 5.2 for the locations of profiles. Vertical exaggeration 3.7X. Based on 1:63,360 scale map with 100-foot contour interval.

formed during a sea-level highstand at about 305 ka.

The long term vertical displacement across the Hope fault zone is estimated to be about $2.5 \pm 0.3$ m/ky. Tectonically induced downcutting below a $29 \pm 2$ ka strath in the terraced piedmont reach 1 km downstream from the Hope fault is $1.3 \pm 0.1$ m/ky. An altitudinal spacing analysis of marine-terrace remnants in the Charwell River drainage basin provides an inferred uplift rate of this part of the Seaward Kaikoura Range of $3.8 \pm 0.2$ m/ky. Long-term vertical displacement across the Hope fault is $3.8 \pm 0.2$ m/ky $- 1.3 \pm 0.1$ m/ky $= 2.5 \pm 0.3$ m/ky, which compares well with the 2.6 m/ky short-term rate estimated by Van Dissen at Sawyers Creek.

A key aspect of tectonically induced downcutting by a stream subject to episodes of aggradation is that each attainment of type 1 dynamic equilibrium will be at a progressively lower level in bedrock. If 5 ky have elapsed since the previous time of type 1 dynamic equilibrium for a reach that is being raised 1 m/ky, then a new major strath should be approximately 5 m below the preceding major strath. By definition a stream has yet to attain type 1 dynamic equilibrium as long as it remains in the aggradation deposits above the youngest strath. Pauses in degradational return to base-level conditions are times of formation of fill-cut and minor strath terraces, which are times of attainment of static equilibrium—the second class of attainment of the base level of erosion (Section 1.3.1).

The Quaternary evolution of the Charwell River basin can be summarized as follows. The inferred uniform uplift rate of about 3.8 m/ky during the past 300 ky, if extended to the uppermost 460 m of the drainage basin, suggests that this part of the Seaward Kaikoura Range is roughly 0.5 my old and that the ''Kaikoura orogeny'' (Cotton, 1974; Suggate, 1965) may have begun since 1 Ma. Evidence for large late Quaternary glaciers is lacking. Instead, deep fluvial canyons are inset between high rugged hillslopes, which were formed by alternation of periglacial and fluvial processes

during full-glacial climate, and by fluvial processes during interglacial times. Denudation of the graywacke and minor argillite has not been spatially uniform. Marine-terrace remnants on the ridgecrests clearly indicate that hillslope denudation rates have been a function of slope steepness. Maximum downcutting rates have been along the valley floors–the avenues of maximum concentration of stream power which has been increased by increase of relief of the watershed reach at rates of $2.5 \pm 0.3$ m/ky. Steep, convex hillslopes (Fig. 5.7) and proximity to a fault capable of generating magnitude $7.2 \pm 0.2$ earthquakes every $200 \pm 50$ years favor landslides as an important hillslope process. The dimensions of this actively degrading fluvial system are summarized in Table 5.2. The piedmont reach has been offset from its source watershed by right-lateral movements of Hope fault at a uniform $33 \pm 3$ m/ky during the past 40 ky. Two older piedmont valleys of the Charwell River with distinctive aggradation surfaces and flights of degradation terraces are preserved at distances of 2 and 4 km west of the present valley of the river.

## 5.2 Aggradation–Degradation Events

### 5.2.1 CHARACTERISTICS

Modes of operation of streams draining the Seaward Kaikoura Range varied with changing late Quaternary climates. Interglacial climates like that of the present were characterized by excess stream power that promoted stream-channel downcutting. Holocene interglacial climates have continued for sufficiently long to enable large streams, whose reaches are being raised tectonically at less than 0.5 to 1.5 m/ky, to attain their long-term base levels of erosion. Cutting of major strath surfaces characterizes these streams at such times of type 1 dynamic equilibrium.

Times of full-glacial climatic conditions were characterized by backfilling of piedmont valleys and local deposition of alluvial fans downstream from the range-bounding Hope fault. Aggradation of 20 to 50 m was common even from fairly small watersheds, such as the 30 km² main branch of the Charwell River. In a general sense, the times of aggradation and degradation reflected global climatic change that locally caused large variations in rates of bedload and water yielded from hillslopes.

Climate-change-induced increases in bedload transport rate resulted in distinctive episodes of fluvial gravel deposition. From oldest to youngest these are named the Quail Downs, Dillondale, Flax Hills, Stone Jug (Fig. 5.5), and Dog Hills aggradation events. Their diagnostic features are summarized here with an emphasis on the Flax Hills and Stone Jug events.

### 5.2.1.1 Flax Hills

Flax Hills aggradation was preceded by extensive tectonic strath beveling during a period when the Charwell River had achieved its base level of erosion for the entire 13-km-long reach from the Seaward Kaikoura Range to the junction with the Conway River. A brief episode of strath cutting appears to have occurred locally at about 40 ka (Table 5.3). Both the river and most piedmont tributary streams had stream power in excess of that needed to transport the bedload; they widened their valley floors by eroding laterally into soft bedrock. Two times of strath cutting suggest that Flax Hills deposition may have occurred in two stages. Their sedimentology and stratigraphy appear to be similar, so they are grouped as a single aggradation event.

Aggradation began with deposition of 2 m (locally 6 m) of alternating beds of well-sorted sand and gravel that locally included lenses of peaty silt and fine-grained sand with fossil reeds and leaves. Transported fossil wood is common in silt, sand, and gravel. Contacts between beds are clear, and bed thickness commonly is 5 to 30 cm. The gravel lenses were deposited by a high energy braided stream; the lenses of cross-bedded sand are similar to sands deposited in depressions scoured in the streambed near cliffy banks of the present braided river. Some silts may be floodplain de-

TABLE 5.2 Morphometric parameters of the Charwell River drainage basin from topographic Map NZMS 1, Charwell, S48; 1:63,360, 100-foot contour interval.

| Parameter | Main Stream | Right Branch |
|---|---|---|
| Basin area | 30.6 km² | 9.6 km² |
| Drainage density[a] | 3.1 km/km² | 3.4 km/km² |
| Bifurcation ratios[b] | | |
| 1st to 2nd order | 4.0 | 5.9 |
| 2nd to 3rd order | 3.4 | 5.8 |
| 3rd to 4th order | 4.5 | |
| Relief ratio[c] | 0.17 | 0.21 |
| Maximum valley side slope[d] | 1.7 | 1.3 |
| Basin elongation ratio[e] | 0.85 | 0.64 |
| Mean basin slope[f] | 0.56 | 0.59 |
| Hypsometric integral[g] | 0.45 | 0.47 |

[a] Average stream length per unit of drainage-basin area (Horton, 1932).

[b] A measure of the degree of branching of a drainage net as described by Strahler (1952) stream orders.

[c] Relative relief of a drainage basin described by the ratio of basin relief to basin height.

[d] Steepest slope along this side of trunk stream over a distance of one-third of the relief between the stream channel and the ridgecrest.

[e] Planimetric shape of a drainage basin described by the ratio of the diameter of a circle with an area equal to the basin area to the horizontal length between the two most distant points in the basin (Cannon, 1976).

[f] Average slope of a drainage basin measured by randomly selected points between contour lines.

[g] The relative volume of landmass below the sequence of contours that describe a drainage basin, expressed as an integration of the percentages of heights and areas of all contour lines (Strahler, 1952).

posits. Fossil tree trunks in growth position are encased by sequences of 2- to 20-mm-thick beds. Each bed typically grades upward from sand to silt to clayey or peaty silt, but some beds have reverse particle-size grading. A modern analog for the depositional environment of the graded beds is not present along the Charwell River. The graded silt beds appear to be low-energy slackwater deposits, each bed recording a flood event. These diverse basal deposits generally are reduced and appear to record a transitional fluvial regime of slow aggradation between the strath cutting of the preceding type 1 dynamic equilibrium period and the subsequent episode of rapid aggradation. A likely depositional environment may have been a braided stream with a sinuous thalweg bordered by a forested floodplain. Slow aggradation of the stream channel would be accompanied by deposition of floodplain sand and silt that buried the lower parts of standing trees.

The overlying aggradation facies consists of 35 to 50 m of massive, silty, water-laid gravels. Indistinct beds are 20 to 100 cm thick. Rare pieces of transported carbonized wood are present in the lower half of this massive facies. Slackwater environments were created in tributary valleys where thick Charwell River gravels were deposited across their mouths. Silty sand lenses are not common, but occur at any position in the aggradation gravels; they record floodplain slackwater deposits and channel infillings by fluid or eolian processes. Other depositional hiatuses are absent, including buried soil profiles, cut-and-fill structures, and different styles of bedding or depositional processes. Prominent, 1- to 2-mm-thick cutans of exceptionally fine clay on gravel clasts are ubiquitous and are one of the best ways to distinguish Flax Hills from younger and older deposits.

Three-color facies for the basal transitional and the overlying aggradation deposits appear to record postdepositional diagenetic groundwater environments. Matrix and surfaces of gravel clasts in the basal deposits are bluish gray. Occasional pockets of yellow-brown silt and sand and bright reddish-brown gravel suggest initial deposition of fluvial materials derived from oxidized hillslope colluvial sources or diagenetic oxidation. Matrix in the upper parts of Flax Hills gravel sections typically is reddish brown silty sand. Discontinuous brownish-black stains (manganese–iron oxyhydroxide–organic matter) typically coat more than half the

surfaces of the gravel clasts. A transitional, variegated, brightly colored facies occurs between the bluish-gray and reddish-brown facies. Silt matrix and clay cutans in the transitional facies range from yellow to yellowish-red, but the underlying surfaces of bluish-gray gravel clasts have minimal yellowish-red staining.

Soil profiles capping fill-terrace treads of Flax Hills age are cumulate, being formed mainly in 1.1 to 1.6 m of loess that accumulated after gravel deposition ceased. Basal soil-profile horizons are dense fragipans with moderate gleying and strong mottling. Upper soil-profile horizons are yellow-brown silty loams with strong subangular blocky structure. Clear to abrupt boundaries between the upper and lower horizons are suggestive of two phases of loess accumulation or of a complex history of horizon development in loess mantles that thickened at variable rates, and whose lower horizons developed fragipan characteristics.

### 5.2.1.2 Stone Jug

Stone Jug aggradation was preceded by less extensive strath beveling than that occurring before the Flax Hills aggradation event. The Charwell River achieved type 1 dynamic equilibrium for only a few kilometers downstream from the Hope fault; well-defined straths characterize this reach (Fig. 5.17). Straths were not cut elsewhere in the piedmont reach because stream power was too small or the base level of erosion was attained for too short a time span, or both. At 8 km downstream from the fault, the degradation event between the Flax Hills and Stone Jug aggradation events consisted only of downcutting into Flax Hills gravels.

Stone Jug aggradation began abruptly and consisted of 15 to 30 m of thick, poorly bedded, yellowish-brown, water-laid silty gravels deposited on underlying major straths or in channels cut into Flax Hills gravels. This uniform oxidized, massive lithology is the only depositional facies. Discontinuous brownish-black stains typically coat less than half of the clast surfaces in the basal gravels and are absent to faint in upper gravels. Silt lenses and all other types of depositional hiatuses are absent. Cutan development on gravel clasts is incipient and consists of silty clay. The lack of reduced fluvial gravels and minimal cutan development suggest that effects of postdepositional groundwater flushing were minor compared with those for Flax Hills deposits.

Soil profiles capping fill-terrace treads of Stone Jug age are cumulate, being formed mainly in 0.5 to 1.1 m of loess that has accumulated since gravel deposition. B soil-profile horizons are yellow-brown silty loams with moderately strong subangular blocky structure. Incipient weak, coarse prismatic structure occurs only at a few sites but suggests initial stages of development of fragipan structure where soil-moisture conditions are suitable.

Both the Stone Jug and Flax Hills are major aggradation events whose fill terraces begin in the valleys of the Seaward Kaikoura Range and continue to the ocean along many streams of the region. Stone Jug aggradation along the Charwell River consists of coalescing alluvial fans adjacent to the mountain front and of fill terraces inset into Flax Hills deposits in downstream reaches. The Stone Jug event along many other streams consists only of fill terraces, because backfilling was insufficient to completely fill the valley and allow an alluvial fan to spread out over Flax Hills surfaces. Threshold times represented by the treads of Flax Hills and Stone Jug aggradation surfaces were synchronous for large drainage basins (Bull, 1990), which were influenced mainly by climatic change, and diachronous for small drainage basins, which were influenced to a greater extent by lithologic controls (Section 5.2.2).

### 5.2.1.3 Dog Hills

The Dog Hills aggradation event is defined as diachronous small alluvial fans and valley fills that were deposited on bedrock or older alluvium immediately downstream from the range-bounding Hope fault. Brief episodes of aggradation characterize these smaller than 1 to 2 km$^2$ fluvial systems draining the crush zone of the Hope fault. The typical section consists of oxidized interbedded water-laid and debris-flow deposits; depositional

hiatuses include incipient buried soil profiles and cut-and-fill structures. The typical soil profile has a 25-cm brownish-black A horizon and an AB horizon that extends to 35 cm, below which is a 15- to 25-cm Bw horizon that is a yellow-brown, pebbly, silty loam.

### 5.2.2 TIMES OF DEPOSITION

The times of the three aggradation events have been estimated with a variety of dating methods. Work in progress provides the following tentative age assignments. Volcanic ash occurs on the east side of the Charwell River as an 8- to 15-cm-thick layer in a colluvial saddle, as concentrations of shards at the lower third position in Stone Jug silty gravels, and in loessial pre-Stone Jug soils. Analysis of major and trace elements by Stephen Weaver of the University of Canterbury indicates that this is another locale of the Kawakawa tephra (Campbell, 1986; Mew et al., 1986) which was derived from a volcanic center 490 km to the north. The Kawakawa tephra is the only late Quaternary volcanic event recorded in the South Island, and its radiocarbon age of $22.6 \pm 0.2$ ka (Wilson et al., 1988) clearly identifies the Stone Jug aggradation event as occurring during the most recent time of full-glacial climate.

Radiocarbon dating of organic materials collected from deposits beneath the terrace treads provides stratigraphic ages, which are quite useful for some purposes (Table 5.3). These ages do not represent times of initial stream-channel downcutting responsible for the overlying stream-terrace tread. The apparent advantage of having a radiocarbon age is offset by the necessity of having to estimate the elapsed time between deposition of the dated material and formation of the terrace tread. A single age does not define an aggradation rate. Where a pair of radiocarbon ages is available, tread ages can only be estimated by assuming uniform aggradation rates. Sections with multiple radiocarbon ages typically reveal rates of valley-fill and alluvial-fan aggradation that increase or decrease (Fig. 4.36) rather than remain uniform.

Initial results of radiocarbon dating of wood within 0.5 to 2.1 m of bedrock provide radiocarbon ages of 48 and 40 ka, which approximate the times of cessation of strath cutting and beginning of deposition of the basal Flax Hills transitional deposits. Radiocarbon ages from other parts of the South Island (Knuepfer, 1984) indicate that the Stone Jug (latest Otiran) aggradation began at about $26 \pm 2$ ka and ended about $14 \pm 2$ ka. Dating of wood and charcoal from Dog Hills deposits provides a variety of ages that include 0.6, 0.8, and 3.3 ka.

A dozen thermoluminescence analyses are being made by Steven Forman of the University of Colorado from a selection of 63 samples collected from Charwell River deposits; this work will be done after 20 radiocarbon samples are dated.

Analyses of cobble weathering rinds (Section 5.2.3.1) on surficial graywacke cobbles is highly useful for estimating ages of stream-terrace treads and alluvial-fan surfaces younger than 20 ka because rind-thickness growth has been calibrated with radiocarbon ages (Whitehouse et al., 1986; Knuepfer, 1988). Weathering rind ages constrain the end of Stone Jug aggradation to older than 13 ka, and provide ages for cessation of deposition of Dog Hills fanheads of $3.8 \pm 0.5$, $6.6 \pm 0.7$ and $8.7 \pm 1.0$ ka.

Establishment of constant fault slip rates (Section 4.2.3.2.2; Weldon & Sieh, 1985) provides a way of estimating the beginning and end of the Flax Hills and Stone Jug aggradation events. Six cobble weathering-rind ages by four different workers indicate that horizontal slip rates of the Hope fault at the Charwell River were $33 \pm 3$ m/ky for time spans ranging from 11 and 4 ka to the present. Ages of leading and trailing edges of tectonically translocated asymmetric alluvial fans of the river and of five other large streams to the east were estimated by dividing the amounts of horizontal tectonic displacement by the assumed uniform 33 m/ky slip rate. The resulting set of calibrated ages indicates that synchronous fan deposition associated with six large fluvial systems occurred as two distinct episodes—between $38 \pm 1$

and $31 \pm 1$ ka (Flax Hills) and between $26 \pm 2$ and $14 \pm 2$ ka (Stone Jug). Two sites have a record of the earlier Flax Hills pulse of deposition between $49 \pm 5$ and $43 \pm 5$ ka.

Reaction and response times to climatic perturbations that initiated aggradation or degradation reflect different sets of interacting variables in each fluvial system. Three factors are particularly worth noting: (1) drainage-basin area is an important control on available stream power, (2) increases in sediment yield may be a function of lithologic variations that include the degree of shearing and crushing by movements along faults of the Hope fault system, and (3) spatial effects include altitudinal variations in geomorphic processes affecting yield of sediment and water. Watersheds for which these three factors are similar tend to have synchronous times of aggradation.

### 5.2.3 TIMES OF DEGRADATION

A key aspect of the Charwell River terrace analysis was provided by the age-control studies made by Peter Knuepfer (1984, 1988). Knuepfer estimated terrace-tread ages from analyses of surficial cobble weathering-rind thicknesses and soil-profile characteristics that were calibrated by radiocarbon ages at many sites in the central and northeastern South Island.

The most desirable datum for age control is the terrace tread. Terrace treads are synchronous or diachronous time lines in fluvial landscapes and represent crossings of the threshold of critical power, periods of equilibrium, or both. In a flight of degradation terraces, a stream-terrace tread and the riser to the next higher tread are a pair of landforms created at the same time. The modern analogy is the river floodplain and its adjacent streambank.

#### 5.2.3.1 Weathering Rinds

Torlesse graywacke is the source of the terrace gravels. Weathering rinds form rapidly on surficial cobbles; rind thicknesses have been measured on rockfalls with ages of less than 0.2 ka. The rate of increase of weathering-rind thickness was initially calibrated at 14 sites (Chinn, 1981; Whitehouse et al., 1980,1986; Whitehouse, 1983; Whitehouse & McSaveney, 1983). The ideal calibration site has abundant material for radiocarbon dating, instantaneous emplacement, and lack of subsequent disturbance of surficial graywacke boulders and cobbles. An example is a forest buried by a rock avalanche. Rind thickness increases as a power function of time and is quite useful as a dating tool for glacial moraines, talus, and rock avalanches with ages of less than 20 ka. Knuepfer (1984,1988) applied weathering-rind dating techniques to stream terraces.

The procedure was to break the tops of 50 to 200 medium-grained graywacke surficial cobbles and boulders collected from the middle of a broad terrace tread that had never been ploughed. The thickness of the light-gray weathering rind (Fig. 5.8) was measured in bright sunlight to a precision of 0.1 mm, using a 7-power comparator with 0.1-mm reticle graduations. The rind-thickness data were tabulated as a bar graph with 0.2-mm thickness classes. These data were smoothed as a weighted overlapping mean (Fig. 5.9) to better identify prominent modes of rind thickness. Data for two of the Charwell River terraces are shown together with data for a young fan surface, whose age is estimated as being slightly younger than a $0.73 \pm 0.043$-ka stratigraphic radiocarbon age (Table 5.3). Distributions of cobble weathering-rind thicknesses for young surfaces tend to be unimodal with pronounced peaks; distributions for older terraces are progressively more dispersed and polymodal. The graph for terrace 1 (Table 5.4) weathering-rind thicknesses has multiple peaks that apparently record a history of surficial events on the terrace tread. These include the time of terrace-tread formation and subsequent episodes of surface disturbance such as uprooting of large numbers of trees during violent windstorms, or episodes of shattering of surficial rocks by frost splitting. Bioturbation by rodents is not a problem because bats were the only terrestrial mammals in New Zealand until colonizations by Polynesians and Europeans. The common-sense approach for old terraces is to

select the oldest prominent mode ($>5\%$ of the sample) in a smoothed graph of rind thicknesses as being indicative of the time of terrace-tread formation.

Factors other than time have a minimal influence on growth rates of weathering rinds. At a few other New Zealand sites Knuepfer had to sample lithologies that are moderately different from the usual medium-grained phrenite-pumpellyite grade graywacke. He found no differences in weathering-rind characteristics. Annual precipitation at the calibration sites ranges from 1200 to 5000 mm, but there is no statistical relationship between annual precipitation and weathering-rind thicknesses of a given age (Chinn, 1981; Whitehouse et al., 1986; Knuepfer, 1984). The range of possible temperature-influenced variations of rind growth are virtually negligible for the wide range of present temperatures at the calibration sites. These conclusions are important for the Charwell River site because (1) gravel-clast lithologies are not identical for any sample, (2) the Charwell age estimates are tied to calibration sites from a larger region with large spatial climatic variations, and (3) large temporal climatic variations have occurred at each stream-terrace sample site during the late Quaternary.

Knuepfer combined his thickness of weathering rind, Wr, data from his calibration sites with those of the initial users of the technique to obtain the power function calibration curve of Figure 5.10, which is in calendric years, t, before a 1980 reference year.

$$t = 966 \pm 66 \ Wr^{1.30 \pm 0.05} \qquad (5.1)$$

Standard errors for the age estimates range from 5 to 20 percent. Knuepfer (1988) used cobble weathering-rind ages to date seven Stone Jug degradation terraces along the Charwell River that range in age from $10.8 \pm 1.8$ to $3.9 \pm 0.5$ ka (Table 5.4).

### 5.2.3.2 Soil-Profile Development

Soil genesis on the Charwell River terrace treads is highly time dependent. Thus soil-profile characteristics can be used for age estimates, although

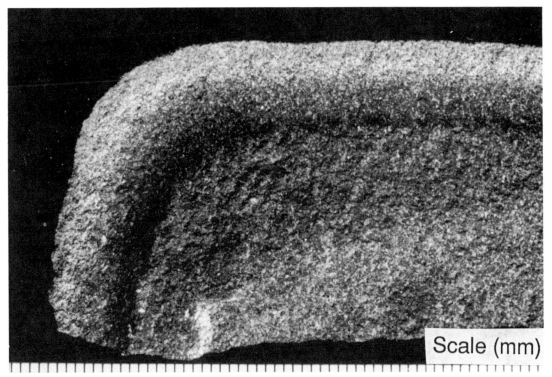

FIGURE 5.8 Photograph of weathering rind in a greywacke cobble. The white part of 50 to 200 rinds are measured for estimates of terrace-tread age. Scale in mm.

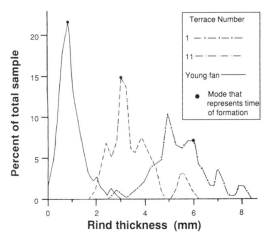

FIGURE 5.9 Changes in the thickness and range of weathering rinds on surficial greywacke cobbles on Charwell River terraces of different ages. The plots are smoothed rind distributions from bar graphs and black dots are assigned modes for time of terrace-tread formation. Modified from Figure C10 of Knuepfer, 1984.

these estimates have larger uncertainties than cobble weathering-rind ages. Potassium, phosphate, magnesium, and calcium are progressively leached from the B soil horizons with the passage of time, and the oxides of iron and aluminum are concentrated (Knuepfer, 1984,1988). Knuepfer quantified his field descriptions of soil morphology by calculating a soil-development index.

Both field and laboratory data show systematic changes with time for the Stone Jug degradation terraces. Both oxalate-extractable iron and alumina tend to accumulate in the 20- to 120-cm depth range, and older B horizons show progressively larger amounts of these illuviated sesquioxides (Fig. 5.11). Soil-profile indices (Section 2.4.1.4.2) are plotted against weathering-rind ages for six terraces (Fig. 5.12). Rapid and internally consistent changes in weathering and soil properties have occurred during the 14 ky since the formation of the Stone Jug aggradation surface.

Knuepfer (1984) examined the internal consis-

TABLE 5.3 Summary of radiocarbon analyses from sites along the south side of the Seaward Kaikoura Range.

| Laboratory Sample Number | Latitude | Longitude | Material | Conventional Date (ka)[a] | $\delta^{13}C$ (‰) | Calendric Age[b] | Comments |
|---|---|---|---|---|---|---|---|
| NZ[c]-R11694/4 | 42°23.83'S | 173°23.59'E | Driftwood in bedded gravel | 0.582 ± 0.036 | −25.4 | 0.638 ± 0.024 ka. 51% probability that date is between 1376 and 1432 A.D. 44% probability that date is between 1302 and 1363 A.D. | [d] |
| NZ-R11694/6 | 42°25.53'S | 173°21.63'E | Charcoal in colluvial wedge on top of soil profile | 0.678 ± 0.048 | −24.9 | 0.667 ± 0.022 ka. 95% probability that date is between 1275 and 1398 A.D. | [d] |
| NZ-R11694/1 | 42°24.33'S | 173°21.90'E | Charcoal in debris-flow bed in small fan cut by the Hope fault | 0.733 ± 0.043 | −25.4 | 0.691 ± 0.015 ka. 72% probability that date is between 1229 and 1322 A.D. | [d]1 m higher in section than A-3536 |
| A[e]-3536 | 42°24.33'S | 173°21.90'E | Charcoal in debris flow bed in small fan cut by the Hope fault | 0.840 ± 0.060 | −26.2 | 0.738 ± 0.054 ka. 88% probability that date is between 1154 and 1261 A.D. | 1 m below sample NZ-R11694/1 |
| NZ-R11694/3 | 42°23.84'S | 173°23.60'E | Driftwood in peat lens beneath paleosol | 3.370 ± 0.043 | −26.3 | 3.59 ± 0.029 ka. 52% probability that date is between 1676 and 1619 B.C. 48% probability that date is between 1736 and 1678 B.C. | [d] |

| Sample | Latitude | Longitude | Description | Age | δ¹³C | Notes |
|---|---|---|---|---|---|---|
| A-5703 | 42°25.10S | 173°19.39'E | Charcoal in debris-flow bed in alluvial-fan deposits | 4.400±0.120 | −24.7 | 4.98±0.115 ka. 73% probability that age is between 3140 and 2910 B.C. 21.85 B.C.[f] |
| NZ-R11752/3 | 42°26.34'S | 173°22.31'E | Peat layer in clayey silt | 20.000±1.0 | −30.1 | [d] |
| NZ-R11752/1 | 42°27.69'S | 173°21.62'E | Driftwood branches in sand lens | >34.300 | −26.9 | [d] |
| A-5697 | 42°27.50'S | 173°21.79'E | Carbonized driftwood in oxidized sand lens on strath 17.5 m above Charwell River | 35.92 +1.25/−1.08 | −28.6 | |
| NZ-R11752/2 | 42°27.42'S | 173°20.99'E | Branch of driftwood in gravel lens | >38.500 | −28.0 | [d] |
| A-5219 | 42°15.30'S | 173°43.35'E | Trunk in debris flow | 39.620 +3.320/−2.340 | −28.0 | Same site but different tree trunk than sample R2 |
| NZ-(R2)[g] | 42°15.30'S | 173°43.35'E | Trunk in debris flow | >41.500 | | [d] |
| A-5700 | 42°28.47'S | 173°19.40'E | Tree trunk in growth position buried by slackwater beds; about 20 m above strath and 21.6 m above Goat Hills Creek | 43.73 +3.62/−2.49 | −26.3 | |
| A-5702 | 42°28.41'S | 173°19.40'E | Driftwood trunk or branch in gravel bed 3.1 m above strath | 46.87 +5.78/−3.33 | −27.3 | |
| A-5699 | 42°26.03'S | 173°21.99'E | Driftwood below deformed lake beds upstream from the now inactive Charwell fault | 46.29 +5.31/−3.17 | −26.2 | |
| A-5695 | 42°25.75'S | 173°21.85'E | Driftwood mat in pond upstream from fault crossing Charwell River | >47.4 | −26.8 | |

(continued)

TABLE 5.3 Summary of radiocarbon analyses from sites along the south side of the Seaward Kaikoura Range. (Continued)

| Laboratory Sample Number | Latitude | Longitude | Material | Conventional Date (ka)[a] | $\delta^{13}C$ (‰) | Calendric Age[b] | Comments |
|---|---|---|---|---|---|---|---|
| A-5698 | 42°27.50'S | 173°21.79'E | Fresh driftwood in re-duced sand lens on strath 15 m above Charwell River | 47.60 +5.64/−3.28 | −28.6 | | |
| A-5704 | 42°26.03'S | 173°19.40'E | Driftwood trunk or branch in gravel bed 1.8 m above strath | >47.94[h] | −29.4 | | |
| A-5701 | 42°28.47'S | 173°19.40'E | Tree trunk in growth position buried by slackwater beds; about 20 m above strath and 18.2 m above Goat Hills Creek | 49.19[i] | −27.2 | | |

[a] Half-life of 5568 years is used; 1 sigma counting errors are shown. Age before 1950 A.D. has been corrected for carbon isotope fractionation using the $\delta^{13}C$ value normalized to −25‰.

[b] Calibrated $^{14}C$ age to account for variations in the specific activity of $^{14}C$ in atmospheric $CO_2$, using the program of Stuiver and Reimer (1986). A. D. Calendric age = 1950 − A.D. calibrated age. B.C. Calendric age = 1949 + B.C. calibrated age.

[c] Radiocarbon laboratory at the New Zealand Institute of Nuclear Sciences.

[d] The author greatly appreciates the assistance of the New Zealand Geological survey in dating this sample.

[e] Radiocarbon laboratory at the University of Arizona.

[f] Calibrated $^{14}C$ age to account for variations in the specific activity of $^{14}C$ in atmospheric $CO_2$, using the comparisons of uranium-thorium ages and radiocarbon ages of Barbados corals (Bard et al., 1990). 1949 A.D. − (20,000 + 3,800) years = 21.85 ka.

[g] R2 is a field sample number.

[h] This sample had a $^{14}C$ activity between 1 $\sigma$ and 1 $\sigma$ above the background activity, and therefore is reported as an "apparent age".

[i] This sample had a $^{14}C$ activity of less than 1 $\sigma$ and 2 $\sigma$ above the background activity.

TABLE 5.4 Age estimates of Charwell River degradation terraces.

| Terrace | Age (ka) |
|---|---|
| Stone Jug aggradation surface | 14 ± 2.0 |
| Stone Jug degradation terraces | |
| 1 | 10.8 ± 1.9 |
| 2 | 8.5 ± 1.4 |
| 3 | 6.7 ± 1.0 |
| 5 | 6.3 ± 0.9 |
| 6 | 5.5 ± 0.8 |
| 9 | 4.7 ± 0.6 |
| 11 | 3.9 ± 0.5 |
| Dog Hills alluvial fan | 0.7 ± 0.05 |

Source: Kneupfer, 1988.

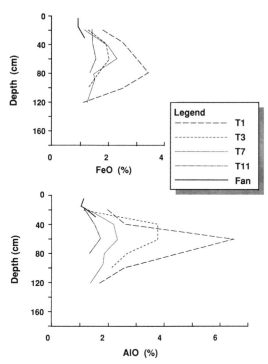

FIGURE 5.11 Oxalate-extractable iron and aluminum, Charwell River terrace soils. Percentages of FeO and AlO were determined by oxalate extraction. From Knuepfer, 1984, Figure 53.

tency of cobble weathering-rind and soil-profile methods by comparing age estimates of terraces 1 and 3 against that of terrace 7, which was assigned a value of 1. The ratios for weathering-rind ages for terraces 1, 3, and 7 are 2.3: 1.4: 1.0; the ratios for the soil-profile indices are 2.2: 1.5: 1.0. This cross-check indicates that the age estimates shown in Table 5.3 are internally consistent within the assigned uncertainties.

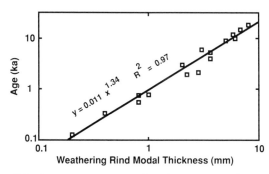

FIGURE 5.10 Increase of surficial greywacke cobble weathering-rind thickness with age, South Island, New Zealand. From Knuepfer, 1988, data set.

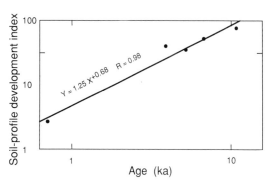

FIGURE 5.12 Profile index versus. weathering-rind age, Charwell River terraces. From Knuepfer, 1984, Figure 40.

## 5.2.4 TIMES OF MAJOR STRATH CUTTING

Attainment of type 1 dynamic equilibrium is re-corded by beveling of major straths that, in the case of the Charwell River, subsequently are bur-ied by the massive gravels of the next aggradation event. In contrast, brief times of attainment of static equilibrium are represented by minor straths of degradation terrace flights typically capped by thin layers of lag gravels that are well sorted and distinctly bedded (Section 5.4.4).

The age control summarized in Sections 5.2.2 and 5.2.3 constrains the times of major strath formation. The dating suggests that strath cutting occurred at about 48 and 40 ka and sometime between 31 and 26 ka, and again from about 4 to 0 ka. These age estimates conform with a model of times of major strath formation that coincide with times of global sea-level highstands. Clearly the prolonged beveling of broad straths that is occurring now coincides with the present sea-level highstand. The major straths beneath the Flax Hills and Stone Jug aggradation gravels are tentatively assigned ages of 53, 40, and 29 ka using the chronology of Chappell and Shackleton (1986), which is fine-tuned by the astronomical clock (Sec-tion 2.1.2.2). Such fine tuning of age estimates can be done for the Charwell fluvial system be-cause it has short reaction times to climatic per-turbations.

Although the Charwell and nearby rivers are close to the ocean (Fig. 5.1), it is highly unlikely that present or past type 1 dynamic equilibrium conditions were direct responses to base-level rises or stable base levels associated with sea-level high-stands. Even base-level rise caused by dam con-struction affects fluvial processes for only a short distance upstream from the perturbation (Leopold & Bull, 1979).

The coincidence between the ages of Charwell River straths and straths elsewhere in New Zealand and North America, and the ages of global marine terraces dated by $^{230}$Th/$^{234}$U analyses of coral from New Guinea (or dated by the astronomical clock) is the result of generally similar timing of climate-

controlled processes and resulting landforms in two markedly different geomorphic systems. Both systems are controlled by fluctuations between glacial and interglacial global climates. Of course, fluvial systems do not respond identically to a climatic perturbation. Nor does the timing of the marine and fluvial equilibria coincide exactly. But, in general, times of rapid aggradation by New Zealand streams seem to have been times of max-imum accumulation of ice on the continents and lowstands of glacioeustatic sea levels. New Zea-land streams are sufficiently powerful, for the erodibility of materials, to downcut rapidly and reestablish briefly type 1 dynamic equilibrium con-ditions during times of interglacial climates, which are times of melting of continental ice masses and attainment of sea-level highstands.

## 5.3 Hillslope Process-Response Models

This section considers possible interactions be-tween variables in the watershed part of the fluvial system that might be responsible for sustained aggradation and subsequent degradation by streams in the piedmont reach. The models are based on data from the Charwell River area supplemented by observations made elsewhere in the north half of the South Island.

Terrace stratigraphy and sedimentology of the piedmont reach may be regarded as outputs of the watershed reach, outputs that contain significant paleoclimatic information. Flax Hills deposits have basal blue gravels (with rare oxidized pockets), oxidized upper gravels, and a transitional facies between the two. Although not common, silt len-ses can occur anywhere in Flax Hills deposits, and fossilized wood is remarkably common in the lower half of many sections. It would seem that the climate immediately preceding and during basal Flax Hills times was wet and mild enough for extensive straths to be beveled and for trees to be common. With the onset of full-glacial climate, mass movement hillslope processes—periglacial and landslides—may have contributed to a large sediment-yield increase that caused the piedmont reach to move strongly to the aggradational side of the threshold of critical power.

Stone Jug deposits are entirely oxidized, have minimal clay and mineral coatings, have no silt lenses (except next to active faults), and are devoid of wood. In contrast to the preceding Flax Hills aggradation event, it seems that the climate during Stone Jug times was dry and cold. Prior to abrupt initiation of aggradation, straths were beveled only near the mountains. At 6 km downstream from the mountain front, streamflow had managed to cut deeply into Flax Hills gravels but stream power was insufficient to degrade into the underlying bedrock and bevel a new major strath below the level of the Flax Hills strath. Fossil wood cannot be found because a depression of tree line of about 800 m left none to be incorporated in Stone Jug gravels. Aggradation may have resulted from both decreased streamflow and increased bedload. Minimal postdepositional infiltration of streamflows and groundwater flushing resulted in minimal development of cutans and coatings, and maintained an oxidizing environment to eliminate fragments of small plants.

Loessial silts are progressively thinner on progressively younger terrace treads (ages ranging from 16 to 1 ka) and are indicative of the relative abundance of loess from regional sources. An abrupt decrease in loess thickness occurs for terraces younger than about 9 ka. Loess accumulation was more rapid during the Pleistocene–Holocene climatic transition than during the Holocene, and may have peaked during times of full-glacial climate. Periglacial processes in much of the Seaward Kaikoura Range probably increased sources of silt for fluvial and eolian processes during each period of full-glacial climate.

The foregoing summary is used to outline three process-response models for fluvial-system behavior during the (1) Flax Hills aggradation event, (2) Stone Jug aggradation event, and (3) Stone Jug degradation event. The first two process-response models are an attempt to explain contrasting rates of change. The Flax Hills event lasted only about 7 ky but resulted in 30 to 40 m of aggradation in all 13 km of the piedmont reach. The Stone Jug event lasted about 12 ky but resulted in only 15 to 30 m of aggradation concentrated in the upstream 6 km of the piedmont reach.

### 5.3.1 FLAX HILLS AGGRADATION EVENT

One possible scenario outlined for Flax Hills aggradation (Fig. 5.13) assumes climatic control of no change in mean annual precipitation and moderate decrease in mean annual temperature, which caused a decrease in the proportion of precipitation falling as rain. Discontinuous permafrost conditions became common on south-facing slopes and at higher altitudes within the Charwell River watershed. Debris flow and landslide activity probably were increased by wet cold conditions and infiltration during a prolonged snowmelt season. Solifluction became an important hillslope process, and landslides may have been more frequent than at present. Stripping of hillslope colluvial mantles that had accumulated during a previous milder climate would have been favored by the presence of an active freeze-thaw layer above seasonally or perennially frozen ground. Freezing of interstitial water during winters would decrease strength of colluvial materials by creating ice lenses and by separating soil particles (French, 1988; Lewkowicz, 1988). Infiltrating water from spring and summer snowmelt would tend to saturate thawed colluvium in the active layer, thereby decreasing grain-to-grain resisting forces through increase of buoyant support and increasing driving forces through development of seepage stresses equal to local head differentials. Such factors would create ideal conditions for creep, gelifluction, and solifluction mudslides on hillslopes that were steep because of tectonically induced downcutting caused by rapid uplift (Fig. 5.7). The resulting bedload increases to streams caused aggradation of valley floors during times of periglacial hillslope regimes in a manner similar to that outlined by Peltier (1950) and Zuener (1959).

Forests probably were replaced by tussock grasses

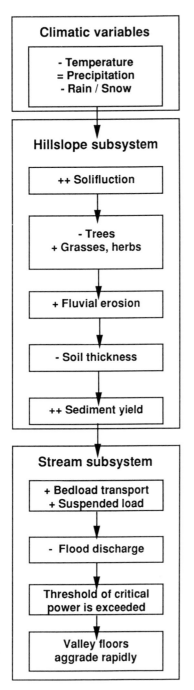

FIGURE 5.13 Process-response model for the hillslopes of the Charwell River basin for a change from humid mesic to humid frigid conditions. Changes resulted in the Flax Hills aggradation event. Symbols for moderate changes are (+) and (−) and for major changes are (+ +) and (− −).

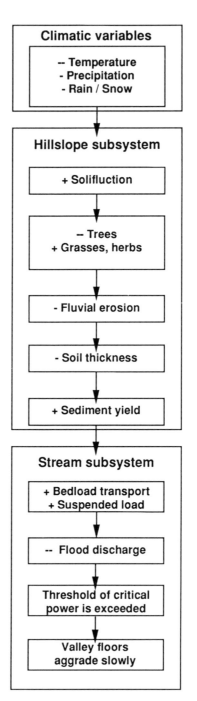

FIGURE 5.14 Process-response model for the hillslopes of the Charwell River basin for a change from humid mesic to semiarid pergelic conditions. Changes resulted in the Stone Jug aggradation event. Symbols for moderate change are (+) and (−) and for major changes are (+ +) and (− −).

and herbaceous plants at the higher altitudes if trees were eliminated by a treeline depression of 400 to 600 m. The density of plant cover and the soil-binding action of roots may not have changed greatly, however. Subalpine scrub and large tussock grasses above the present treeline in the Southern Alps are dense; they provide many roots to bind soil and intercept virtually all raindrops. Fluvial erosion is assumed to have increased moderately, because summer rains would fall on soil that had been disrupted by landslide and solifluction processes during snowmelt. Net depletion of the hillslope sediment reservoir decreased soil thickness, which in turn reduced plant growth.

These postulated changes would have had profound effects on the stream subsystem, mainly through massive increase in bedload transport rate. Peak flood discharge would decrease moderately because snowmelt generally creates smaller peak discharges than does a rainfall regime with the same mean annual precipitation (Costa & Baker, 1981, Fig. 12–19). Prolonged high flows during the snowmelt season could have transported huge amounts of gravel to the piedmont reach. The lack of evidence for glaciation in the Charwell River watershed suggests virtually complete melting of the snowpack during most years. The stream power/resisting power ratio decreased to much less than 1.0 (Table 5.5). The capacity of the watershed reach to supply both large amounts of bedload and the water required to transport it resulted in extremely large bedload transport rates and extensive rapid aggradation of the piedmont reach. The threshold of critical power was crossed again after about 7 ky, and a few degradation terraces were cut in 2 or 3 ky. The brief period of downcutting was terminated by the next aggradation event.

## 5.3.2 STONE JUG AGGRADATION EVENT

A considerably different scenario is needed to explain the Stone Jug aggradation event (Fig. 5.14), which occurred during relatively cold, dry, glacial times. A moderate decrease in mean annual precipitation and a major decrease in temperature seem to have been the main climatic changes. The

rain/snow ratio decreased more than for the Flax Hills scenario.

Assuming these and other (Section 5.1.3) climatic controls, the hillslopes would be affected by a moderate increase in periglacial processes, but not to the same extent as during Flax Hills times. Continuous permafrost conditions probably were widespread, but the amount of available water was the limiting factor for both solifluction mudslides (moderate increase) and fluvial erosion (moderate decrease). Soil thicknesses, already depleted by multiple episodes of glacial conditions during the prior 50 ky, decreased again as accelerated depletion of the hillslope sediment reservoir substantially increased sediment yields.

The most likely effect on the stream subsystem of these changes in watershed processes was an increase in amount of bedload being supplied to the Charwell River concurrent with a major decrease in stream discharge to transport bedload. Stream power decreased as a result of greater importance of snowmelt processes and lesser annual runoff during this period of dryer climate. The lack of transitional basal Stone Jug beds suggests an abrupt transition from strath cutting to aggradation of massive silty gravel. Basal Flax Hills beds suggest sufficient stream power for transport and sorting of gravel. Basal Stone Jug massive silty gravels suggest a slug of gravel derived from reworked diamicts that moved downstream only as rapidly as the limited stream power would allow. Depositional hiatuses did not occur as streams reworked gravel temporarily stored in bars and as new increments of silty gravel were supplied from the watershed reach. The net effect of moderate increase of sediment yield and moderate decrease of streamflows for transporting gravel was slow aggradation of a much smaller total volume of valley fill than during the Flax Hills aggradation event.

## 5.3.3 STONE JUG DEGRADATION EVENT

The third process-response model outlines a likely scenario for the Stone Jug degradation event. It is assumed that:

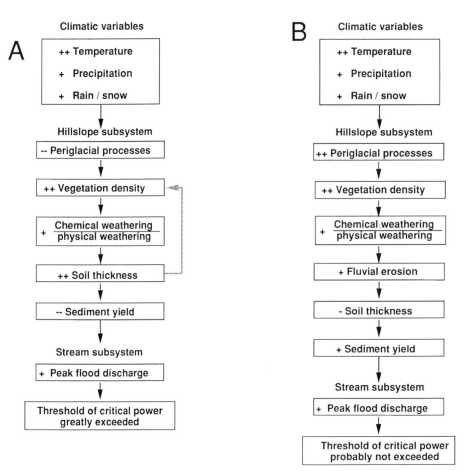

FIGURE 5.15 Process-response model for the hillslopes of the Charwell River basin for a change from semiarid pergelic to humid mesic conditions. Changes resulted in the Stone Jug degradation event. Symbols for moderate change are (+) and (−) and for major change are (++) and (−−). A. Watershed from the basin mouth to the present treeline, 450 to 1200 m. Self-enhancing feedback mechanism is shown by the dotted line with arrow. B. Watershed from the present treeline to highest part of the basin, 1200 to 1600 m.

1. Full-glacial conditions had ceased by 14 ± 2 ka.

2. Recolonization of the hillslopes by bushes and grasses occurred during a 14- to 10-ka climatic transition when precipitation and temperature were increasing.

3. Podocarp-hardwood forests were reestablished shortly after onset of Holocene climates at 10 ka.

4. Treeline may have been higher than at present during a 2- to 3-ky mid-Holocene period that was wetter and 1°C warmer than at present.

More than one Holocene process-response model should be used for the Charwell basin despite its monolithologic nature and moderate relief of only 1100 m. The present treeline at an altitude of 1200 m is a useful basis for defining the two models

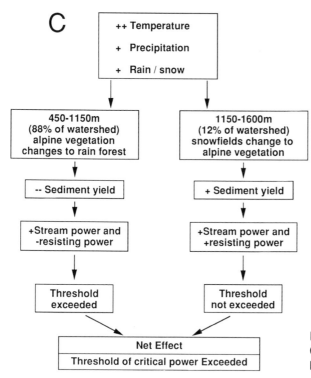

C

Figure 5.15 *(continued)*
C. Entire watershed from the basin mouth to the highest summit, 450 to 1600 m.

diagrammed in Figures 5.15A and 5.15B. Vegetation is such an important geomorphic variable that the presence or absence of forest can be used to define two process-response models for the watershed. The key climatic changes from latest Pleistocene full-glacial climate to milder Holocene climate are assumed to have been increase of temperature and precipitation and the proportion of precipitation occurring as rain instead of snow. Above 1200 to 1300 m, areas of presumed semi-permanent snowfields changed to alpine vegetation (the shrubs and tussock grasses that presently grow there). The more important changes occurred within the 450– to 1200–m altitude range where presumed subalpine vegetation was replaced by tall shrubs and then by beech forest.

The most likely effects of the Pleistocene–Holocene climatic change are used to construct a process-response model for the 90 percent of the Charwell basin area lying between 450 and 1200

m (Fig. 5.15A) for an assumed change in climate from semiarid frigid to humid mesic. The mass-movement hillslope regime associated with periglacial processes became a fluvial-process regime. Landslides continued to be important, particularly where tectonically induced valley downcutting created convex valley sides with steep, unstable footslopes.

The regional and local data presently available suggest the following scenario. Major increases in vegetation density occurred as bushes, and then trees, returned to form a protective vegetative cloak on the hillslopes. The warmer and wetter climate, together with the abundant organic acids produced by tree, such as beech (*Nothofagus*), may have substantially increased rates of chemical weathering of hillslope colluvium and bedrock. Return of trees would initiate increases in soil thickness, which acted as a self-enhancing feedback mechanism that further favored expansion of forests, thus increasing storage in the hillslope sediment reservoir. The net effect was a progressive and substantial decrease in sediment yield to the Charwell River.

The latest Pleistocene-early Holocene decreases in sediment yield (Fig. 5.15A) represent a major decrease in resisting power that came at a time of climatically induced increase in stream power. An important factor may have been increase in storm runoff. Sustained high discharge is possible during summer melting of snowpacks, but increases in the proportion of precipitation that fell as rain probably resulted in substantially larger peak flood discharges, especially during large rainfalls on snowpacks.

At present the Seaward Kaikoura Range receives occasional intense rainfall from tropical storms during the late summer and early fall. Floods generated by Cyclone Alison devastated much of the area in 1975. Thus the presumed return of tropical storms during the Holocene may have caused episodes of markedly greater stream power.

A different process-response model (Fig. 5.15B) is used for the 12 percent of the Charwell River basin lying above 1150 m, where stone stripes and talus formation are still active. The suggested climatic change for this part of the hillslope subsystem is from semiarid pergelic-frigid to humid frigid-mesic. Snow mantled the terrain during more of the year at times of full-glacial climate. Denudation processes beneath permanent snowfields may be minimal. Reduction of the area blanketed by snowfields would increase freeze-thaw frequency and associated periglacial mass movements, and would expose the hillslopes to occasional intense summer rainfall. Thus both periglacial processes and fluvial erosion would increase with the onset of Holocene climates. The density of vegetation below the altitude of the present tree line would change from bare ground beneath snow to alpine and subalpine plant communities. Increase of fluvial and periglacial erosion would tend to increase sediment yield and an increase in vegetative cover would tend to decrease yield. The overall effect

of changes in these variables may have been a modest increase in bedload sediment yield. Such a presumed increase in resisting power would be largely offset by increase in stream power because a larger proportion of the precipitation fell as rain instead of as snow. The effect of these offsetting changes in variables is indeterminate, but exposure of slopes to erosional processes may have resulted in a modest decrease in the stream power/resisting power ratio for the 10 percent of the drainage basin lying above 1200 m. Figure 5.15B suggests that this minor component of the total hillslope system may have had little effect on the stream subsystem in the piedmont reach.

The combined effects of climatic changes on the 90 percent (Fig. 5.15A) and the 10 percent (Fig. 5.15B) of the hillslope subsystem would be for slow degradation during times of transitional latest Pleistocene climates and rapid degradation during Holocene climates. The net effect (Fig. 5.15C) was the Stone Jug degradation event, the details of which are recorded by the flight of complex-response terraces in the piedmont reach.

The combined effects of concurrent increases in stream power and decreases in resisting power resulted in the threshold of critical power being greatly exceeded during the early and middle Holocene. The rate of stream-channel downcutting was largely a function of how far removed the stream system was from equilibrium (Table 5.5). Total downcutting was a function of how much a given reach of the river had been elevated by regional and local uplift since the time of type 1 dynamic equilibrium associated with the latest major strath beveling (just prior to the start of Flax Hills aggradation), and by deposition of silty gravel on the strath. Valley-floor degradation provided a new source of bedload, but it was insufficient to offset the major decrease of hillslope sediment yield.

### 5.3.4 TECTONIC CONTROLS

Relatively steep longitudinal profiles increase stream power and facilitate bedload transport. Faster uplift

TABLE 5.5 Summary of geomorphic responses to relative changes of late Quaternary climate in Charwell River fluvial system.

| Feature (from 0 ka, top, to 40 ka, bottom) | Climate[b] | | Hillslope Yields | | Stream Power/ Resisting Power | Comments |
|---|---|---|---|---|---|---|
| | Temperature | Precipitation | Sediment | Water | | |
| Present fluvial system | Warm | Wet | Small | Large | 1.0 | Basis for comparisons with prehistorical conditions |
| Stone Jug[a] degradation, middle Holocene | Warm | Wet | Small | Large | 2.0 | Frequent large floods |
| Stone Jug degradation, latest Pleistocene | Cool | Fairly dry | Fairly large | Fairly small | 1.1 | Slow downcutting |
| Stone Jug threshold | Cool | Fairly dry | Fairly large | Fairly small | 1.0 | 2-ky-long equilibrium |
| Stone Jug aggradation | Cold | Dry | Large | Small | 0.7 | Too dry for rapid aggradation |
| Stone Jug strath | Fairly cold | Fairly dry | Fairly small | Fairly large | 1.0 | Too dry for optimal strath erosion |
| Flax Hills degradation | Cool | Fairly wet | Fairly large | Fairly large | 1.5 | Only enough time for 2 or 3 terraces |
| Flax Hills aggradation | Fairly cold | Fairly wet | Very large | Fairly large | 0.5 | Optimal climate for periglacial processes |
| Basal Flax Hills beds | Cool | Fairly wet | Fairly small | Large | 0.9 | Slow deposition during transition |
| Flax Hills strath | Cool | Wet | Small | Large | 1.0 | Extensive straths |

[a] See Section 5.2.1 for definitions of names.
[b] Relative to the present climate.

257

of the watershed reach than the piedmont reach favored bedload transport in the Seaward Kaikoura Range and deposition of gravel downstream from the Hope fault for both the Flax Hills and Stone Jug aggradation events. A watershed reach that is rising three times as fast as the piedmont reach is much less likely to attain the base level of erosion. As a result of rapid uplift, the valley floor of the Charwell River is much narrower and five times steeper upstream than downstream from the Hope fault (aerial view of Figure 5.4). Aggradation in the downstream part of the watershed reach probably began later and stopped sooner than in the adjacent piedmont reach because tectonic controls tended to cause a larger stream power/resisting power ratio upstream from the Hope fault. Conversely, lesser piedmont gradients favored rapid deposition of thick gravel derived from the mountains. The amount of tectonically induced downcutting needed to achieve new base levels of erosion varied with uplift rates of different parts of

the piedmont reach. These rates range from 0.4 to 1.3 m/ky.

## 5.4 Responses of the Charwell River to Climatic Change

Process-response models for the stream subsystem can be evaluated now that we have preliminary ages for stream-terrace treads and for deposits in the piedmont reach, and have outlined possible interactions between geomorphic processes in the watershed during aggradation and degradation events.

Useful landforms (Fig. 5.16) for evaluating responses of the Charwell River—and many other streams—to late Quaternary climatic changes include:

1. Straths capped by massive gravels—referred to as major straths; this stream-terrace landform

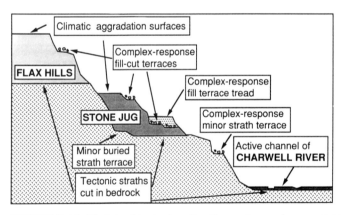

FIGURE 5.16 Sketch of tectonic, climatic, and complex-response terraces of the Charwell River. Late Quaternary geomorphic responses to climatic change and tectonic perturbations, together with complex-response adjustments, are recorded as stream terraces (instead of only as a stratigraphic record) because of the large vertical space provided by long-term tectonic and episodic depositional elevation of the stream channel. The Flax Hills and Stone Jug aggradation surfaces are treads of major fill terraces that record the ends of climatically-induced aggradation events. Complex responses are preserved as flights of minor strath, fill-cut, and fill terraces that record pauses in degradation from the level of an aggradation surface to the subsequent tectonic strath terrace. Each tectonic strath forms during a period of dynamic equilibrium after tectonically induced downcutting lowers the longitudinal profile of the river to a new base level of erosion.

records the end of degradation events and represents time of attainment of type 1 dynamic equilibrium. Major straths are the fundamental tectonic stream-terrace landform because their spacing is controlled by magnitudes of tectonically induced downcutting between times of attainment of the base level of erosion.

2. Aggradation surfaces that occur as major fill terraces and as alluvial fans. Such surfaces are the fundamental climatic stream-terrace landform because they mark the end of climatic-change-induced deposition of massive gravels.

3. Flights of degradation terraces—referred to as minor straths and fill-cut terraces. These complex-response terraces generally mark pauses in stream-channel downcutting between the end of an aggradation event and the next time of major strath cutting.

The Charwell River repeatedly passes through equilibrium-aggradation-degradation cycles. Equilibrium ends when a major strath is buried by gravel. Cessation of deposition leaves an aggradation surface as the stream crosses the threshold of critical power and degrades until the base level of erosion is attained again (Fig. 5.16). Each successive major strath is lower than the preceding one by an amount equal to the intervening tectonically induced downcutting. In this way tectonic inputs are modulated by dominant climatic controls.

Information and concepts from the preceding sections can be used to summarize the geomorphic responses to climatic change. Aggradation occurred during times of full-glacial global climate, and major strath cutting occurred near the end of times of interglacial global climate. Cutting of major straths seems to have occurred at about 50 to 48, 41 to 39, 30 to 28, and 4 to 0 ka. Synchronous deposition of alluvial fans and valley-floor aggradation occurred in the piedmont reach during about 48 to 43 and 38 to 31 ka (Flax Hills) and 26 to 14 ka (Stone Jug). Flights of degradation terraces were cut between 31 and 29 ka (two terraces) and between 14 and 4 ka (10 terraces).

Deposition of the diachronous Dog Hills alluvium occurred throughout the late Quaternary as random brief episodes of aggradation from small unstable watersheds in the crush zone of the Hope fault.

This section is primarily concerned with the Stone Jug degradation event. It emphasizes factors that affect the magnitudes and rates of renewed downcutting after the Stone Jug aggradation event. Magnitudes of stream-channel downcutting were a function of (1) the thickness of valley fill that accumulated during the aggradation event and (2) the amount of uplift that occurred since cutting of the Stone Jug major strath.

### 5.4.1 CLIMATE-CHANGE INTERRUPTIONS OF TECTONICALLY INDUCED DOWNCUTTING

Aggradation events temporarily reverse the tectonically induced trend of progressive erosional lowering of valley floors. Degradation was delayed 7 and 12 ky in the piedmont reach by the Flax Hills and Stone Jug aggradation events. Renewed downcutting started at a higher altitude than that of the preceding base level of erosion because the streambed had been raised by an amount equal to the depositional base-level rise. Deposits of 15 to 50 m thick above the Flax Hills and Stone Jug straths are functions of the strengths and types of climatic perturbations and of the time spans during which watershed hillslopes furnished gravel in amounts that exceeded the bedload transport rate in piedmont reaches.

Large variations in the altitude of the valley floor during the past 29 ky can be illustrated by a single Stone Jug outcrop (Fig. 5.17). After the Stone Jug major strath was beveled, the active channel rose about 30 m, after which it declined about 70 m. About 23 m of uniformly massive, medium-grained, silty gravel above the Stone Jug major strath is capped by 1 to 2 m of well-sorted and bedded bouldery gravel of a degradation terrace that was cut into the alluvial fan of the Stone Jug aggradation event. Degradation removed about 7 m of Stone Jug alluvial-fan deposits before the time of static equilibrium, represented by the boul-

T1 tread & soil

lower 23 m of 36 m of aggradation gravels

Strath

Soft bedrock

River on strath

FIGURE 5.17 View of an 11-ka fill-cut terrace in the main branch of the Charwell River, 1 km upstream from the junction with the right branch. About 23 m of gravel lies on a major strath believed to have formed at 29 ± 2 ka. The 39 m between the buried and present straths reflects tectonically induced river downcutting during the past 29 ky (inferred uplift rate of 1.3 m/ky).

der bed on the tread of the fill-cut terrace. The well-sorted and bouldery deposits of the fill-cut terrace represent a much different streamflow regime than that of the underlying deposits. Latest Pleistocene and Holocene degradation downstream from the Hope fault totaled 65 to 75 m; about 26 to 36 m was through Stone Jug aggradation gravels, and 39 m was through soft sheared sandstone beneath the 29–ka strath.

Holocene degradation continued far below the level of the Stone Jug strath. It reveals the pervasive influence of long-term uplift on stream-channel downcutting as the Charwell River adjusted to the effects of the climatic perturbation and reestablished its base level of erosion. About 39 m of tectonically induced downcutting between the two times of attainment of type 1 dynamic equilibrium, indicated by the 29 ka and 0 ka straths, implies a long-term uplift rate for this reach of 1.3 m/ky.

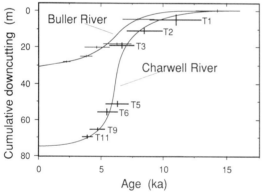

FIGURE 5.18 Generalized downcutting curves for the Buller and Charwell rivers. The Buller River site is 80 km northwest of the Charwell River and is west of the Alpine fault. The crosses mark the positions of degradation terraces formed during brief pauses in downcutting when static equilibrium was attained; the lines through the crosses show the estimated degree of uncertainty as to terrace ages and positions below the aggradation surface. Modified from Figures 23 and 24 of Knuepfer, 1984, using Knuepfer, 1988, data set.

Aggradation events also interrupted of degradational processes in the watershed reach; the stream could not continue to downcut in narrow canyons or erode fault scarps. Surface ruptures along the Hope fault during Stone Jug valley backfilling ruptured both bedrock and the overlying alluvium. Streamflows quickly eradicated scarps formed in loose gravel, but the underlying suballuvial bedrock fault scarp became progressively higher until subsequent Holocene degradation exposed bedrock once more. The time span between the onset of aggradation and the time at which the river had degraded the valley fill sufficiently to expose bedrock again was about 17 ky (roughly 26 to 9 ka). Cumulative displacement by the Hope fault during these 17 ky probably was large:

$$17 \pm 3 \text{ ky} \times 2.5 \pm 0.3 \text{ m/ky} = 42.5 \text{ m} + 13.5 \text{ m} (5.2)$$
$$1 - 11.5 \text{ m}$$

Thus a prominent bedrock waterfall was exhumed during the Stone Jug degradation event which, after exposure, could retreat upstream from the Hope fault as a nickpoint or nickzone. A presently anomalous reach upstream from the Hope fault departs from a smooth longitudinal profile by about 40 m (Bull & Knuepfer, 1987), which is in agreement with the values of mountain uplift rates and delay in bedrock fault-scarp degradation used in Equation 5.2.

### 5.4.1.1 Catching Up to the Base Level of Erosion

Ages and heights of Stone Jug degradation terraces have been used to reconstruct the degradation history since 14 ka as the Charwell River caught up with the total amount of post-29-ka uplift of the piedmont reach. Downcutting curves for the Charwell and Buller rivers (Fig. 5.18) are generalized because each terrace tread used for age control represents a brief pause in degradation (Section 5.4.4). Initially slow degradation accelerated rapidly by the middle Holocene and then decreased during the late Holocene as a new base level of erosion was approached. Total downcut-

FIGURE 5.19 Charwell River near Flax Hills farm beveling a major strath 17 m below the Flax Hills strath. The highest terrace is the tread of the 14-ka Stone Jug aggradation surface, which is 62 m above the active channel. The 11-ka cut terrace is at upper right, below which are several other Stone Jug fill-cut and strath terraces.

ting was about 65 to 75 m for the Charwell River and about 32 m for the Buller River.

The middle Holocene period of maximum downcutting rates appears to coincide with a climate that may have been 1°C warmer than at present (Table 5.1). Two factors seem to have been important during this climatic optimum. The incidence of intense rainfalls derived from tropical moisture sources may have been even greater than at present. Treeline probably was raised. Even dense plant cover can not reduce runoff during extreme storm events (Costa & Baker, 1981), but it can reduce sediment yields. Larger and more intense rainfalls together with increased plant cover may have caused the mid-Holocene to be a time

of maximum departure from the threshold of critical power.

### 5.4.2 SYNCHRONOUS AND DIACHRONOUS DEGRADATION TERRACES

The first Stone Jug degradation terrace (Figs. 5.17, 5.19) is synchronous along the Charwell River but is diachronous along the tributary piedmont streams. The terrace is easy to identify because it is well preserved and is the first terrace below the Stone Jug aggradation surface. Cobble weathering rind ages along the Charwell River range from 10.8 to 11.8 ka, or about 11.3 ± 0.5 ka, which is within the dating uncertainty level of ± 1.9 ka.

The first degradation terrace tread on Dillon Creek, 1 km upstream from the junction with the Charwell, has a cobble weathering rind age of $5.3 \pm 0.7$ ka. This age clearly is diachronous when compared with the 10.8 to $11.8 \pm 2.0$-ka ages of terrace treads in the same topographic position in the flight of the Stone Jug degradation terraces along the river.

Both times and processes of terrace-tread incisement may be different for tributary streams than for the trunk stream. Stone Jug degradation terraces could not form along tributary streams until after degradational base-level fall had occurred along the trunk stream. Degradation by Dillon Creek into Stone Jug gravels deposited at its mouth would then migrate upstream from the junction with the Charwell. It would proceed slowly because of the small stream power produced by this minor tributary watershed. For these reasons the effects of the Charwell River erosional base-level fall were delayed roughly 6 ky for the reach 1 to 2 km upstream from the mouth of Dillon Creek. Response times to base-level fall at the basin mouth are so slow in the Dillon Creek system that degradation terraces have yet to form in the headwater's third of the low relief watershed.

### 5.4.3 STATIC EQUILIBRIUM DEGRADATION TERRACES

A dozen Stone Jug degradation terraces (defined as fill-cut terraces where formed in alluvium and as minor strath terraces where formed in bedrock) are inset below the 14-ka aggradation surface in the piedmont reach (Figs. 5.19–5.22). These are paired minor complex-response (Section 1.7) terraces. They are much different from the primary climatic and tectonic terrace landforms of aggradation surfaces and of major straths beveled at times of attainment of type 1 dynamic equilibrium. Each degradation terrace tread represents brief attainment of static equilibrium followed by resumed downcutting. This section explores the question of how equilibrium is achieved before

the river has attained type 1 dynamic equilibrium conditions.

Pauses in degradation may result from many factors that change stream power or resisting power or both. During the Holocene the stream subsystem tended to be strongly to the degradational side of the threshold of critical power, but temporary perturbations could decrease the stream power/critical power ratio and thereby move the subsystem closer to equilibrium. Decrease in rainfall would decrease runoff and stream power. Brief periods of colder climate might temporarily increase sediment yield as periglacial processes again became moderately important in the highest parts of the drainage basin. Coseismic landslides might introduce such large volumes of additional bedload into the stream subsystem that any pre-landslide excess of stream power would be needed to transport additional bedload rather than doing degradational work. Although such perturbations cannot be eliminated as possible causes for one or more of the minor strath degradation terraces, the field evidence suggests a more general explanation that does not rely on perturbations external to the fluvial system.

Cycles of internal adjustment within the stream subsystem–complex responses–can account for flights of static equilibrium degradation terraces. Each degradation terrace may have been associated with self-arresting feedback mechanisms that increased resisting power. Poor sorting of the gravels during aggradation suggests rapid deposition by braided streams that were unable to winnow and sort the sediment. Degradational terrace gravels represent a much different depositional environment. Although particle size varies between locations, most remnants are capped by gravel that is substantially coarser, and better sorted, than the underlying massive silty gravel. The capping gravel bed at the Figure 5.17 locality has abundant 0.3- to 1-m-boulders. Mosaics of surficial boulders on this and virtually all other degradation terrace treads are best interpreted as being interlocking streambed armor. Remnants of streambed armor

FIGURE 5.20 Stream terraces at the mouth of the Charwell River basin that are offset by right-lateral and vertical movements along the Hope fault. Long-term lateral offset rates are 33 ± 3 m/ky, which is an order of magnitude larger than the 2.5 ± 0.3 m/ky uplift rate of the drainage basin reach relative to the piedmont reach.

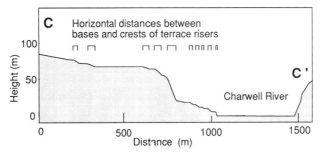

FIGURE 5.21 Cross-valley profile of the piedmont reach of the Charwell River 200 m downvalley from the Hope fault. Sloping surface at upper left is the trailing edge of a Stone Jug alluvial fan; lower surfaces are the cut and strath terraces of the Stone Jug degradation event. A broad major strath is 0.2 to 1.5 m beneath the active channel of the Charwell River. Location of the C–C' line is shown on Figure 5.2. From Figure 22 of Knuepfer, 1984.

264

FIGURE 5.22 Vertical aerial photograph of the mountain–piedmont junction of part of the Seaward Kaikoura Range. The presently entrenched main branch of the Charwell River (upper right) and right branch (center) widen abruptly and are beveling straths downstream from the Hope fault. These reaches have attained type 1 dynamic equilibrium. Distance from the mountain front to the junction of the two forks is 1.9 km.

on degradation terraces attest to an environment of ample stream power during a period of decreasing yield of gravel from hillslope sediment reservoirs.

Winnowing of alluvial gravels during periods of degradation of valley fill by small to moderate streamflows would be accompanied by input of additional cobbles and boulders introduced from upstream reaches. Sand and fine gravel would be

transported through the reach, but large gravel clasts would become concentrated on the streambed. This streambed armor would greatly affect bedload transportation and channel degradation rates (Gomez, 1983; Brayshaw, 1985; Dietrich et al., 1989). Resisting power would be increased by (1) the increase in size of material to be entrained by streamflows from the streambed, (2) closer spacing of large particles, and (3) increases in hydraulic roughness. Visual roughness estimates of $0.025 \pm 0.004$ for the moderately smooth present streambed (which is a post-Cyclone Alison feature) are less than the estimated roughness of $0.035 \pm 0.004$ for remnants of paleostreambed armor. Studies of the effects of the spacing of large particles on streambeds (Leopold et al., 1966; Leopold & Emmett, 1976) indicate that stream power at least one order of magnitude greater is required to entrain cobbles touching one another compared with cobbles spaced more than 8 diameters apart. An additional self-arresting feedback mechanism would be concurrent growth of riparian vegetation as parts of the active channel became stable; this would further increase hydraulic roughness and strength of streambed materials.

It may be hypothesized that winnowing of streambed gravels during degradation occurs during times of moderate-sized streamflows. Self-arresting feedback mechanisms increase resisting power until this power equals stream power for the spectrum of discharges available to transport bedload. Decrease of degradation rates is accompanied by increasing rates of lateral cutting. A small cut surface or strath is formed which becomes a stream-terrace tread upon renewed stream-channel downcutting.

Renewed degradation may have been initiated by a flow event sufficiently large to disrupt and destroy streambed armor and riparian vegetation. This may have been the infrequent tropical storm, such as Cyclone Alison in 1975. Renewed degradation after major flow events would be accompanied by concurrent winnowing of a new streambed armor. Closely spaced major storm events would not allow sufficient time for streambed armor to form and downcutting would continue at a rate commensurate with the magnitude of excess stream power relative to resisting power. Storm events with return periods longer than 0.5 ky would allow sufficient time for formation of streambed armor and lateral cutting by the stream.

Thus, without tectonic perturbations, secular climatic changes, or seismogenic landslides, a flight

of minor strath and fill-cut terraces could be formed before the Charwell River had attained type 1 dynamic equilibrium. In part, this is an example of lithologic control of geomorphic processes. Few, if any, terraces would have been formed if the hillslope subsystem had supplied nothing coarser than sand, or if large boulders were so abundant that they were always common on streambeds. The Stone Jug degradation terraces partially illustrate the variety of adjustments of variables—complex responses—that allow attainment of static equilibrium in streams. Hydraulic roughness is sufficiently important that it can control the times of formation of minor fill-cut and strath terraces. Slope, as defined by the base level of erosion at times of type 1 dynamic equilibrium, is an important controlling variable for major strath terrace formation.

### 5.4.4 CONCLUSIONS

#### 5.4.4.1 Late Quaternary Variations of Stream Power and Resisting Power

Data and conceptual models for the Charwell River provide an opportunity to estimate and compare relative magnitudes of stream power and resisting power for the piedmont reach since 31 ka (Fig. 5.23). This time span includes four chapters of Stone Jug fluvial history—major strath cutting, aggradation event, degradation event, and major strath cutting. Vertical separations between the two plots provide a visual impression of how far removed the piedmont reach of the stream was from equilibrium or threshold conditions during times of aggradation and degradation. Stream power and resisting power plots are horizontal and equal during times of type 1 dynamic equilibrium and major strath cutting at approximately 30 to 28 ka and 4 to 0 ka. The higher position for the 4 to 0 ka equilibrium period acknowledges the presumed greater stream power at present (Table 5.5). A single threshold of critical power is indicated where the two plots cross in a 2-ky-long equilibrium period at about 14 ka.

Nonequilibrium stream behavior in the pied-

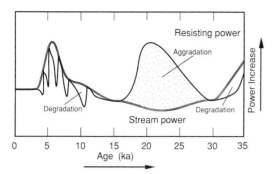

FIGURE 5.23 Changes in the relative magnitudes of stream power and resisting power of the Charwell River during the past 30 ky.

mont reach (Fig. 5.23) prevailed during an aggradational mode of operation that persisted from approximately 26 to 14 ka (calendric ages of approximately 29.0 to 16.6 ka using the corrections of radiocarbon ages described by Bard et al. [1990]). Stream power decreased with the onset of a cold dry climate that caused full-glacial conditions. Reductions in annual precipitation and in the proportion of precipitation falling as rain combined to reduce streamflow peak discharges substantially—and thus stream power. Concurrent late Pleistocene increases in resisting power were due mainly to increases in bedload derived from hillslopes characterized by moderately active periglacial processes. Maximum aggradation rates occurred during full-glacial times. This is not a definitive type of parameter that can be easily recognized in the stratigraphic record.

The transition to warmer and wetter climates during the latest Pleistocene reversed the decrease in stream power, but the aggradational domain continued until stream power exceeded resisting power at a crossing of the threshold of critical power during the 16- to 14-ka time span. Sediment yields also decreased with the onset of transitional latest Pleistocene climates.

The threshold crossing associated with the fill-terrace tread—the Stone Jug aggradation surface—occurred during the latest Pleistocene, by which time as much as 30 m of aggradation gravels had accumulated downstream from the Hope fault. Deposition of alluvial fans ceased adjacent to the Hope fault, but terminal aggradation rates are unknown. The threshold crossing appears to have been gradational. The evidence for a gradational threshold comes from a reach 5 to 7 km downstream from the mountain front. The Charwell River impinged on a hillside of massive, soft sandstone and cut a 150-m-wide strath capped only by 1 to 2 m of Stone Jug deposits and soil profile. The strath-terrace tread is at the same height above the active channel as the Stone Jug aggradation surface on the opposite side of the valley. Such an unusual strath could be cut only if neither aggradation nor degradation prevailed for about $2 \pm 1$ ky at the transition between the aggradational and degradational domains.

The post-14-ka domain has a more complex pattern that reflects about 65 to 75 m of intermittent channel downcutting. Stream power remained high during times of transitional and Holocene climate relative to that of the late Pleistocene. Stream power became larger because of increased precipitation, and because more of the precipitation fell as rain instead of snow. The mid-Holocene warm period between 8 and 5 ka represented a climatic optimum that further accentuated the differences between stream power and resisting power. The time span between about 7 and 6 ka is especially noteworthy because more than 30 m of degradation occurred during this millennium. One can speculate that such conditions may have been associated with an increased incidence of major storm runoffs associated with tropical moisture sources. The overall trend during the Holocene may have been for progressively decreasing sediment yield as detritus in the hillslope sediment reservoir tended to thicken, thereby favoring increase in density of forest vegetation.

The magnitude of the difference between stream power and resisting power since 14 ka is directly proportional to the slopes of the degradation curves for the Charwell River (Fig. 5.18). Stream power and resisting power were the same during the equilibrium period of threshold crossing at about

16 to 14 ka, and stream power probably did not greatly exceed resisting power between 14 and 10 ka. Stream power was much greater than resisting power between about 7 and 6 ka; the river downcut 32 m during this millennium. Downcutting rates progressively decreased during the late Holocene and the streambed has been lowered only about 5.0 m during the past 3.9 ky—a mean rate of 1.3 m/ky, which is the same as the long-term estimated uplift rate for the piedmont reach near the Hope fault (Fig. 5.17). We do not know if downcutting rates were uniform; for example, they may have been 2 m/ky from 3.9 to 2 ka and then zero since 2 ka.

Overall decrease of bedload being supplied to the stream was the primary cause of lesser resisting power during the Holocene, but resisting power fluctuated sharply. It increased during times of streambed armoring and riparian plant growth, which increased hydraulic roughness and the shear stress needed to entrain streambed materials. Each postulated major flood discharge that disrupted streambed armor caused an abrupt decrease in resisting power. Resisting power gradually increased again during the next period of renewed degradation and concurrent renewal of streambed armor.

### 5.4.4.2 Climatic and Tectonic Controls on Stream Behavior in the Piedmont Reach

The Charwell River fluvial system is sensitive to tectonic and climatic perturbations and has numerous complex-response terraces (Fig. 5.24). Termination of each episode of valley-floor alluvial backfilling created a climatic stream terrace—an aggradation surface. It is only near the mountain front of the Seaward Kaikoura Range that the spatially limited Stone Jug aggradation rose high enough on some streams to spread as alluvial fans over Flax Hills surfaces. The stream in the piedmont reach spends most of its time aggrading or in catching up with new base levels of erosion because of a combination of rapid uplift and large amounts of intermittent aggradation. In order to occasionally reach a new base level of erosion the stream had to degrade through the valley fill, and then through an increment of bedrock equal to the amount of uplift since the last time that stream had attained the base level of erosion. It barely had enough time to bevel a new tectonic strath after attaining the base level of erosion, because of the onset of a new aggradation event.

Flights of degradation (complex-response) terraces formed after initial incisement of the aggradation surfaces. The degradation part of the Flax

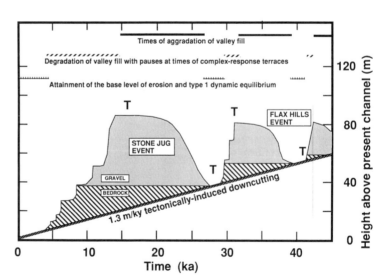

FIGURE 5.24 Summary of combined influence of tectonic and climatic controls on the late Quaternary behavior of the Charwell River as reflected by changes in streambed altitude.

Hills event was so short that only two or three cut terraces were formed; in Figure 5.24 these are inferred from downstream reaches. The degradation part of the Stone Jug event continued for 10 ky, which allowed time for a dozen fill-cut and strath terraces to form. Only the ones dated by Knuepfer (1988) are shown. The types, ages, and heights of complex-response terraces do not appear to correlate with flights of degradation terraces in nearby watersheds.

Times of fill-terrace tread formation, such as at 31 ka, typically are abrupt thresholds separating aggradation from degradation. Aggradation rates may decrease but the switch to the degradational mode tends to be abrupt. For minor pulses of aggradation by small streams during historical times, crossing of the threshold of critical power has been associated with stream-channel incisement during a flood. Flood events at key times may (1) reduce the availability of bedload from the hillslope sediment reservoir to a point where an aggradational mode of operation can no longer be maintained and (2) confine future streamflows sufficiently that a self-enhancing feedback mechanism is established that promotes continued degradation (Baker et al., 1988; Bull, 1988). It probably is unusual for an aggradation event to end as an equilibrium period, as happened during about 2 ky from 16 to 14 ka.

Geomorphic responses to relative changes of late Quaternary climate and uplift in the Charwell River fluvial system are sufficiently complex that only general conclusions are made in Table 5.5. Only two of many climatic factors are estimated, and then only in general terms. It would be preferable to also consider climatic variables such as windiness, frosts and length of growing season, and topographic controls affecting accumulation and melting of snow. Inferences about water and sediment yield are a step further from being measured directly. Considering the generality of these estimates of geomorphic variables, it would be unwise to attempt quantitative estimates of climate-change-induced variations of bedload transport rate, discharge of water, and hydraulic roughness. In a general sense, fluvial system behavior—in both humid and arid regions—is so indeterminate as to preclude all but general predictions of future trends or approximate speculations of past behavior. Such discussions attempt to focus on the most likely or reasonable interactions of system variables. Fortunately the threshold of critical power is defined in terms of combinations of key variables whose changes are observable as landscape changes. The stream power/resisting power ratio provides a numerical index for expressing opinions about how far removed the stream subsystem was from equilibrium or threshold conditions (a value of 1.0). Maximum departure from threshold or equilibrium conditions occurred during Flax Hills aggradation (a suggested ratio of 0.5) and during mid-Holocene Stone Jug degradation (a suggested ratio of 2.0). At other times the fluvial system was close to threshold or equilibrium conditions: barely degrading during the latest Pleistocene climatic transition and during the late Holocene, and slowly aggrading during the initial phases of Flax Hills aggradation.

This concludes the discussion of geomorphic responses to climatic change in four study areas whose present climate ranges from strongly seasonal hyperthermic and extremely arid to moderately seasonal mesic and humid. The conceptual models of Chapter 1—especially the threshold of critical power—have been used to evaluate the markedly different geomorphic responses to climatic change described in Chapters 2 through 5. The stage is now set for comparison of the geomorphic responses to climatic change in these four diverse settings.

# 6

# Different Responses of
# Arid and Humid Fluvial Systems

The diverse impacts of latest Pleistocene–Holocene–climatic changes on fluvial systems were discussed in Chapters 2, 3, 4, and 5 in a sequence from relatively simple to complex geomorphic settings. Most drainage basins in the Mojave Desert (Chapter 2) are monolithologic, are tectonically inactive, and have a small enough altitude range that only one climatic vegetation zone need be considered in hillslope process-response models. The Nahal Yael basin (Chapter 3) is much the same except that it provides an opportunity to examine the effects of climatic change in a drainage basin underlain by three rock types with different sensitivities to such change. Some watersheds in the San Gabriel Mountains (Chapter 4) are multilithologic, but geomorphic responses to climatic change in this lofty range are complex because of multiple climatic vegetation zones and rapid uplift. Although the Charwell River basin (Chapter 5) seems to be underlain by one petrographic rock type, responses of the fluvial system to climatic change are complex because of (1) shearing and fracturing of bedrock in the wide crush zone of the Hope fault, (2) different rates of uplift in the watershed and piedmont reaches,

(3) multiple climatic vegetation altitude zones, and (4) rapid horizontal offset of the watershed reach from the piedmont reach by the highly active Hope fault. The intent of this chapter is to compare and contrast the geomorphic responses to climatic change in these four diverse study areas. First, let's compare differences in the types of climatic change.

The Pleistocene–Holocene transition (see Table 1.2 for definitions of temporal terms) was a time of shifting airmass circulation boundaries. In the Charwell area the maritime airflow remained about the same, but in the other study areas airmass circulations underwent changes.

The Mojave Desert presently receives storms from a winter precipitation regime from the north that is directed by the prevailing westerlies and from a summer precipitation regime from the south that is a combination of monsoonal-type airflow and tropical Pacific storms. In much of North America, late Pleistocene insolation changes affected climatic change only after modulation from the effects of continental ice sheets. During full-glacial times only the winter regime was present, but summer monsoonal rains returned 3 to 4 ky before the onset of Holocene aridity. The fluvial systems responded with two aggradation events; one during the latest Pleistocene and the other during the early Holocene.

In much of North Africa and the Middle East, times of wetter climates occurred during the early Holocene, presumably because of increases in both monsoonal and winter rains. Holocene expansion of African, Mid-East, and Indian lakes probably was a sensitive index of intensity of monsoonal airmass circulation, which is controlled directly by astronomically induced changes in solar insolation (Wright, 1984). The time of subsequent valley-floor aggradation at Nahal Yael was associated with climate-induced reductions in plant cover that occurred during the mid-Holocene.

Changes of late Quaternary airmass circulation patterns in the San Gabriel Mountains may have favored more tropical storms during the mid-Holocene as nearby oceans became warmer, but variations in the strength of the winter storm regime may have been the most important climatic element throughout the late Quaternary.

## 6.1 Weathering and Soil-Formation Rates

Increase in areas of exposed bedrock on hillslopes undergoing accelerated erosion seems to have been a function of degree of aridity as well as of joint abundance and uplift rates. In order of increasing tendency to expose bedrock as a result of climatic change, the four study sites are Charwell River, San Gabriel Mountains, Mojave Desert, and Nahal Yael. Extremely arid hillslopes tend to have thin, patchy soils, and humid (see Table 2.1 for definitions of climatic terms) hillslopes have a tendency to develop thick and extensive colluvial mantles. The volume of material that must be removed for outcrops to be completely exposed probably is an order of magnitude less in extremely arid than in most humid settings. Volume of detritus stored in hillslope sediment reservoirs also increases with increase in joint spacing and decrease in uplift rates.

Weathering and soil-profile formation rates are much more rapid in humid than in arid regions. Weathering rinds on surficial basaltic boulders are useful for dating some arid alluvial geomorphic surfaces in the Mojave Desert that formed during the past 700 ky (Section 2.4.1.2.2). Weathering rinds on surficial graywacke cobbles on the humid terrace treads of the Charwell River form so rapidly that their usefulness is limited to the past 20 ky (Section 5.2.3.1). Rapid rates of cobble weathering-rind development were essential for dating the details of the Charwell degradation-terrace flight, and slow rates were valuable for evaluating frequencies of aggradation events since the early Pleistocene in the Mojave Desert. The key to useful age control is to select dating methodologies that are appropriate for the problem being studied.

In most study areas a dating methodology for stream terraces can be devised that suits the range of ages of interest to the investigator, especially since refinement of techniques utilizing cosmogenic isotopes and thermoluminescence for dating of terraces (Pavich et al., 1986; Pavich, 1987; Philips et al., 1986, 1990; Cerling, 1990; Forman, 1989). Different soils parameters were used for mapping alluvial geomorphic surfaces in the San Gabriel Mountains (iron oxyhydroxides), the Mojave Desert (calcium carbonate), and the Charwell River (aluminum oxides). Rock varnish is excellent for dating terrace treads in hot deserts, but cobble weathering rinds are more appropriate in the humid realm.

Although the soil-profile index was devised with the intent of providing numerical comparisons of field properties of soils chronosequences (Harden, 1982), it may also be influenced by climate. Markedly different increases in rates of soil development for the past 20 ky are summarized by graphs of soil-profile development indices in Figure 6.1. For extremely humid mesic Franz Josef valley, index values are high after only 0.1 ka but then increase comparatively slowly. For the arid hyperthermic Whipple Mountains piedmont, index values are initially low but then increase comparatively rapidly. The curves for intermediate cli-

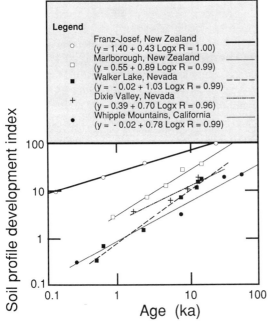

FIGURE 6.1 Variations of soil-profile indices with climate on the gravelly treads of late Quaternary stream terraces. The parent material and present climate for each of the five chronosequences are: Franz Josef—schist, weakly seasonal extremely humid and moderately seasonal mesic; Marlborough—greywacke, moderately seasonal subhumid and moderately seasonal mesic to frigid; Dixie Valley and Walker Lake—granitic, moderately seasonal arid and strongly seasonal mesic; Whipple Mountains—gneiss and schist, strongly seasonal arid and strongly seasonal hyperthermic. Data are from the following sources: New Zealand, Knuepfer (1984, 1988); Dixie Valley, Hecker (1985); Walker Lake, Demsey (1987); and Whipple Mountains, Figure 2.33 in this book.

matic settings occupy intermediate positions and may merge for soils older than 100 ka. The family of curves reveals major differences in both coefficients and exponents of the power functions. Spatial climatic variations of the relation between age and the soil-profile index imply that temporal variations in rates of soil-profile development occurred at sites where climate changed during the

late Quaternary. Climatically controlled differences in weathering rates also influence the availability of sediment that can be yielded from hillslopes in both a temporal and spatial sense.

## 6.2 Climatic and Tectonic Stream Terraces

Aggradation and degradation—these are the processes by which streams depart from equilibrium conditions and initiate processes that lead to the creation of stream terraces. Some of the complexities of stream terraces can be understood through application of the concepts of Chapter 1—tectonically induced downcutting, base level of erosion, complex response, threshold of critical power, diachronous and synchronous response times, and static and dynamic equilibrium. With neither climatic nor tectonic perturbations, streams would cut down to only a single base level of erosion and few, if any, terraces would form. Changes in climatic and tectonic controls make streams far more interesting. These changes in independent variables are responsible for paired climatic and tectonic stream terraces, whose formation sets the stage for the formation of flights of paired or unpaired complex-response terraces. This discussion begins with a summary of these fundamental types of stream terraces and then considers the influences of several climatic settings on genesis of terraces.

Streams may be portrayed as passing through a sequence of stages in response to tectonic or isostatic uplift, or to climatic perturbations. Where stream power, uplift rates, and erodibility of materials permit, streams degrade, achieve, and maintain type 1 dynamic equilibrium. Lateral erosion of bedrock becomes the dominant process and may continue long enough to create extensive beveled surfaces beneath active channels. With sufficient time it may even lead to the formation of pediments. Such major straths are the fundamental tectonic stream-terrace landform.

A stream that maintains type 1 dynamic equilibrium passes through infinite base levels of erosion.

Ideally, no terraces would form, but in the real world slight variations in short-term uplift and stream-channel downcutting rates may result in several low (1- to 5-m-high) strath terraces. These low terraces may be considered tectonic terraces where heights and ages of strath surfaces indicate a mean rate of tectonically induced downcutting that equals the long-term uplift rate. Conditions at each site will dictate whether such terraces should be grouped as a single irregular base level of erosion (a tectonic strath) or split out as separate small tectonic terraces.

Beveling continues until climatic or tectonic perturbations upset the prevailing type 1 dynamic equilibrium. For example, local faulting or folding that raises an equilibrium reach of a stream causes accelerated stream-channel downcutting, which continues upstream from the base-level fall until a new base level of erosion is established (Fig. 4.24). Remnants of the former base level of erosion comprise a major strath terrace that is mantled with a thin blanket of gravelly cutting tools. A climate-change-induced aggradation event may also terminate major strath formation, but in this case the tectonic landform is buried beneath aggradation gravels (Fig. 4.15).

Aggradation events represent brief reversals of long-term trends of tectonically induced downcutting. Aggradation generally is caused by inability of a stream to transport all of its bedload due to decrease in stream discharge, increase in amount and size of gravel, or both. Thicknesses of the valley-floor alluvium provide relative measures of the strength of successive climatic perturbations on a fluvial system, or can be used to compare the same event in adjacent drainage basins. Such geomorphic responses to climatic change are modulated by the sensitivity of the rock types that underlie the source watersheds (Chapter 3). Aggradation of climate-change-induced deposition of valley-floor fills typically accelerates and then decelerates before ceasing when the threshold of critical power is crossed (Fig. 4.29). Initiation of degradation leaves an aggradation surface, this type of fill-terrace tread is the fundamental climatic

stream-terrace landform. Aggradation surfaces may be buried by subsequent episodes of deposition unless intervening tectonically induced downcutting lowers the active channel sufficiently for younger aggradation surfaces to form below older surfaces. Flights of fill terraces with parallel treads are common. Such parallelism of longitudinal profiles is suggestive of similar hydraulic conditions during times of maximum valley-floor aggradation. The combinations of interacting variables may have been different, but each parallel tread records a similar net effect or behavior at the times of crossing of the critical power threshold.

Space for formation of flights of complex-response terraces (Section 1.7) is created either by climatically induced aggradation or by uplift. Either local base-level process may raise an active channel above the previous base level of erosion. The space for complex-response terraces is equal to the difference between the altitude of the tread of the tectonic or climatic terrace and the longitudinal profile of the stream defined by type 1 dynamic equilibrium after completion of the degradation event. The space is equal to the thickness of valley fill along tectonically inactive streams, and is equal to fill thickness plus tectonically induced downcutting where streambeds are being raised. These relatively minor terraces may be young or old, low or high, and limited or extensive. Pauses as streams degrade intermittently are the times of formation of complex-response fill-cut, strath, or fill terraces.

Plentiful good exposures and opportunities for dating deposits and terrace treads help distinguish static equilibrium complex-response terraces, such as those described in Section 5.4.3, from type 1 dynamic equilibrium tectonic straths formed when the stream is at the base level of erosion. Equilibrium terraces formed in alluvium deposited on top of the most recent tectonic strath clearly are complex-response terraces, because fill-cut terraces are above the altitude of the stream at the time of initiation of the most recent aggradation event.

Strath terraces above a modern tectonic strath most likely are complex-response strath terraces if they are of limited areal extent, but additional information about rates of processes is needed in order to be sure. Rates of degradation determined by ages and altitudinal separations of degradation terrace treads are especially useful. Strath terraces may be regarded as complex-response terraces where rates of degradation between times of terrace-tread formation (Fig. 5.18) exceed the long term uplift rate for the reach based on ages and differences in altitude of tectonic terraces. In such cases type 1 dynamic equilibrium had yet to be attained at the times of formation and the strath surfaces should be regarded as minor (complex response) terraces.

Relative abundances of climatic, tectonic, and complex-response terraces vary with climate in the four study areas, primarily as a function of stream power. Major fill terraces and an occasional complex-response terrace are typical of deserts in the American southwest and the Middle East. Total stream power per 1,000 years in these hot deserts is limited, so most streams cut through valley fills but generally lack the time or size to bevel obvious tectonic straths in response to the short term climatic fluctuations of the past 130 ky. Exceptions occur in two types of situations. One is along large rivers such as the Colorado River where it flows through the Mojave Desert. The other is where soft bedrock is present such as on the east side of the Sinai Peninsula where soft Nubian Sandstone is abraded easily by bedload consisting of granitic boulders. Tectonic stream terraces may be common in arid regions but generally require at least a million years to form in hard rocks: one or more levels of pediments record prolonged attainment of the base level of erosion.

The San Gabriel Mountains and the Charwell River have remarkably similar climatic and tectonic stream terraces because total stream power per 1,000 years is very large, watershed lithologies are sensitive to climatic change, and streambed bedrock is fairly soft. The Charwell River has more complex-response terraces than any other site because of a combination of factors that probably include (1) an appropriate range of clast sizes

in the bedload to permit episodic streambed armoring, (2) the highest total stream power of the four sites, and (3) tectonic translocation that preserves stream terraces.

## 6.3 Stream Power and Resisting Power

Both resisting power and stream power underwent changes conducive to aggradation in the four study areas. Increases in discharge of sediment, and particularly in amount and size of bedload, can be considered the most important change favoring aggradation. Increases in bedload size were highly variable and controlled largely by lithologic sensitivity to the impacts of climatic change on hillslopes. Lithologic controls on sediment yield variations were substantial for the graywacke of the Charwell watershed, but sensitive rocks in Nahal Yael caused a major increase in amount and size of bedload sediment from granitic hillslopes that weather mainly to grus and boulders.

Decrease in mean annual streamflow, and possibly total stream power, seems to have promoted aggradation in all four areas, but for different reasons. Onset of cold dry climates that ushered in the Stone Jug event in the Charwell watershed resulted in less annual precipitation and a decrease in the amount of precipitation occurring as rain. For both of these reasons the Charwell was the most likely stream of any in the four study areas to have undergone a true decrease in peak streamflow discharges at the time of a major increase of bedload being supplied from hillslope sediment reservoirs. Prehistorical, but presently unused, trails of grazing animals convincingly show that the Nahal Yael area underwent major late Holocene decreases in mean annual precipitation. It is more difficult to argue that there were also decreases in magnitudes and frequencies of flood discharges. Although large storms probably were much less frequent, the progressively more barren and stripped hillslopes were highly conducive for generation of flashy and powerful streamflow events.

A similar argument can be made for the drainage

basins of the Mojave and Sonoran deserts except that we are not as sure regarding a possible Holocene decrease in storm runoff. Latest Pleistocene and early Holocene debris-flow and water-flood paleohydrologic information suggests that intense summer rainfalls were larger or more common than now. Increases in temperature during the early and middle Holocene eventually were important in decreasing vegetative cover on hillslopes. Then thunderstorms might have caused still flashier flood discharges, even with a moderate decrease in abundance of monsoonal rains.

Holocene climatic changes in the San Gabriel Mountains probably decreased in the importance of runoff from the winter snowpack and increased the relative importance of winter storm rainfalls and autumn tropical storms. All three changes would have tended to increase total annual stream power and the potential of the fluvial system for doing work.

Sediment yields in each of the study areas vary with annual amounts of work done by the fluvial systems (Schumm, 1965). Nahal Yael has a total (60% suspended, 40% bedload) measured sediment yield of 260 m³/km²/yr (Asher Schick, written communication, January 26, 1986). Less is known about the drainage basins in the Mojave Desert. Volumes of Holocene fan deposits provide estimates of bedload sediment yield that range from 50 to 200 m³/km²/yr for watersheds underlain by granitic and metamorphic rocks. Total sediment yield may be on the order of 400 m³/km²/yr. Substantial data have been collected regarding the sediment yields from granitic and metamorphic rocks in the San Gabriel Mountains because of the construction of debris basins at the mouths of virtually every canyon to reduce flood damage to suburban areas. The typical modern sediment yield is high–1000 m³/km/yr. Modern sediment yields in the Seaward Kaikoura Range (O'Loughlin & Pearce, 1982) may be at the upper end of those reported from moist temperate watersheds (Gregory & Walling, 1973). Thomson and McArthur (1969) reported a sediment yield of 1500 to 2000 m³/km²/yr for a basin with properties similar to

those of the Charwell basin. Griffiths (1979) reported a mean annual specific suspended sediment yield of 1300 tonnes/km²/yr for the Waiau River, which drains a large area of the Southern Alps southwest of the Charwell River. It is interesting to note that despite such high sediment yields, the New Zealand rivers have been actively downcutting during the Holocene and the larger streams have reestablished type 1 dynamic equilibrium conditions. Sediment yields during full-glacial times of rapid aggradation may have exceeded 3000 m³/km²/yr. It is clear that present sediment yields for the four study areas vary by 10-fold.

## 6.4 Aggradation Events

Fluvial systems were overwhelmed by Pleistocene–Holocene climatic change. Short-term climate-induced increases in hillslope yields of bedload, or decreases in water yield, or both, cause strong shifts to the aggradational side of the threshold of critical power. This is true for tectonically inactive and highly active settings, and for systems whose present climates range from extremely arid to humid, from hyperthermic to frigid, and from extremely to slightly seasonal.

The net effect of concurrent tectonic and climatic perturbations in the San Gabriel Mountains and Seaward Kaikoura Range has been for aggradation to occur, even where uplift on range-bounding faults is greater than 2 m/ky. Prominent fill terraces that extend far upstream from rising mountain fronts are clear evidence of the relatively greater impact of late Quaternary climatic change. Tectonic elevation of a reach tends to shift the mode of operation to the degradational side of the threshold of critical power and favors long-term tectonically induced downcutting. Aggradation events may be shorter in such reaches than in tectonically inactive reaches. Magnitudes of climatic perturbations needed to initiate aggradation probably are larger in tectonically steepened reaches of eqivalent stream discharge. Westward decreases in uplift rates of the Seaward Kaikoura-Amuri

Range from about 6 to less than 3 m/ky are accompanied by increases in upstream extent of fill terraces. Thicknesses of valley fill and lengths of time spans of aggradation apparently decrease with increase in uplift rate.

## 6.4.1 VEGETATION

All the effects of climatic change on hills and streams need to be considered in a context of changes in vegetation. Changes in watershed plant communities in each study region largely determined the times, magnitudes, and rates of aggradation. Decreases of hillslope plant cover released more detritus from hillslope sediment reservoirs, which caused streams to aggrade. In the mountains of the Mojave and Sonoran deserts, change from juniper-pinyon woodland to desert shrubs was accompanied by marked decreases in plant density to protect hillslope sediment reservoirs that had accumulated detritus during the full-glacial climate of late Pleistocene. In Nahal Yael virtually all plant growth was restricted to watercourses by the late Holocene. Although minor to moderate decreases in density of vegetative cover may have occurred in the San Gabriel Mountains, the more important change was in plant species composition to a highly flammable chaparral plant. This change resulted in fires and consequent barren hillslopes. Changes in treeline altitude apparently were rapid and large during times of climatic change in the Charwell River study area and resulted in major changes in plant species composition.

Climatic controls on plant communities are much different in mesic humid mountains than in hot deserts. Vegetation in thermic arid watersheds is limited mainly by deficiencies in amount of precipitation rather than by low temperatures. Where annual precipitation approaches zero, temperature decreases do little to increase effective precipitation for plant growth. On humid hillslopes vegetation tends to be continuous below the uppermost altitude of plant growth. Minor variations in annual or seasonal precipitation may not affect the percentage of plant cover protecting the soils from

raindrop and fluvial erosion, but can alter plant species composition. Decreases in temperature and windiness can be important; they may change forests to alpine scrub and tussock grasslands, or to arctic-alpine dwarf plant communities. Thus temperature is a more important limiting control than precipitation for plants and associated geomorphic processes in frigid and pergelic humid regions. Plant communities in Nahal Yael are dependent on availability of moisture, but plants in the Charwell area are dependent on changing temperatures in environments of abundant soil moisture.

## 6.4.2 TIMING

Drainage basins in the rugged Seaward Kaikoura Range of New Zealand behaved differently from the arid fluvial systems of the Middle East and southwestern North America. During times of full-glacial and subsequent transitional climates, hillslope sediment reservoirs in hot deserts accumulated colluvium. This was stripped during the Holocene to aggrade valley floors. In marked contrast, hillslopes in the humid mesic mountain ranges had maximum sediment yields during times of full-glacial climates and hillslope sediment reservoirs became more stable during the Holocene.

Highly diverse impacts of Pleistocene and Holocene climatic changes resulted in aggradation events at different times in the four study areas (Fig. 6.2). Pulses of valley-floor aggradation occurred in the Charwell River system with each new onset of full-glacial climate. The piedmont reach is being elevated tectonically at 0.5 to 1.3 m/ky, but stream power is sufficient to allow attainment of type 1 dynamic equilibrium between aggradation events. In such equilibrium reaches, relatively small perturbations are required to initiate the next aggradation event compared with those required for rapidly downcutting reaches. The reach upstream from the Hope fault has a fivefold steeper gradient than the piedmont reach. The result is a greater unit stream power that may delay or prevent much of the watershed reach from switching to the aggradational side of the critical-power threshold.

The fluvial systems of the other study areas behaved differently. Streams in the Mojave Desert aggraded when their hillslopes, which were stable during the late Pleistocene, were partially stripped by latest Pleistocene–early Holocene monsoonal rains, and again during times of change to arid mid-Holocene interglacial climates. Two distinct pulses of aggradation occurred. Mid-Holocene was a time of aggrading valleys in both extremely arid Nahal Yael in the Middle East and the North Fork of the San Gabriel River in the Transverse Ranges of southern California. Drier or warmer mid-Holocene climates were associated with pulses of valley-floor aggradation. Both watersheds seem to have been sensitive to climatic changes because fluvial responses occurred quickly.

The range of possible responses to climatic change shown in Figure 6.2 indicates that one should expect substantial areal variations in the times of aggradation in the Basin and Range Province of the western United States. In southeastern Arizona and in adjacent Sonora, Mexico, there appears to be only one strong climatic-change-induced episode of Holocene aggradation. It occurred at about 8 ka, and is presumed to be the result of initiation of hotter drier climates that resulted in the replacement of woodlands with the plant communities of the Sonoran and Chihuahuan Deserts. The lower Colorado River region had two distinct episodes of aggradation, one associated with the return of the monsoonal rains during the latest Pleistocene and the second with the transition from woodland to desert scrub about 8 ka. Much farther to the north in Idaho, the time of piedmont aggradation was during the full-glacial climate of the latest Pleistocene, and the Holocene was characterized by entrenchment of valley fills and piedmont fans. It is presumed that the full-glacial climates were conducive to vigorous peri-glacial (and locally glacial) processes in these northern mountains. Thus appropriate process-response models might be those of Figures 5.14 and 5.15 (Charwell River) instead of those of Figure

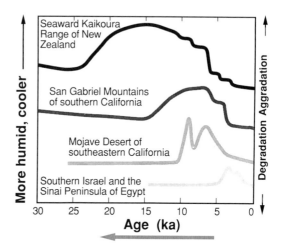

FIGURE 6.2 Generalized comparisons of the times of aggradation and degradation and of equilibrium periods for the study regions of Chapters 2, 3, 4, and 5. Equilibrium periods are horizontal for the tectonically inactive Mojave and Sinai deserts but are gently sloping for the tectonically active San Gabriel and Kaikoura mountains, to reflect long-term rates of tectonically induced downcutting.

2.38 (Mojave Desert). Between the northern and southern parts of the Basin and Range province there should be an interesting transition zone where some fluvial systems aggraded mainly during times of full-glacial climates, some aggraded mainly during times of interglacial climates, and others had prolonged, diachronous aggradation events in the style of Cajon Creek (Section 4.4.2).

Because of their great heights the Panamint Range in the Basin and Range Province and the San Gabriel Mountains in the Transverse Ranges of southern California may be considered altitudinal climatic transition zones that are similar to latitudinal climatic transition zones. Large basins in lofty mountains may have aggraded during both full-glacial and interglacial climates, in response to partial stripping of detritus in hillslope sediment reservoirs in different altitude zones of the watershed. The North Fork hillslopes did not respond to the climatic changes of the latest Pleistocene and early Holocene, but yielded huge volumes of sediment during a $3 \pm 1$ ky mid-Holocene aggradation event. Cajon Creek aggraded slowly over $11 \pm 2$ ky, apparently in response to a combination of full-glacial and Holocene sediment-yield increases derived from different altitude ranges of the drainage basin. Considering both the temporal and spatial contrasts, we should expect to find marked variations in the times of aggradation in different parts of lofty mountains in semiarid and subhumid regions.

The formation of both tectonic strath and climatic aggradation surfaces is influenced by global climatic change but the two may differ in degree of synchroneity. The present happens to be a time of attainment of the base level of erosion for many powerful streams, especially those flowing on soft materials. Modern straths that extend for many kilometers beneath stream-channel gravels of powerful rivers are clear evidence for straths being a synchronous fluvial landform. Thus major straths are synchronous landforms regardless of whether strath genesis was diachronous. As an equilibrium tectonic landform, synchronous major straths in coastal areas may be graded to marine-terrace shore platforms formed during the sea-level highstand of the past 6 ky. Strath formation may end at the beginning of an aggradation event, which buries the beveled bedrock surface, or because of stream-channel incisement into the surface.

Aggradation surfaces mark times of switching from aggradational to degradational modes of operation, a change that commonly occurs at different times along a stream. Aggrading valley floors may extend for many kilometers, but unlike in strath genesis, the level of streamflow is rising rather than remaining at about the same level. Rates of aggradation typically increase and then decrease as the threshold of critical power is approached (Fig. 4.29).

Aggradation surfaces of climatic terraces commonly are diachronous in arid and semiarid regions where available stream power is small and amounts of bedload are large. Such controls severely limit the rates at which streams can respond to either

climatic or tectonic perturbations, and large spatial variations in times of initiation of aggradation can occur (Fig. 4.36, 4.37).

Aggradation surfaces are diachronous for long rivers where a slug of bedload resulting from a climatic change shifts downstream over a time span of 1 to 4 ky (Jackson et al., 1982; Fisk, 1944). Such climatic change terraces are more likely to by synchronous for short powerful streams of humid regions.

It is important to make the distinction between synchroneity of time of formation of a terrace landform within a single fluvial system and synchroneity of time of formation of stream terraces on a regional basis. Regional synchroneity involves the important topic of terrace correlations. Tectonic stream terraces would seem to have the best regional synchroneity, especially where graded to marine terraces. This is because of times, like the present, that favor attainment of the base level of erosion. Climatic stream terraces—aggradation surfaces—may vary greatly in their degree of regional synchroneity. One should expect diachronous terraces (1) in regions characterized by strongly seasonal arid to semiarid climate and (2) where drainage basins have a large range of area, relief, and rock types. In a general sense all the diverse basins in such a region will respond to a given climatic perturbation but times of formation of aggradation surfaces for individual watersheds may vary by 3 to more than 10 ky. Conversely, fluvial systems with similar properties and processes will tend to have synchronous climatic terraces.

Aggradation surfaces of suites of adjacent drainage basins in humid regions can be synchronous. An example is the fluvial systems of the Seaward Kaikoura Range (Section 5.4.2). Such terraces seem to be synchronous in regard to their times of genesis, both within a given drainage basin and in other drainage basins of the climatic-lithologic-topographic province.

Complex-response terraces tend to be regionally diachronous because they result from local adjustments within individual fluvial systems. Examples of geomorphic processes that may operate on local instead of regional scales to create regionally diachronous complex-response terraces include (1) local floods that disrupt stream-channel gravels and renew degradation that leads to the creation of new fill-cut or strath terraces, (2) landslides that cause temporary aggradation of valley floors, and (3) local faulting that initiates a minor pulse of aggradation downstream from the fault zone as headcutting migrates upstream, and terrace-tread incisement upstream from the fault zone that is concurrent with the headcutting.

Ultimately, the earth's climate is a function of solar insolation, so the combined effects of variations in the precise cycles of earth's orbital parameters not only may be a fundamental cause of climatic change, but also provide the ultimate geochronological tool—an astronomical clock. Geomorphic utilization of the astronomical clock is beset by the need to understand the time lags of response to astronomical perturbations that are associated with diachronous or synchronous behavior of fluvial systems.

### 6.4.3 FREQUENCY

The frequency of late Pleistocene aggradation events generally was much lower than the number of climate-change-induced sea-level highstands or lowstands. The marine record (Chappell & Shackleton, 1986) is remarkable in the way that it reflects the climatic effects of astronomical perturbations of the earth's orbital parameters (Figs. 2.4, 6.3). The clearly similar times of insolation maxima in the northern hemisphere and the $^{230}Th/^{234}U$ ages of corals that provide ages for sea-level highstands underscore the relevance of the astronomical clock for the sea-level change system, which appears to behave synchronously with minimal response time to changes in solar insolation. Eleven marine terraces formed at times of sea-level highstands during the past 130 ky, but far fewer aggradation events have been recognized in southwestern North America during the same time span. Times of late Quaternary aggradation in the San Gabriel Mountains and in the adjacent Mojave

Desert appear to be the same—roughly 125, 55, and 10 ka. All three times coincide with rising sea level but this is only one-third of the potential times of aggradation suggested by the marine record of climatic fluctuations. The times of aggradation events in the Mojave Desert and San Gabriel Mountains coincide were times of cutting of major strath terraces along the Charwell River. The Charwell River aggradation events occurred during times of full-glacial climate between the intervals of strath cutting, but about half of each interval between times of strath cutting was a time of a degradation event. Cosmogenic chlorine dating of surficial boulders in the glacial moraines along the east side of the Sierra Nevada by Phillips and colleagues (1990) provides ages of four major glacial advances that are in good agreement with the process-response models proposed for the Charwell River and the Mojave Desert and San Gabriel Mountains study areas. Sierra Nevada glacial advances occurred just prior to aggradation in the other California study areas and at the same time as some of the New Zealand aggradation events. The remarkably good agreement between the sea-level record and the geomorphic responses to climatic changes in the four study areas underscores the pervasive influence of global climatic changes.

Why have the times of decrease of effective precipitation during the late Quaternary not resulted in more aggradation events in the Mojave Desert and San Gabriel Mountains? Perhaps responses to climatic change in fluvial (or glacial) systems are considerably more complex than simple sea-level rises caused by melting of glacial ice in response to variations in solar radiation.

Indeed, the answer lies in the complex behavior of the hillslope subsystems. Major aggradation of mountain valleys and piedmonts occurs only when large amounts of bedload are rapidly eroded from fully stocked hillslope sediment reservoirs. A major aggradation event is unlikely now in places like Nahal Yael because insufficient sediment remains to be stripped from the barren slopes. Hillslope soils and colluvium first must be renewed to

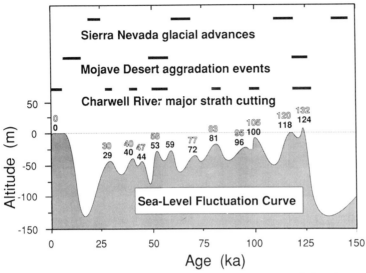

FIGURE 6.3 Comparison of times of insolation maxima during the past 150 ky of solar insolation maxima (solid numbers), $^{230}$Th/$^{234}$U ages for sea-level highstands (open numbers) from Chappell and Shackleton (1986), chlorine 36 ages of major glacial advances in the Sierra Nevada (Phillips et al., 1990), times of aggradation events in the Mojave Desert (Chapter 2), and times of major strath cutting along the Charwell River, New Zealand (Chapter 5). Times of straths younger than 40 ka are dated; older ages are estimated from calibrated rates of long-term tectonically induced downcutting.

supply sediment for the next aggradation event. Renewal may require more than 40 ka on many hillslopes, which might explain why only three of the four prolonged dry periods during the past 140 ky in the Searles Lake stratigraphic record (Smith, 1979) are represented by aggradation events in the fluvial-system record of climatic change.

Replenishment of the hillslope sediment reservoir is as important as erosion in the production of an aggradation event. Conditions that favor rapid and progressive increases in hillslope plant and soil cover may be infrequent or may require long time spans. A lack of prolonged droughts would seem to be an essential ingredient. Presumably both the plant communities and weathering and soil-forming processes were much different during times of increased storage of hillslope detritus than at present.

Replenishment requires a set of self-enhancing feedback mechanisms that are opposite those de-

scribed in Figure 2.38. Another, and possibly opposite, type of climatic change has to occur that is of sufficient strength to reverse hillslope processes and allow progressive increases in hillslope plant and soil cover. It almost seems that the number of Quaternary aggradation events is related to the frequency of climatic changes sufficiently strong to permit reversals of self-enhancing feedback mechanisms that promote either progressive aggradation or degradation.

The Seaward Kaikoura Range is different from southwestern North America in that aggradation events occur frequently. Synchronous aggradation occurred between 26 and 14 ka, 38 and 31 ka, and 48 and 43 ka, and possibly after times of major strath formation at about 80, 100, and 120 ka. The main contrast with the desert study areas seems to be in the rates of adjustment of hillslope processes. In humid regions such as the Seaward Kaikoura Range with potentially unlimited sources of bedload, fluctuations of hillslope erosion may occur as rapidly as sea-level fluctuations. In such areas there seems to be a better correlation between marine and continental records of responses to climatic change.

In summary, fluvial aggradation events appear to be driven by self-enhancing feedback mechanisms that deplete hillslope sediment reservoirs. Where sources of bedload are limited, such as in Nahal Yael, the Mojave Desert, and possibly parts of the San Gabriel Mountains, accumulation of hillslope detritus must occur before another aggradation event can occur. Times of storage of detritus in hillslope sediment reservoirs are characterized by types of self-enhancing feedback mechanisms opposite those present during episodes of stripping. Eventually the system becomes sensitive again to climatic perturbations, which trigger the next aggradation event, but the climatic perturbation has to be strong enough to reverse the self-enhancing feedback mechanisms once again. Aggradation events are few in many locales (1) because of difficulty in reversing self-enhancing feedback mechanisms, (2) because of potentially long times needed to weather and accumulate hillslope detritus in arid regions, and (3) because many climatic changes may represent perturbations of insufficient strength to initiate and maintain an aggradation event. All three factors depend on lithologic controls.

## 6.5 Degradation and Reattainment of Equilibrium

Streams in both arid and humid regions tend to return to type 1 dynamic equilibrium after aggradation has raised the streambed. Aggradation along equilibrium reaches of streams in tectonically inactive terrains is followed by a degradation event that reestablishes the same longitudinal profile that was present before the aggradation event. Most New Zealand and San Gabriel Mountains streams are in tectonically active terrains. They also tend to downcut to their respective equilibrium profiles,

which, because of continuing uplift, are below major straths beveled prior to aggradation.

Holocene channel downcutting rates seem to be directly related to size of flood discharge and total annual stream power available in each study area. Degradation rates are not constant. Degradation accelerates to a maximum value and then decelerates as a new base level of erosion is approached. Estimated ranges of downcutting rates for Nahal Yael, Mojave Desert, San Gabriel Mountains, and Charwell River are 0.2 to 2, 1 to 3, 2 to 7, and 4 to 32 m/ky, respectively. Brief periods of attainment of static equilibrium are represented by pauses in degradation for the Charwell River and many streams in the San Gabriel Mountains.

## 6.6 Time Lags of Responses to Late Quaternary Climatic Change

Availability of stream power also plays a role in determining the time lags of response of fluvial systems to climatic change. Only generalizations are possible in estimating the time lags of response for fluvial systems as diverse as those discussed in Chapters 2 through 5. Reaction time is the time between initiation of a climate-change-induced perturbation on a hillslope and the first indication of response of a given reach of the stream subsystem to that perturbation. It depends primarily on the magnitude of the perturbation, the distance between the reach and the hillslope source of the perturbation, and how far removed the reach of the stream was from equilibrium or threshold conditions. Reaction times are in large part a function of the capacity of a stream to transport bedload. They also are a function of bedload size; they should be shorter for transport of sand and pebbles than for transport of boulders. Rivers of humid regions have more available power than do ephemeral streams such as those of Nahal Yael, whose flows often fail to reach the piedmont. Thus treads of major fill terraces, or alluvial-fan deposition, should tend to be more synchronous for powerful perennial streams and more diachronous for ephemeral small streams of extremely arid regions. Times of initiation and ending of aggradation events for streams of the Seaward Kaikoura Range with basins larger than 30 km$^2$ are synchronous within the limits of dating methods. The more arid climatic settings of Hanuapah Canyon in the Panamint Range and Cajon Creek in the San Gabriel Mountains may be responsible for response times that are so long that aggradation events overlap one another.

The outcrop-area-controlled self-enhancing feedback mechanism makes hillslopes sensitive to climatic perturbations. Bedrock outcrops are conspicuous on mountains in hot deserts. In arid regions colluvial soil mantles are thin, and in Nahal Yael only a decimeter or two of colluvium mantled granitic slopes during times of semiarid climate. These hillslope sediment reservoirs are stripped rapidly because of the small volumes of material to be eroded and because progressive increases in areas of bedrock outcrops are a powerful self-enhancing feedback mechanism (Fig. 2.38). Fluvial systems in the eastern Mojave Desert had sufficient time to aggrade and then recross the threshold of critical power between the two times of different climate in the period 13 to 7 ka. The late Quaternary stratigraphy and geomorphology of the region records a prominent hiatus within the Holocene aggradation event. The lack of hiatuses in the stratigraphic record of aggradation events of the Charwell River may result in part because fractured graywacke is not conducive to the formation of massive outcrops that rise above adjacent colluviated slopes.

Hillslope subsystems of the San Gabriel Mountains probably had much longer reaction and relaxation times to impacts of climatic change than did those of Nahal Yael. Bedrock outcrops are sparse now and may have been virtually absent on many hillslopes prior to the most recent aggradation event. The preceding 40 to 100 ky had produced several meters of weathered hillslope detritus. The time needed to completely empty hillslope sediment reservoirs in the San Gabriel Mountains—if it even occurs—may be two orders

of magnitude longer than for Nahal Yael, despite 10-fold greater annual precipitation.

## 6.7 Different Responses in the Mojave Desert and San Gabriel Mountains

Similarities between the arid thermic to hyperthermic Mojave Desert and adjacent subhumid to humid mesic San Gabriel Mountains study regions are surprisingly few. Pleistocene–Holocene climatic change in both areas has resulted in partial stripping of granitic hillslope colluvium, which furnished bedload to form thick fill terraces. The contrasting landscape changes in response to changing climates of the two study regions are the subject of this brief section.

Some contrasts may be attributed to the marked differences in soil moisture between arid and subhumid climates; soil leaching indices are about 10 mm in the Mojave Desert and 260 to 600 mm in the San Gabriel Mountains. The result is thin calcic soils in the hot desert and thick noncalcic soils in the cool Transverse Ranges. Early and mid-Holocene soils have argillic horizons in the San Gabriel Mountains but only cambic or thin calcic horizons in the drier Mojave Desert. Another contrast is the rate of hillslope sediment production. Rates of rock weathering tend to be less than soil-erosion rates in deserts. Conversion of rock to colluvial materials probably is two orders of magnitude slower in the Mojave Desert than in the subhumid parts of the Transverse Ranges. Hillslope sediment reservoirs continue to be depleted in the desert and detritus now occurs only as patches of thin colluvium between expanding areas of outcrops. In contrast, accelerated erosion of colluvial mantles in much of the San Gabriel Mountains may have occurred only during the mid-Holocene. The late Holocene return to cooler and moister conditions occurred before the slopes could be stripped; vegetation cover became denser and hillslope erosion rates decreased. Self-enhancing feedback mechanisms were strong enough to continue the mode of hillslope stripping in the Mojave

Desert (and Nahal Yael), but apparently not in the San Gabriel Mountains where colluvium is commonly thicker than 1 m and buries most of the bedrock.

The tectonically inactive nature of the lower Colorado River region contrasts with the highly active San Gabriel Mountains region. This has several implications. Tectonically steepened stream gradients result in relatively greater unit stream power for a given discharge. Instead of being erosional remnants of small mountain ranges with small watersheds, the lofty San Gabriel Mountains have 100- to 1000-km$^2$ drainage basins. The much larger altitude range of the tectonically active range results in a great variety of geomorphic processes; for example, splitting of rocks by freeze-thaw processes is an important process at the higher altitudes. The numbers and ages of Holocene stream terraces are similar in most drainage basins of the eastern Mojave Desert, but even adjacent drainage basins in the San Gabriel Mountains may have different Holocene terraces.

Both areas have thick fill terraces, but the aggradation and incisement processes of fill-terrace formation are substantially different. Aggradation of desert valleys occurred because of rapid stripping of a thin hillslope sediment reservoir after a change to markedly less vegetation cover or an increase in intense summer-type precipitation events, or both. Thick fills were deposited in the valleys of the San Gabriel Mountains because the volume of material stripped from the hillslopes increased after postulated change to a climate with fewer snowfalls, more intense rainfalls, and more flammable types of plants. Strengths of the self-enhancing feedback mechanisms as outlined in Figure 2.38 may be so different in the two regions that additional, more detailed process-response models may be needed for evaluation of differences in fluvial system behavior (for example, see Figures 5.13, 5.14, and 5.15).

Stream terraces reflect major differences in the operation of fluvial systems between the Mojave and adjacent deserts and the San Gabriel Mountains. In the comparatively simpler Mojave Desert,

drainage basins in a 100,000 km$^2$ area generally have the same number of terrace-forming events. Even casual visitors to the San Gabriel Mountains note marked differences in abundance and heights of stream terraces from canyon to canyon, and closer inspection reveals equal diversity of ages of terrace-tread soils.

An intriguing question is, Why is there only one major Holocene aggradation event in the valley of the North Fork of the San Gabriel River? Although the North Fork has not been tectonically quiescent, impacts of recent uplift have been much less than at the range front. Thus the effects of regional perturbations on landscape changes, such as the two climatic changes that occurred in the Mojave Desert during the late Quaternary, should be readily apparent. They are not. Only one prominent Holocene fill terrace is present. In contrast, the Q3a and Q3b alluvial geomorphic surfaces of the Mojave Desert attest to two major late Quaternary aggradation events.

The answer to this apparent paradox of dissimilar numbers of Holocene fill terraces in the two study regions may lie in differences of threshold conditions that determined whether a hillslope remained stable or was subject to accelerated erosion. Protective plant cover probably was marginal on desert hillslopes even during the latest Pleistocene. The first transition from pinyon-juniper woodlands to juniper-Joshua tree plant communities was sufficient for the threshold separating aggrading from degrading hillslope sediment reservoirs to be crossed. The change to desert scrub at about 8 ka simply resulted in marked acceleration of hillslope erosion.

In contrast, the change from late Pleistocene climate to early Holocene climate in the San Gabriel Mountains may not have greatly changed hillslope plants. Even if changes in plant species composition did occur, the lack of an early Holocene valley-fill aggradation period indicates that the vegetation cover remained sufficiently dense to protect the hillslopes from accelerated erosion. Even today, summer monsoonal rains rarely penetrate over the lofty Transverse Ranges to the watersheds on the Pacific Ocean side of the mountains. So increase of monsoonal rains at 12 to 9 ka in the San Gabriel Mountains may have been minimal compared with the increase in the eastern Mojave Desert, which is much closer to the sources of summer monsoonal moisture. Rapid mid-Holocene valley aggradation suggests major changes occurred in plant density, plant type, or both. These changes made San Gabriel Mountains slopes susceptible to accelerated erosion during intense rainfalls. Warmer, drier mid-Holocene climates presumably were important in promoting both a change to flammable chaparral vegetation and to increased brush fires.

\*  \*  \*  \*  \*  \*  \*

Studies of the timing of geomorphic responses to climatic change in even a few areas reveal profoundly different numbers and times of geomorphic events (Fig. 6.3). Aggradation events occur during full-glacial times in humid mesic climates and during interglacial times in arid thermic climates. Numbers of aggradation events vary greatly from region to region, perhaps reflecting climatic and lithologic controls of the time needed to replenish hillslope sediment reservoirs so the next stripping event can send a large surge of detritus through the fluvial system. The record of major straths is more likely to be complete in humid than in arid regions, especially where soft rocks allow powerful streams to degrade to new base levels of erosion between closely spaced episodes of aggradation.

Climatic change: past, present, and future–how suddenly it seems to occur, and how little we know about its impact! This book has explored how climatic change has affected geomorphic processes on hills and streams and in soil profiles in many tectonic and climatic settings. Landscapes in much of the world have changed dramatically in response to climate change from ice ages to the present. The timing of climatically induced aggradation and degradation events seems to be in step with variations in the earth's orbital parameters—the astronomical clock. Lessons from histories of landscape change provide insight into future impacts of natural or human-induced climatic change on storm patterns, flood frequency, landslides, stability of valley floors, and agricultural productivity.

# References Cited

Ahlborn, A.O., 1982, Santa Ana river basin flood hazard: San Bernardino County Museum Association Quarterly, v. 29, 95 pp.

Ahnert, F., 1970, Functional relationships between denudation, relief, and uplift in large mid-latitude drainage basins: American Journal of Science, v. 268, pp. 243–263.

Allen, C.C., 1978, Desert varnish of the Sonoran Desert—Optical and electron probe microanalysis: Journal of Geology, v. 86, pp. 743–752.

Allen, J.R.L., 1974, Reaction, relaxation and lag in natural systems; general principles, examples and lessons: Earth Science Reviews, v. 10, pp. 263–342.

Allis, R.G., 1981, Continental underthrusting beneath the Southern Alps of New Zealand: Geology, v. 9, pp. 303–307.

Amit, R., and Gerson, R., 1986, The evolution of Holocene Reg (gravelly) soils in deserts; an example from the Dead Sea region: Catena, v. 13, pp. 59–79.

Amundson, R.G., Chadwick, O.A., Sowers, J.M., and Doner, H.E., 1989a, The stable isotope chemistry of pedogenic carbonates at Kyle Canyon, Nevada: Soil Science Society of America Journal, v. 53, pp. 201–210.

Amundson, R.G., Chadwick, O.A., Sowers, J.M., and Doner, H.E., 1989b, Soil evolution along an altitudinal transect in the eastern Mojave Desert, U.S.A.: Geoderma, v. 43, pp. 349–371.

Amundson, R.G., Sowers, J.M., and Chadwick, O.A., 1989c, Influence of time and climate on pedogenesis in a desert alluvial fan system, in T.J. Rice, Jr. (editor), Soils geomorphology relationships in the Mojave Desert, California, Nevada: Field Tour Guidebook for the 1989 Soil Science Society of America Annual Meeting Pre-Meeting Tour, October 12–14, 1989, pp. 90–129.

Andrews, E.D., 1979, Hydraulic adjustment of the East Fork River, Wyoming, to the supply of sediment, in D.D. Rhodes and G.P. Williams (editors), Adjustments of the fluvial system: Proceedings of the 10th Annual Binghamton Geomorphology Symposia Series, Dubuque, Iowa, Kendall/Hunt, pp. 69–94.

Andrews, E.D., 1983, Entrainment of gravel from naturally sorted riverbed material: Geology Society of America Bulletin, v. 94, pp. 1225–1231.

Antevs, E., 1948, The Great Basin, with emphasis on glacial and post-glacial times: University of Utah Bulletin, v. 38, pp. 168–191.

Antevs, E., 1954, Climate of New Mexico during the last glacial-pluvial: Journal of Geology, v. 62, pp. 182–191.

Antevs, E., 1955, Geologic-climatic dating in the west: American Antiquity, v. 20, pp. 317–335.

Arkley, R.J., 1963, Calculations of carbonate and water movement in soil from climatic data: Soil Science, v. 96, pp. 239–248.

Bachman, G.O., and Machette, M.N., 1977, Calcic soils and calcretes in the southwestern United States: U.S. Geological Survey Open-File Report 77–794, 162 pp.

Bagnold, R.A., 1973, The nature of saltation and of bedload transport in water: Proceedings, Royal Society of London, ser. A., v. 332, pp. 473–504.

Bagnold, R.A., 1977, Bed-load transport by natural rivers: Water Resources Research, v. 13, pp. 303–312.

Baker, V.R., 1973, Paleohydrology and sedimentology of Lake Missoula flooding in eastern Washington: Geological Society of America Special Paper 144.

Baker, V.R., 1988, Cataclysmic processes in geomorphological systems: Zeitschrift für Geomorphologie, Supplementband 67, pp. 25–32.

Baker, V.R., and Costa, J. E., 1987, Flood power, in Mayer, Larry and Nash, David (editors), Catastrophic Flooding: London, Allen and Unwin.

Baker, V.R., Kochel, R.C., and Patton, P.C., 1988, Flood Geomorphology: New York, John Wiley, 503 pp.

Baker, V.R., and Penteado-Orellana, M.M., 1977, Adjustment to Quaternary climatic change by the Colorado River in central Texas: Journal of Geology, v. 85, pp. 395–422.

Baker, V.R., and Penteado-Orellana, M.M., 1978, Fluvial sedimentation conditions by Quaternary climatic change in central Texas: Journal of Sedimentary Petrology, v. 48, pp. 433–451.

Bard, E., Hamelin, B., Fairbanks, R.G., and Zindler, A., 1990, Calibration of the $^{14}C$ timescale over the past 30,000 years using mass spectrometric U-Th ages from Barbados corals: Nature, v. 345, pp. 405–410.

Barrell, J., 1917, Rhythms and measurement of geologic time: Geological Society of America Bulletin, v. 28, pp. 745–904.

Barshad, L., 1966, The effect of a variation in precipitation on the nature of clay mineral formation in soils from acid and basic igneous rocks: Proceedings, International Clay Conferences, Jerusalem, v. 1, pp. 167–173.

Bar Yosef, O., and Phillips, J.L. (editors), 1977, Prehistoric investigations in Gebel Maghara, northern Sinai, Qedem, v. 7: Jerusalem.

Bateman, P., and Wahrhaftig, C., 1966, Geology of the Sierra Nevada, in Geology of northern California: California Division of Mines Bulletin 190, pp. 107–172.

Begin, Z.B., 1975, Structural and lithologic constraints on stream profiles in the Dead Sea region: Journal of Geology, v. 83, pp. 97–111.

Begin, Z.B., Broecker, W., Buchbinder, B., Druckman, Y., Kaufman, A., Magaritz, M., and Neev, D., 1985, Dead Sea and Lake Lisan levels in the last 30,000 years, a preliminary report: Israel Geological Survey, GSI/29/85, 18 pp.

Begin, Z.B., Ehrlich, A., and Nathan, Y., 1974, Lake Lisan: The Pleistocene precursor of the Dead Sea: Geological Survey of Israel Bulletin No. 63, 30 pp.

Begin, Z.B., and Schumm, S.A., 1984, Gradational thresholds and landform singularity; significance for Quaternary studies: Quaternary Research, v. 21, pp. 267–274.

Benson, L.V., and Paillet, F.L., 1989, The use of total lake-surface area as an indicator of climatic change: examples from the Lahontan basin: Quaternary Research, v. 32, pp. 262–275.

Berger, A., 1979, Insolation signatures of Quaternary climatic changes: Nuovo and Sim, v. 2C, pp. 63–87.

Berger, A., 1980, The Milankovitch astronomical theory of paleoclimates; a modern review: Vistas Astronomical, v. 24, pp. 103–122.

Berger, A., Imbrie, J., Hays, J., Kukla, G., and Saltzman, B. (editors), 1984, Milankovitch and climate, understanding the response to astronomical forcing: Boston, D. Reidel, 895 pp.

Birkeland, P.W., 1982, Subdivision of Holocene glacial deposits, Ben Ohau Range, New Zealand, using relative-dating methods: Geological Society of America Bulletin, v. 93, pp. 433–449.

Birkeland, P.W., 1984a, Soils and geomorphology: New York, Oxford University Press, 372 pp.

Birkeland, P.W., 1984b, Holocene soil chronofunctions, Southern Alps, New Zealand: Geoderma, v. 34, pp. 115–134.

Birkeland, P.W., 1990, Soil geomorphic analysis and chronosequences—a selective overview, *in* P.L.K. Knuepfer and L.D. McFadden (editors), Soils and Landscape Evolution, 1990 Binghamton Geomorphology Symposium: Geomorphology, v. 3, pp. 207–224.

Birkeland, P.W., Machette, M.N., and Haller, K.M., 1990, Soils as a tool for applied Quaternary geology, manual for a short course: Salt Lake City, Utah Geological and Mineral Survey, Utah Department of Natural Resources, Miscellaneous Publication Series.

Bishop, D.G., Bradshaw, J.D., and Landis, C.A., 1985, Provisional terrane map of South Island, New Zealand, *in* D.G. Howell (editor), Tectonostratigraphic terranes of the circum-Pacific region: Circum-Pacific Council for Minerals and Energy Earth Sciences, series 1, pp. 515–521.

Blackwelder, E., 1948, Historical significance of desert lacquer: Geological Society of America Bulletin, v. 59, p. 1367.

Blanford, H.F., 1884, On the connection of the Himalayan snowfall with dry winds and seasons of drought in India: Proceedings, Royal Society of London, v. 37, pp. 3–32.

Blom, R., Elachi, C., and Evans, D., 1982, SIR-A radar images of sand dunes and volcanic fields: IEEE IGARSS, Munich, West Germany, pp. 9.1–9.6.

Bloomfield, C., 1981, The translocation of metals in soils, *in* D.J. Greenland and M.H.B. Hayes (editors), The chemistry of soil processes: London, John Wiley & Sons, Ltd., pp. 463–504.

Boettinger, J.L., and Southard, R.J., 1989, Granitic pediments of the western Mojave Desert, *in* T.J. Rice (editor), Soils geomorphology relationships in the Mojave Desert, California, Nevada: Field Tour Guidebook for the 1989 Soil Science Society of America Annual Meeting Pre-Meeting Tour, October 12–14, 1989, pp. 1–39.

Bowman, D., 1971, Geomorphology of the shore terraces of the late Pleistocene Lisan Lake, Israel: Palaeogeography, Palaeoclimatology, Palaeoecology, v. 9, pp. 183–209.

Bowman, D., 1974, Geomorphology of river terraces on the western bank of the Dead Sea—geomorphology of the shore terraces of the late Pleistocene Lisan Lake, *in* R. Gerson and M. Inbar

(editors), Field study program for the international symposium on geomorphic processes in arid environments, pp. 40–51.

Bowman, D., 1978, Determination of intersection points within a telescopic alluvial fan complex: Earth Surface Processes and Landforms, v. 3, pp. 265–276.

Bowman, D., 1988, The declining but non-rejuvenating base level—the Lisan Lake, the Dead Sea area, Israel: Earth Surface Processes and Landforms, v. 13, pp. 239–249.

Bradley, R.S., 1985, Quaternary paleoclimatology, methods of paleoclimatic reconstruction: Boston, Allen and Unwin, 472 pp.

Bradley, W.C., Hotton, J.T., and Twidale, C.R., 1978, Role of salts in development of granitic tafoni, South Australia: Journal of Geology, v. 86, pp. 647–654.

Bradshaw, J.D., Adams, C.J., and Andrews, P.B., 1980, Carboniferous to Cretaceous on the Pacific margin of Gondwana; the Rangitata phase of New Zealand, *in* M.M. Cresswell and P. Vella (editors), Proceedings of the fifth International Gondwana Symposium: Rotterdam, A.A. Balkema, pp. 217–221.

Brakenridge, G.R., 1978, Evidence for a cold, dry, full-glacial climate in the American Southwest: Quaternary Research, v. 9, pp. 22–40.

Brakenridge, G.R., 1981, Late Quaternary floodplain sedimentation along the Pomme de Terre River, southern Missouri: Quaternary Research, v. 15, pp. 62–76.

Brayshaw, A.C., 1985, Bed microtopography and entrainment thresholds in gravel-bed streams: Geological Society of America Bulletin, v. 96, pp. 218–223.

Broecker, W.S., 1984, Terminations, *in* A. Berger, J. Imbrie, J. Hays, G. Kukla, and B. Saltzman (editors), Milankovitch and climate, understanding the response to astronomical forcing: Boston, D. Reidel, pp. 687–698.

Brookes, I.A., 1989, Early Holocene basinal sediments in the Dakleh Oasis region, south central Egypt: Quaternary Research, v. 32, pp. 139–152.

Brunsden, D., 1980, Applicable models of long term landform evolution: Zeitschrift für Geomorphologie, Supplementband 36, pp. 16–26.

Brunsden, D., and Thornes, J.B., 1979, Landscape

sensitivity and change: Transactions of the Institute of British Geographers, v. 4, pp. 463–484.

Bryson, R.A., and Lowry, W.P., 1955, Synoptic climatology of the Arizona summer precipitation singularity: Bulletin of the American Meteorological Society, v. 36, pp. 329–339.

Bryson, R.A., and Swain, A.M., 1981, Variations of monsoon rainfall in Rajasthan: Quaternary Research, v. 16, pp. 135–145.

Bryson, R.A., and Wendland, W.M., 1967, Tentative climatic patterns for some late glacial and postglacial episodes in central North America, in W.J. Mayer-Oakes (editor), Life, land, and water: Winnipeg, University of Manitoba Press, pp. 277–278.

Budel, J., 1957, Die "Doppelten Einebnungsflachen" in den feuchten Tropen: Zeitschrift für Geomorphologie, Neue Folge, 1, pp. 201–228.

Budel, J., 1982, Climatic geomorphology: Princeton, New Jersey, Princeton University Press, 443 pp.

Bull, W.B., 1974a, Effects of Holocene climate on arid fluvial systems, Whipple Mountains, California: American Quaternary Association, 3rd Biennial Meeting, Discussant Paper, p. 64.

Bull, W.B., 1974b, Geomorphic tectonic analysis of the Vidal region, in Information concerning site characteristics, Vidal Nuclear Generating Station: Los Angeles, Southern California Edison Company, Appendix 2.5B, Amendment 1, 66 pp.

Bull, W.B., 1975, Allometric change of landforms: Geological Society of America Bulletin, v. 86, pp. 1489–1498.

Bull, W.B., 1976a, Sensitivity of fluvial systems in hot deserts to climatic change: American Quaternary Association 4th Biennial Meeting, Discussant Paper, pp. 42–43.

Bull, W.B., 1976b, Landforms that do not tend toward a steady state, in W.N. Melhorn and R.C. Flemal (editors), Theories of landform development: State University of New York at Binghamton, Publications in Geomorphology, 6th Annual Meeting, pp. 111–128.

Bull, W.B., 1977a, The alluvial-fan environment: Progress in Physical Geography, v. 1, pp. 222–270.

Bull, W.B., 1977b, Tectonic geomorphology of the Mojave Desert: U.S. Geological Survey Contract Report 14–08–001–G–394; Office of Earth-

quakes, Volcanoes, and Engineering, Menlo Park, California, 188 pp.

Bull, W.B., 1978, Geomorphic tectonic activity classes of the south front of the San Gabriel Mountains, California: U.S. Geological Survey Contract Report 14–08–001–G–394; Office of Earthquakes, Volcanoes, and Engineering, Menlo Park, California, 59 pp.

Bull, W.B., 1979, Threshold of critical power in streams: Geological Society of America Bulletin, v. 90, pp. 453–464.

Bull, W.B., 1980, Geomorphic thresholds as defined by ratios, in D. Coates and J. Vitek (editors), Thresholds in geomorphology: London, Allen and Unwin Ltd., pp. 259–263.

Bull, W.B., 1984, Tectonic geomorphology: Journal of Geological Education, v. 32, pp. 310–324.

Bull, W.B., 1985, Correlation of flights of global marine terraces, in M. Morisawa and J. Hack (editors), Tectonic geomorphology: Proceedings of the 15th Annual Geomorphology Symposium, State University of New York at Binghamton; Hemelhempstead, England, George Allen and Unwin, pp. 129–152.

Bull, W.B., 1988, Floods–degradation and aggradation, in V.R. Baker, R.C. Kochel, and P.C. Patton (editors), Flood Geomorphology: New York, John Wiley, pp. 157–165.

Bull, W.B., 1990, Stream-terrace genesis–implications for soil development, in P.L.K. Knuepfer and L.D. McFadden (editors), Soils and Landscape Evolution, 1990 Binghamton Geomorphology Symposium: Geomorphology, v. 3, pp. 351–368.

Bull, W.B., and Cooper, A.F., 1986, Uplifted marine terraces along the Alpine fault, New Zealand: Science, v. 234, pp. 1225–1228.

Bull, W.B., and Knuepfer, P.L.K., 1987, Adjustments by the Charwell River, New Zealand to uplift and climatic changes: Geomorphology, v. 1, pp. 15–32.

Bull, W.B., Menges, C.M., and McFadden, L.D., 1979, Stream terraces of the San Gabriel Mountains, California: U.S. Geological Survey Contract Report 14–08–001–G–394; Office of Earthquakes, Volcanoes, and Engineering, Menlo Park, California, 139 pp.

Bull, W.B., and Schick, A.P., 1979, Impact of climatic change on an arid watershed: Nahal Yael, southern Israel: Quaternary Research, v. 11, pp. 153–171.

Burke, R.M., and Birkeland, P.W., 1979, Reevaluation of multiparameter relative dating techniques and their application to the glacial sequence along the eastern escarpment of the Sierra Nevada, California: Quaternary Research, v. 11, pp. 21–51.

Burrows, C.J., 1977, Forest vegetation, in C.J. Burrows (editor), Case history and science in the Cass District, Canterbury, New Zealand: University of Canterbury, Christchurch, pp. 233–257.

Burrows, C.J., 1979, A chronology for cool-climate episodes in the Southern Hemisphere 12,000–1,000 years B.P.: Palaeogeography, Palaeoclimatology, Palaeoecology, v. 27, pp. 287–347.

Burrows, C.J., Chinn, T.J.H., and Kelly, M., 1976, Glacial activity in New Zealand near the Pleistocene-Holocene boundary in light of new radiocarbon dates: Boreas, v. 5, pp. 57–60.

Burrows, C.J., and Russell, J.B., 1975, Moraines of the upper Rakaia Valley: Royal Society of New Zealand Journal, v. 5, pp. 463–477.

Butzer, K.W., 1975, Patterns of environmental change in the Near East during late Pleistocene and early Holocene times, in F. Wendorf and A.E. Marks (editors), Problems in prehistory; North Africa and the Levant: Dallas, Southern Methodist University Press, pp. 389–410.

Campbell, I. B., 1986 New Occurrences and distribution of the Kawakawa tephra in South Island, New Zealand: New Zealand Journal of Geology and Geophysics, v. 29, pp. 425–435.

Campbell, J.D., and Coombs, D.S., 1966, Murihiku Supergroup (Triassic to Jurassic) of Southland and South Otago: New Zealand Journal of Geology and Geophysics, v. 9, pp. 393–398.

Cannon, P.J., 1976, Generation of explicit parameters for a quantitated geomorphic study of the Mill Creek drainage basin: Oklahoma Geology Notes, v. 36, pp. 3–17.

Cerling, T.E., 1990, Dating geomorphologic surfaces using cosmogenic $^3$He: Quaternary Research, v. 33, pp. 148–156.

Cerling, T.E., Quade, J., Wang, Y., and Bowman, J.R., 1989, Carbon isotopes in soils and palaeosols as ecology and palaeoecology indicators: Nature, v. 341, pp. 138–139.

Chadwick, O.A., Brimhall, G.H., and Hendricks, D.M., 1990, From a black to a gray box—a mass balance approach to understanding soil processes, in P.L.K. Knuepfer and L.D. McFadden (editors), Soils and Landscape Evolution, 1990 Binghamton Geomorphology Symposium: Geomorphology, v. 3, pp. 369–390.

Chadwick, O.A., and Davis, J.O., 1990, Soil-forming intervals caused by eolian sediment pulses in the Lahontan Basin, northwestern Nevada: Geology, v. 18, pp. 243–246.

Chadwick, O.A., Hecker, S., and Fonseca, J., 1984, A soils chronosequence at Terrace Creek; studies of late Quaternary tectonism in Dixie Valley, Nevada: U.S. Geological Survey Open-File Report 84–90, 29 pp.

Chadwick, O.A., Sowers, J.M., and Amundson, R.G., 1988, Morphology of calcite crystals in clast coatings from four soils in the Mojave Desert region: Soil Science Society of America Journal, v. 52, pp. 211–219.

Chandler, R.J., 1972, Periglacial mudslides in Vestspitzbergen and their bearing on the origin of fossil solifluction shears in low angled clay slopes: Quarterly Journal of Engineering Geology, v. 5, pp. 223–241.

Chappell, J., 1983, A revised sea-level record for the last 300,000 years on Papua New Guinea: Search, v. 14, pp. 99–101.

Chappell, J., and Shackleton, N.J., 1986, Oxygen isotopes and sea level: Nature, v. 324, pp. 137–140.

Chase, C.G., and Wallace, T.C., 1986, Uplift of the Sierra Nevada, California: Geology, v. 14, pp. 730–733.

Chase, C.G., and Wallace, T.C., 1988, Flexural isostasy and uplift of the Sierra Nevada, California: Journal of Geophysical Research, v. 93, pp. 2795–2802.

Chinn, T.J.H., 1975, Late Quaternary snowlines and cirque moraines within the Waimakariri watershed: University of Canterbury M.S. Thesis, Christchurch, New Zealand, 213 pp.

Chinn, T.J.H., 1981, Use of rock weathering-rind thickness for Holocene absolute age-dating in New Zealand: Arctic and Alpine Research, v. 13, pp. 33–45.

Chorley, R.J., 1962, Geomorphology and the general systems theory: U.S. Geological Survey Professional Paper 500–B, 10 pp.

Chorley, R.J., and Kennedy, B.A., 1971, Physical geography, a systems approach: London, Prentice-Hall International, 370 pp.

Chorley, R.J., Schumm, S.A., and Sugden, D.E., 1984, Geomorphology: London, Methuen, 605 pp.

Christenson, G.E., and Purcell, C., 1985, Correlation and age of Quaternary alluvial-fan sequences, Basin and Range province, southwestern United States, in D.E. Weide (editor), Soils and Quaternary geology of the southwestern United States: Geological Society of America Special Paper 203, pp. 115–122.

Church, M., and Slaymaker, O., 1989, Disequilibrium of Holocene sediment yield in glaciated British Columbia: Nature, v. 337, pp. 452–454.

COHMAP, 1988, Climatic changes of the last 18,000 years: Observations and model simulations: Science, v. 241, pp. 1043–1052.

Cole, K.L., 1982, Late Quaternary environments in the eastern Grand Canyon: Vegetational gradients over the last 25,000 years: University of Arizona, Ph.D. Dissertation.

Cole, K.L., 1985, Past rates of change, species richness, and a model of vegetational inertia in the Grand Canyon, Arizona: American Naturalist, v. 125, pp. 289–303.

Cole, K.L., 1986, The Lower Colorado River valley; a Pleistocene desert: Quaternary Research, v. 25, pp. 392–400.

Colman, S.M., 1982, Chemical weathering of basalts and andesites; evidence from weathering rinds: U.S. Geological Survey Professional Paper 1246, 51 pp.

Colman, S.M., and Pierce, K.L., 1981, Weathering rinds on andesitic and basaltic stones as a Quaternary age indicator, western United States: U.S. Geological Survey Professional Paper 1210, 56 pp.

Colman, S.M., Pierce, K. L., and Birkeland, P. W., 1987, Suggested terminology for Quaternary dating methods: Quaternary Research, v. 28, pp. 314–319.

Cooke, R.U., 1970, Stone pavements in deserts: Association of American Geographers Annals, v. 60, pp. 560–577.

Cooke, R.U., 1984, Geomorphological hazards in Los Angeles: London, George Allen and Unwin, 206 pp.

Cooke, R.U., and Warren, A., 1973, Geomorphology in deserts: London, Batsford Ltd., 394 pp.

Costa, J.E., 1986, A history of paleoflood hydrology in the United States, 1800–1970: EOS, v. 67, pp. 425, 428–430.

Costa, J.E., and Baker, V.R., 1981, Surficial geology, building with the earth: New York, John Wiley, 498 pp.

Cotton, C.A., 1974, Bold coasts: Wellington, A.W. Reed, 354 pp.

Cotton, C.A., and Te Punga, 1955, Solifluction and periglacially modified landforms at Wellington, New Zealand: Royal Society of New Zealand Transactions, v. 82, pp. 1000–1031.

Craig, R.G., Roberts, B.L., and Singer, M.P., 1984, Climates and lakes of the Death Valley drainage system during the last glacial maximum: Kent State University Report to Battelle Memorial Institute, Pacific Northwest Laboratories, 157 pp.

Crampton, J.S., 1985, The geology of the Monkey Face area, Marlborough, with special reference to the Mesozoic strata: B.Sc. Honors thesis, University of Otago, Dunedin, New Zealand.

Crook, R., Jr., 1986, Relative dating of Quaternary deposits based on P-wave velocities in weathered granitic clasts: Quaternary Research, v. 25, pp. 281–292.

Crook, R., Jr., and Kamb, R., 1980, A new method of alluvial age dating based on progressive weathering, with application to the time-history of fault activity in southern California: U.S. Geological Survey Open-File Report 80–1144.

Crook, R., Jr., Kamb, B., Allen, C.R., Payne, C.M., and Proctor, R.J., 1978, Quaternary geology and seismic hazard of the Sierra Madre and associated faults, western San Gabriel Mountains, California: Final Technical Report to U.S. Geological Survey Contract No. 14–08–0001–15258.

Dan, J., 1981, Soils of the Arava Valley, in J. Dan, R. Gerson, H. Koyundjisky, and D. Yaalon (editors), Aridic soils of Israel, properties, genesis, and management: International Conference on Aridic Soils, Israel Agricultural Research Organization Institute of Soils and Water, pp. 297–349.

Dan, J., Gerson, R., Koyundjisky, H., and Yaalon, D. (editors), 1981, Aridic soils of Israel: Agricultural Research Organization, Bet Dagan Special Publication 190, 353 pp.

Dan, J., and Yaalon, D.H., 1971, On the origin and nature of the paleopedological formations in the coastal desert fringe areas of Israel, *in* D.H. Yaalon (editor), Paleopedology: Israel University Press, pp. 245–260.

Dan, J., Yaalon, D.H., Moshe, R., and Nissim, S., 1982, Evolution of Reg soils in southern Israel and Sinai: Geoderma, v. 28, pp. 173–202.

Danin, A., 1986, Patterns of biogenic weathering as indicators of paleoclimates in Israel: Proceedings of the Royal Society of Edinburg, v. 89B, pp. 243–253.

Danin, A., Gerson, R., Garty, J., and Marton, K., 1982, Patterns of limestone and dolomite weathering by lichen and blue-green algae and their palaeoclimatic significance: Palaeogeography, Palaeoclimatology, Palaeoecology, v. 37, pp. 221–233.

Dansgaard, W.S., Johnsen, S.J., Moller, J., and Langway, C.C., 1969, One thousand centuries of climatic record from Camp Century on the Greenland Ice Sheet: Science, v. 166, pp. 377–381.

Davis, O.K., 1984, Multiple thermal maxima during the Holocene: Science, v. 244, pp. 617–619.

Davis, O.K., Anderson, R.S., Fall, P.L., O'Rourke, M.K., and Thompson, R.S., 1985, Palynological evidence for early Holocene aridity in the southern Sierra Nevada of California: Quaternary Research, v. 24, pp. 322–332.

Davis, O.K., Sheppard, J.C., and Robertson, S., 1986, Contrasting climatic histories for the Snake River Plain, Idaho, resulting from a multiple thermal maxima: Quaternary Research, v. 26, pp. 321–339.

Davis, W.M., 1889, The rivers and valleys of Pennsylvania: National Geographic Magazine, v. 1, pp. 183–253.

Davis, W.M., 1899, The geographical cycle: Geographical Journal, v. 14, pp. 481–501.

Davis, W.M., 1902, Base-level, grade, and peneplain: Journal of Geology, v. 10, pp. 77–111.

Demsey, K., 1987, Holocene faulting and tectonic geomorphology along the Wassuk Range, west-central Nevada: University of Arizona, Geosciences Department, M.S. prepublication manuscript, 64 pp.

Denny, C.S., 1965, Alluvial fans in the Death Valley region, California and Nevada: U.S. Geological Survey Professional Paper 466, 62 pp.

Derbyshire, E., 1973, Climatic geomorphology: London, McMillan, 296 pp.

Derbyshire, E., 1976, Geomorphology and climate: New York, John Wiley and Sons.

Dickey, D.D., Carr, W.J., and Bull, W.B., 1980, Geologic map of the Parker NW, Parker, and parts of the Whipple Mountains SW and Whipple Wash Quadrangles, California and Arizona: U.S. Geological Survey Miscellaneous Investigations Series Map I-1124.

Dietrich, W.E., Kirchner, J.W., Ikeda, H., and Iseya, F., 1989, Sediment supply and the development of the coarse surface layer in gravel-bedded rivers: Nature, v. 340, pp. 215–217.

Doell, R.R., Dalrymple, G.V., and Cox, A., 1966, Geomagnetic polarity epics, Sierra Nevada data, part 3: Journal of Geophysical Research, v. 71, pp. 531–541.

Dorn, R.I., 1983, Cation-ratio dating: A new rock varnish age-determination technique: Quaternary Research, v. 20, pp. 49–73.

Dorn, R.I., 1984a, Speculations on the cause and implications of rock varnish microchemical laminations: Nature, v. 310, pp. 767–770.

Dorn, R.I., 1984b, Geomorphological interpretation of rock varnish in the Mojave Desert, *in* J.C. Dohrenwend (editor), Surficial geology of the eastern Mojave Desert, California, Field Trip 14: 97th Annual Meeting of the Geological Society of America, pp. 69–87.

Dorn, R.I., 1988, A rock varnish interpretation of alluvial-fan development in Death Valley, California: National Geographic Research, v. 4, pp. 56–73.

Dorn, R.I., 1989a, Accelerator mass spectrometry radiocarbon dating of rock varnish: Geological Society of America Bulletin.

Dorn, R.I., 1989b, Cation-ratio dating; a geographical perspective: Progress in Physical Geography, v. 13, p. 559–596.

Dorn, R.I., Bamforth, D.B., Cahill, T.A., Dohrenwend, J.C, Turrin, B.D., Donahue, D.J., Jull, A.J.T., Long, A., Macko, M.E., Weil, E.B., Whitley, D.S., and Zabel, T.H., 1986, Cation-ratio and accelerator radiocarbon dating of rock varnish on Mojave artifacts and landforms: Science, v. 231, pp. 830–833.

Dorn, R.I., and DeNiro, M.J., 1985, Stable carbon isotope ratios of rock varnish organic matter: A new paleoenvironmental indicator: Science, v. 227, pp. 1472–1474.

Dorn, R.I., DeNiro, M.J., and Ajie, H.O, 1987c,

Isotopic evidence for climatic influence on alluvial-fan development in Death Valley, California: Geology, v. 15, pp. 108–110.

Dorn, R.I., Jull, A.J.T., Donahue, D.J., Linick, T.W., and Toolin, L.J., 1989, Accelerator mass spectrometry radiocarbon dating of rock varnish: Geological Society of America Bulletin, v. 101, pp. 1363–1372.

Dorn, R.I., and Oberlander, T.M., 1981, Microbial origin of desert varnish: Science, v. 213, pp. 1245–1247.

Dorn, R.I., and Oberlander, T.M., 1982, Rock varnish: Progress in Physical Geography, v. 6, pp. 317–367.

Dorn, R.I., Tanner, D., Turrin, B.D., and Dohrenwend, J.C., 1987a, Cation-ratio dating of Quaternary materials in the east-central Mojave Desert, California: Physical Geography, v. 8, pp. 72–81.

Dorn, R.I., Turrin, B.D., Jull, A.J.T., Linick, T.W., and Donahue, D.J., 1987b, Radiocarbon and cation-ratio ages for rock varnish on Tioga and Tahoe morainal boulders of Pine Creek, eastern Sierra Nevada in California, and paleoclimatic implications: Quaternary Research, v. 28, pp. 38–49.

Douglas, A.V., 1976, Past air-sea interactions over the eastern north Pacific Ocean as revealed by tree-ring data: University of Arizona Ph.D. dissertation, 196 pp.

Douglas, A.V., 1981, On the influence of warm equatorial conditions in the central Pacific on climatic patterns in the United States, 1977–1980: Proceedings of the 5th Annual Climate Diagnostics Workshop, October 22–24, 1980, Seattle, Washington, U.S. Department of Commerce, National Oceanic and Atmospheric Administration, pp. 239–250.

Douglas, I., 1967, Man, vegetation and sediment yields of rivers: Nature, v. 215, pp. 925–928.

Eckis, R.P., 1928, Alluvial fans of the Cucamonga district, southern California: Journal of Geology, v. 36, pp. 224–247.

Edwards, R.L., Chen, J.H., Ku, T.L., and Wasserburg, G.J., 1987, Precise timing of the last interglacial period from mass spectrometric determination of thorium-230 in corals: Science, v. 236, pp. 1547–1552.

Elvidge, C.D., 1982, Reexamination of the rate of desert varnish formation reported south of Barstow, California: Earth Surface Processes, v. 7, pp. 345–348.

Elvidge, C.D., and Moore, C.B., 1979, A model for desert varnish formation: Geological Society of America Abstracts with Programs, v. 11, p. 271.

Emmett, W.W., 1974, Channel aggradation in western United States as indicated by observations at Vigil Network sites: Paper presented to the International Symposium on Geomorphic Processes in Arid Environments, Jerusalem, Israel.

Emmett, W.W., 1976, Bedload transport in two large gravel-bed streams, Idaho and Washington: Proceedings of the Third Federal Inter-Agency Sedimentation Conference, 4.101–4.113.

Enzel, Y., 1989, Hydrology of a large closed arid watershed as a basis for paleohydrological studies in the Mojave River drainage system, Soda Lake and Silver Lake Playas, southern California: University of New Mexico Ph.D. dissertation.

Enzel, Y., Brown, W.J., Anderson, R.Y., and Wells, S.G., 1988, Late Pleistocene-early Holocene lake stand events recorded in cored lake deposits and shore features, Silver Lake Playa, eastern Mojave Desert, southern California: Geological Society of America Abstracts with Programs, Cordilleran Section, v. 20, p. 158.

Evenari, J., Yaalon, D.H., and Gutterman, Y., 1974, Note on soils with vesicular structure in deserts: Geomorphology, v. 18, pp. 162–172.

Eybergen, F.A., and Imeson, A.C., 1989, Geomorphic processes and climatic change: Catena, v. 16, pp. 307–319.

Farr, T. G., 1985, Age-dating volcanic and alluvial surfaces with multipolarization data: in NASA/JPL Aircraft SAR Workshop Proceedings: JPL Publication 85–39, pp. 31–36.

Fisk, H.N., 1944, Geological investigation of the alluvial valley of the lower Mississippi River: Vicksburg, Mississippi, Mississippi River Commission.

Flint, R.F., 1971, Glacial and Quaternary geology: New York, John Wiley and Sons.

Forman, S.L., 1989, Application and limitations of thermoluminescence to date Quaternary sediments: Quaternary International, v. 1, pp. 47–59.

Forman, S.L., Jackson, M.E., McCalpin, J., and Maat,

P., 1988, The potential of using thermoluminescence to date buried soils in colluvial and fluvial sediments from Utah and Colorado, U.S.A., preliminary results: Quaternary Science Reviews, v. 7, pp. 68–77.

Forman, S.L., Machette, M.N., Jackson, M.E., and Maat, P., 1989, Evaluation of thermoluminescence dating of paleoearthquakes on the American Fork segment, Wasatch fault zone, Utah: Journal of Geophysical Research, v. 94, pp. 1622–1630.

Free, E.E., 1911, Desert pavements and analogous phenomena: Science, v. 33, p. 355.

French, H.M., 1988, Active layer processes, in M.J. Clark (editor), Advances in periglacial geomorphology: New York, John Wiley, pp. 151–177.

Frye, J.C., 1961, Fluvial deposition and the glacial cycle: Journal of Geology, v. 69, pp. 600–603.

Galloway, R.W., 1970, The full-glacial climate in the southwestern United States: Annals of the Association of American Geographers, v. 60, pp. 245–256.

Gat, J.R., and Magaritz, M., 1980, Climatic variations in the eastern Mediterranean Sea area: Naturwissenschaften, v. 67, pp. 80–87.

Gerson, R., 1981, Geomorphic aspects of the Elat Mountains, in J. Dan, R. Gerson, H. Koyundjisky, and D. Yaalon (editors), Aridic soils of Israel, properties, genesis, and management: International Conference on Aridic Soils, Israel Agricultural Research Organization Institute of Soils and Water, pp. 279–296.

Gerson, R., 1982a, Talus relicts in deserts: A key to major climatic fluctuations: Israel Journal of Earth-Sciences, v. 31, pp. 123–132.

Gerson, R., 1982b, The Middle East; landforms of a planetary desert through environmental changes: Striae, v. 17, pp. 52–78.

Gerson, R., and Amit, R., 1987, Rates and modes of dust accretion and deposition in an arid region: the Negev, Israel, in L. Fostick and I. Reid (editors), Desert sediments: ancient and modern: Geological Society Special Publication 35, pp. 157–169.

Gerson, R., Amit, R., and Grossman, S., 1985, Dust availability in deserts; a study in the deserts of Israel and the Sinai: Institute of Earth Sciences, Hebrew University of Jerusalem Contract Report DAJA45–83–C-0041 for the U.S. Army Research, Development and Standardization Group.

Gerson, R., Bull, W.B., Fleischhauer, L.H., McHargue, L.H., Mayer, L., Shih, E.H.H., Tucker, W.C., 1978, Origin and distribution of gravel in stream systems of arid regions: U.S. Air Force Office of Scientific Research Contract Report No. F49–620–77–C-0115.

Gerson, R., and Yair, A., 1975, Geomorphic evolution of some small desert watersheds and certain paleoclimatic implications, Santa Catherina area, southern Sinai: Zeitschrift für Geomorphologie, v. 19, pp. 66–82.

Gilbert, G.K., 1879, Geology of the Henry Mountains (Utah): U.S. Geographical and Geological Survey of the Rocky Mountain Region, Washington, D.C., U.S. Government Printing Office, 170 pp.

Gile, L.H., 1975, Holocene soils and soil-geomorphic relations in an arid region of southern New Mexico: Quaternary Research, v. 5, pp. 321–360.

Gile, L.H., Hawley, J.W., and Grossman, R.B., 1981, Soils and geomorphology in the Basin and Range area of southern New Mexico—Guidebook to the Desert Project: New Mexico Bureau of Mines and Mineral Resources Memoir 39, 222 pp.

Gile, L.H., Peterson, F.F., and Grossman, R.B., 1965, The K horizon—master soil horizon of carbonate accumulation: Soil Science, v. 99, no. 2, pp. 74–82.

Gile, L.H., Peterson, F.F., and Grossman, R.B., 1966, Morphological and genetic sequences of carbonate accumulation in desert soils: Soil Science, v. 101, pp. 347–360.

Gile, L.H., Peterson, F.F., and Grossman, R.B., 1979, The Desert Project soil monograph: U.S. Department of Agriculture, Washington, D.C., U.S. Soil Conservation Service, 984 pp.

Gillespie, A.R., 1982, Relative dating of moraines using acoustic wave speeds in boulders, in Quaternary glaciation and tectonism in the southeastern Sierra Nevada, Inyo County, California: Ph.D. dissertation, California Institute of Technology, Pasadena, California, pp. 53–189.

Gillespie, A.R., Kahle, A.B., and Palluconi, F.D., 1984, Mapping alluvial fans in Death Valley, California, using multi-channel thermal infrared images: Geophysical Research Letters, v. 11, pp. 1153–1156.

Goldberg, P., 1977, Late Quaternary stratigraphy of Gebel Maghara, in O. Bar Yosef and J.L. Phillips (editors), Prehistoric investigations in Gebel Maghara, northern Sinai, Qedem, v. 7: Jerusalem, pp. 11–31.

Goldberg, P., and Bar-Yosef, O., 1982, Environmental and archeological evidence for climatic change in the southern Levant, in J.L. Bintliff and W. van Zeist (editors), Palaeoclimates, palaeoenvironments and human communities in the eastern Mediterranean region in later prehistory: British Archaeological Reports International Series (no. 133), Oxford, pp. 399–414.

Gomez, B., 1983, Temporal variations in bedload transport rates, the effect of progressive armoring: Earth Surface Processes and Landforms, v. 8, pp. 41–54.

Goodfriend, G.A., 1987, Radiocarbon age anomalies in shell carbonate of land snails from semi-arid areas: Radiocarbon, v. 29, pp. 159–167.

Goodfriend, G.A., 1988, Mid-Holocene rainfall in the Negev Desert from $^{13}$C of land snail shell organic matter: Nature, v. 333, pp. 757–760.

Goodfriend, G.A., in press, Rainfall in the Negev Desert during the middle Holocene, based on $^{13}$C of organic matter in land snail shells: Quaternary Research.

Goodfriend, G.A., and Magaritz, M., 1988, Palaeosols and late Pleistocene rainfall fluctuations in the Negev Desert: Nature, v. 332, pp. 144–146.

Goodfriend, G.A., Magaritz, M., and Carmi, I., 1986, A high stand of the Dead Sea at the end of the Neolithic period: paleoclimatic and archeological implications: Climatic Change, v. 9, pp. 349–356.

Goudie, A.S., 1978, Dust storms and their geomorphological implications: Journal of Arid Environments, v. 1, pp. 291–310.

Gould, S.J., 1966, Allometry and size in ontogeny and phylogeny: Cambridge Philosophical Society Biological Review, v. 41, pp. 587–640.

Graf, W.L., 1979, Catastrophe theory as a model for change in fluvial systems, in D.D. Rhodes and G.P. Williams (editors), Adjustments of the fluvial system: Proceedings of the 10th Annual Binghamton Geomorphology Symposia Series, Dubuque, Iowa, Kendall/Hunt, pp. 13–32.

Graf, W.L., 1982a, Distance decay and arroyo development in the Henry Mountains region, Utah:

American Journal of Science, v. 282, pp. 1541–1554.

Graf, W.L., 1982b, Spatial variations of fluvial processes in semiarid lands, in C.E. Thorn (editor), Space and time in geomorphology: Proceedings of the 12th Annual Binghamton Symposium, Boston, Allen and Unwin, pp. 193–217.

Graf, W.L., 1983a, Variability of sediment removal in a semiarid watershed: Water Resources Research, v. 19, pp. 643–652.

Graf, W.L., 1983b, Downstream changes in stream power, Henry Mountains: Association of American Geographers Annals, v. 73, pp. 373–387.

Gregory, K.J., and Walling, D.E., 1973, Drainage basin form and process: New York, John Wiley, 456 pp.

Griffiths, G.A., 1979, High sediment yields from major rivers of the western Southern Alps, New Zealand: Nature, v. 282, pp. 61–63.

Griffiths, G.A., and McSaveney, M.J., 1983, Distribution of mean annual precipitation across some steepland regions of New Zealand: New Zealand Journal of Science, v. 26, pp. 197–209.

Gvirtzman, G., 1976, Late Wurm temperature depression in the Middle East—15 C; evidence from fossil snowlines on Mount Hermon and Jebel Catharina, Sinai, in Geography in Israel, a collection of papers offered to the 23rd International Geographical Congress, U.S.S.R., July-August 1976, pp. 364–372.

Hack, J.T., 1960, Interpretation of erosional topography in humid temperate regions: American Journal of Science (Bradley Volume) 258–A, pp. 80–97.

Hack, J.T., 1965a, Geomorphology of the Shenandoah Valley, Virginia and West Virginia, an origin of the residual ore deposits: U.S. Geological Survey Professional Paper 484, 84 pp.

Hack, J.T., 1965b, Postglacial drainage evolution in the Ontonagan area, Michigan: U.S. Geological Survey Professional Paper 504–B, pp. 1–40.

Hack, J.T., 1973, Stream-profile analysis and stream-gradient index: U.S. Geological Survey Journal of Research, v. 1, pp. 421–429.

Hales, J.E., 1974, Southwestern United States summer monsoon source—Gulf of Mexico or Pacific Ocean?: Weatherwise, v. 27, pp. 148–155.

Hamilton, W., 1964, Geologic map of the Big Maria Mountains, NE Quadrangle, Riverside Co., California and Yuma Co., Arizona: U.S. Geological Survey, Map GQ-350.

Hanes, T.L., 1971, Succession after fire in the chaparral of southern California: Ecological Monographs, v. 41, pp. 27–52.

Hanes, T.L., 1977, Chaparral, in Terrestrial vegetation of California: New York, John Wiley and Sons.

Harden, J.W., 1982, A quantitative index of soil development from field descriptions; examples from a chronosequence in central California: Geoderma, v. 28, pp. 1–28.

Harden, J.W., Matti, J.C., and Terhune, C., 1986, Late Quaternary slip rates along the San Andreas fault near Yucaipa, California, derived from soil development on fluvial terraces: Geological Society of America Abstracts with Programs, v. 18, p. 113.

Harden, J.W., and Taylor, E.M., 1983, A quantitative comparison of soil development in four climatic regimes: Quaternary Research, v. 28, pp. 342–359.

Harrington, C.D., and Whitney, J.W., 1987, Scanning electron microscope method for rock-varnish dating: Geology, v. 15, pp. 967–970.

Harrison, J.B.J., McFadden, L.D., and Weldon, R.J., III, 1990, Spatial soil variability in the Cajon Pass chronosequence: implications for the use of soils as a geochronological tool, in P.L.K. Knuepfer and L.D. McFadden (editors), Soils and Landscape Evolution, 1990 Binghamton Geomorphology Symposium: Geomorphology, v. 3, pp. 399–419.

Haynes, C.V., Jr., 1968, Geochronology of late Quaternary alluvium, in R.B. Morrison and H.E. Wright (editors), Means of correlation of Quaternary successions: Salt Lake City, University of Utah Press, pp. 591–631.

Haynes, C.V., Jr., 1982, Great sand sea and Selima sand sheet, eastern Sahara: geochronology of desertification: Science, v. 217, pp. 629–633.

Haynes, C.V., Jr., 1987, Holocene migration rates of the Sudano-Sahelian wetting front, Araba'in, eastern Sahara, in A.E. Close (editor), Prehistory of arid north Africa: Dallas, Southern Methodist University Press, pp. 69–84.

Haynes, C.V., Jr., and Mead, A.R., 1987, Radiocarbon dating and paleoclimatic significance of subfossil Limicolaria in northwestern Sudan: Quaternary Research, v. 28, pp. 86–99.

Hecker, S., 1985, Timing of Holocene faulting in part of a seismic belt, west-central Nevada: University of Arizona, Geosciences Department, M.S. prepublication manuscript, 42 pp.

Hicken, E.J., 1983, River channel changes: retrospect and prospect, in J.D. Collinson and J. Lewin (editors), Modern and ancient fluvial systems: International Association of Sedimentologists Special Publication 6, pp. 61–83.

Hooke, R.LeB., and Lively, R.S., 1979, Dating of Quaternary deposits and associated tectonic events uranium-thorium methods, Death Valley, California: Report for NSF Grant EAR-7919999.

Horton, R.E., 1932, Drainage basin characteristics: American Geophysical Union Transactions, v. 13, pp. 350–361.

Howard, A.D., 1959, Numerical systems of terrace nomenclature; a critique: Journal of Geology, v. 67, pp. 239–243.

Howard, A.D., 1965, Geomorphological systems–equilibrium and dynamics: American Journal of Science, v. 263, pp. 302–312.

Huesser, L., 1978, Pollen in the Santa Barbara Basin, a 12,000 year record: Geological Society of America Bulletin, v. 89, pp. 673–678.

Hume, T.M., Sherwood, A.M., and Nelson, C.S., 1975, Alluvial sedimentology of the Upper Pleistocene Hinuera Formation, Hamilton Basin, New Zealand: Journal of the Royal Society of New Zealand, v. 5, pp. 421–462.

Hunt, C.B., 1962, Stratigraphy of desert varnish: U.S. Geological Survey Professional Paper 424–B, pp. 194–195.

Hunt, C.B., and Mabey, D.R., 1966, Stratigraphy and structure of Death Valley, California: U.S. Geological Survey Professional Paper 494–A, 162 pp.

Huntington, E., 1907, Some characteristics of the glacial period in non-glaciated regions: Geological Society of America Bulletin, v. 18, pp. 351–388.

Huntington, E., 1914, The climatic factor as illustrated in arid America: Carnegie Institute of Washington Publication 192, Washington, D.C., 341 pp.

Hutchinson, J.N., 1974, Periglacial solifluxion; an approximate mechanism for clayey soils: Geotechnique, v. 24, pp. 438–443.

Imbrie, J., Hays, J.D., Martinson, D.G., McIntyre, A., Mix, A.C., Morley, J.J., Pisias, N.G., Prell,

W.L., and Shackleton, N.J., 1984, The orbital theory of Pleistocene climate; support from a revised chronology of the marine k$^{18}$O record, *in* A. Berger, J. Imbrie, J. Hays, G. Kukla, and B. Saltzman (editors), Milankovitch and climate, understanding the response to astronomical forcing: Boston, D. Reidel, pp. 269–305.

Imbrie, J., and Imbrie, K.P., 1979, Ice ages; solving the mystery: Short Hills, New Jersey, Enslow.

Israel Meteorological Service, 1967, Series A, Meteorological Notes, no. 21, Climatological standard normals of rainfall, 1931–1960.

Israel Meteorological Service, 1977, Map of average annual rainfall (2nd edition): Israel Meteorological Service, Bet Dagan.

Israel Meteorological Service, 1983, Series A, Meteorological Notes, no. 41, v. 1: Averages of temperature and relative humidity, 1964–1979.

Issar, A., and Eckstein, Y., 1969, The lacustrine beds of Wadi Feiran, Sinai; their origin and significance: Israel Journal of Earth Sciences, v. 18, pp. 21–28.

Jackson, L.E., Jr., MacDonald, G.M., and Wilson, M.C., 1982, Paraglacial origin for terraced river sediments in Bow Valley, Alberta: Canadian Journal of Earth Sciences, v. 19, pp. 2219–2231.

Janda, R.J., Nolan, K.M., Harden, D.R., and Colman, S.M., 1975, Watershed conditions in the drainage basin of Redwood Creek, Humboldt County, California, as of 1973: U.S. Geological Survey Open-File Report 75–568, Menlo Park, California, 257 pp.

Jane, G.T., 1986, Wind damage as an ecological process in mountain beech forests of Canterbury, New Zealand: New Zealand Journal of Ecology, v. 9, pp. 25–39.

Jennings, J.N., 1968, Tafoni, *in* R.W. Fairbridge (editor), Encyclopedia of geomorphology: New York, Reinhold, pp. 1103–1104.

Jenny, H., 1980, The soil resource—origin and behavior: New York, Springer-Verlag, 377 pp.

Jessup R. W., 1960, The Stony Tableland soils of the Australian arid zone and their evolutionary history: Soil Science, v. 11, pp. 188–196.

Johnson, A.M., 1970, Physical processes in geology: Freeman and Cooper, San Francisco, 577 pp.

Johnson, D.L., Keller, E.A., and Rockwell, T.K., 1990, Dynamic pedogenesis: new views on some key soil concepts, and a model for interpreting Quaternary soils: Quaternary Research, v. 33, pp. 306–319.

Johnson, R.G., 1982, Brunhes-Matuyama magnetic reversal dated at 790,000 years B.P. by marine-astronomical correlations: Quaternary Research, v. 17, pp. 135–147.

Jones, P.D., Wigley, T.M.L., and Wright, P.B., 1986, Global temperature variations between 1861 and 1984: Nature, v. 322, pp. 430–434.

Jung, H.J., and Bach, W., 1985, GCM-derived climatic change scenarios due to a $CO_2$-doubling applied for the Mediterranean area: Arch. Met. Geoph. Biocl., Ser. B 35, pp. 323–339.

Kahle, A.B., Shumate, M.S., and Nash, D.B., 1984, Active airborne infrared laser system for identification of surface rock and minerals: Geophysical Research Letters, v. 11, pp. 1149–1152.

Kaufman, A., 1971, U-series dating of Dead Sea basin carbonates: Geochimica et Cosmochimica Acta, v. 35, pp. 1269–1281.

Keigwin, L. D., and Jones, G. A., 1988, Global evidence for abrupt climate events in the marine record of deglaciation: American Geophysical Union Transactions, v. 69, p. 299.

Keller, E.A., 1979, Tectonic geomorphology of the central Ventura basin: Semiannual Technical Report 14–08–0001–17678, Office of Earthquake Studies, U.S. Geological Survey, Menlo Park, California.

Keller, E.A., Zepeda, R.L., Laduzinsky, D.M., Seager, D.B., and Zhao, X., 1985, Late Pleistocene: Final Technical Report, U.S. Geological Survey Contract 14–08–0001–21829, 120 pp.

Kerr, R.A., 1989, The global warming is real: Science, v. 243, p. 603.

Kerr, R.A., 1990, New greenhouse report puts down dissenters: Science, v. 249, p. 481.

King, T.J., 1976, Late Pleistocene–early Holocene history of coniferous woodlands in the Lucerne valley region, Mojave Desert, California: Great Basin Naturalist, v. 36, pp. 227–238.

Knighton, D., 1984, Fluvial forms and processes: London, Edward Arnold, 218 pp.

Knox, J.C., 1972, Valley alluviation in southwestern Wisconsin: Association of America Geographers Annals, v. 62, pp. 401–410.

Knox, J.C., 1976, Concept of the graded stream, *in* W. Melhorn and R. Flemal (editors), Theories of landform development: Binghamton, State University of New York, Publications in Geomorphology, pp. 168–198.

Knox, J.C., 1983, Responses of river systems to Holocene climates, in H.E. Wright, Jr. (editor), Late-Quaternary environments of the United States, v. 2, The Holocene: Minneapolis, University of Minnesota Press, pp. 26–41.

Knox, J.C., 1984, Fluvial responses to small scale climate changes, in J.E. Costa and P.J. Fleischer (editors), Developments and applications of geomorphology: Berlin, Springer Verlag, pp. 318–342.

Knox, J.C., 1985, Responses of floods to Holocene climatic change in the upper Mississippi Valley: Quaternary Research, v. 23, pp. 287–300.

Knox, J.C., 1987, Historical valley floor sedimentation in the upper Mississippi Valley: Association of America Geographers Annals, v. 77, pp. 224–244.

Knox, J.C., McDowell, P.F., and Johnson, W.C., 1978, Holocene fluvial stratigraphy and climatic change in the Driftless area, Wisconsin, in W.C. Mahaney (editor), Quaternary soils: University of East Anglia, Norwich, England, Geo Abstracts Ltd.

Knuepfer, P.L.K., 1984, Tectonic geomorphology and present-day tectonics of the Alpine shear system, South Island, New Zealand: Tucson, University of Arizona, Ph.D. dissertation, 508 pp.

Knuepfer, P.L.K., 1988, Estimating ages of late Quaternary stream terraces from analysis of weathering rinds and soils: Geological Society of America Bulletin, v. 100, pp. 1224–1236.

Kochel, R.C., 1988, Geomorphic impact of large floods; review and new perspectives on magnitude and frequency, in V.R. Baker, R.C. Kochel, and P.C. Patton (editors), Flood Geomorphology: New York, John Wiley, 503 pp.

Kochel, R.C., Baker, V.R., and Patton, P.C., 1982, Paleohydrology of southwestern Texas: Water Resources Research, v. 18, pp. 1165–1183.

Koons, D., 1955, Cliff retreat in the southwestern United States: American Journal of Science, v. 253, pp. 44–52.

Ku, T.L., Bull, W.B., Freeman, S.T., and Knauss, K.G., 1979, $Th^{230}$-$U^{234}$ dating of pedogenic carbonates in gravelly desert soils of Vidal Valley, southeastern California: Geological Society of America Bulletin, Part 1, v. 90, pp. 1063–1073.

Ku, T.L., Ivanovich, M., and Luo, S., 1990, U-series dating of late interglacial high sea stands: Barbados revisited: Quaternary Research, v. 33, pp. 129–147.

Kutzbach, J.E., 1983, The changing pulse of the monsoon, in J.S. Fein and P.L. Stephens (editors), Monsoons: New York, John Wiley, 632 pp.

Kutzbach, J.E., and Guetter, P.J., 1984, Sensitivity of monsoon climates in orbital parameter changes for 9000 yr B.P.; experiments with the NCAR general circulation model, in A. Berger, J. Imbrie, J. Hays, G. Kukla, and B. Saltzman (editors), Milankovitch and climate, understanding the response to astronomical forcing: Boston, D. Reidel, pp. 801–820.

Kutzbach, J.E., and Guetter, P.J., 1986, The influence of changing orbital parameters and surface boundary conditions on climatic simulations for the past 18,000 years: Journal of Atmospheric Science, v. 43, pp. 1726–1759.

Kutzbach, J.E., and Otto-Bleisner, B.L., 1982, The sensitivity of the African-Asian monsoon climate to orbital parameter changes for 9000 yr B.P. in a low-resolution general circulation model: Journal of Atmospheric Science, v. 39, pp. 1177–1188.

Lajoie, K.R., Kern, J.P., Wehmiller, J.F., Kennedy, G.L., Mathiesen, S.A., Sarna-Wojcicki, A.M., Yerkes, R.F., and McCrory, P.F., 1979, Quaternary marine shorelines and crustal deformation, San Diego to Santa Barbara, California, in P.L. Abbott (editor), Geological excursions in the southern California area: San Diego State University, Department of Geological Sciences, pp. 3–15.

Lajoie, K.R., Sarna-Wojcicki, A.M., and Yerkes, R.F., 1982, Quaternary chronology and rates of crustal deformation in the Ventura area, California: Geological Society of America Guidebook 78th Annual Meeting of the Cordilleran Section, pp. 43–51.

LaMarche, V.C., Jr., 1973, Holocene climatic variations inferred from treeline fluctuations in the White Mountains, California: Quaternary Research, v. 3, pp. 632–660.

LaMarche, V.C., Jr., 1974, Paleoclimatic inferences from long tree-ring records: Science, v. 183, pp. 1043–1048.

Lamb, H.H., 1972, Climate; present, past and future, vol. 1, Fundamentals and climate now: London, Methuen.

Lamb, H.H., 1977, Climate; present, past and future, vol. 2: London, Methuen, 835 pp.

Langbein, W.B., and Leopold, L.B., 1964, Quasi-equilibrium states in channel morphology: American Journal of Science, v. 262, pp. 782–794.

Langbein, W.B., and Leopold, L.B., 1968, River channel bars and dunes—theory of kinematic waves: U.S. Geological Survey Professional Paper 422–L, 19 pp.

Langbein, W.B., and Schumm, S.A., 1958, Yield of sediment in relation to mean annual precipitation: American Geophysical Union Transactions, v. 39, pp. 1076–1084.

Langbein, W.B., and others, 1959, Annual runoff in the United States: U.S. Geological Survey Circular 52.

Lattman, L.H., 1973, Calcium carbonate cementation of alluvial fans in southern Nevada: Geological Society of America Bulletin, v. 84, no. 9, pp. 3013–3028.

Lensen, C.J., 1962, Sheet 16—Kaikoura: New Zealand Geological Survey, Geological Map of New Zealand, scale 1:250,000.

Leopold, L.B., and Bull, W.B., 1979, Base level, aggradation, and grade: Proceedings, American Philosophical Society, v. 123, pp. 168–202.

Leopold, L.B., and Emmett, W.W., 1976, Bedload measurements, East Fork River, Wyoming: National Academy of Science Proceedings, v. 73, pp. 1000–1004.

Leopold, L.B., Emmett, W.W., and Myrick, R.M., 1966, Channel and hillslope processes in a semiarid area, New Mexico: U.S. Geological Survey Professional Paper 352–G, pp. 193–253.

Leopold, L.B., and Langbein, W.B., 1962, The concept of entropy in landscape evolution: U.S. Geological Survey Professional Paper 500–A, 20 pp.

Leopold, L.B., and Maddock, T., Jr., 1953, The hydraulic geometry of stream channels and some physiographic implications: U.S. Geological Survey Professional Paper 252.

Leopold, L.B., and Miller, J.P., 1954, A post-glacial chronology for some alluvial valleys in Wyoming: U.S. Geological Survey Water-Supply Paper 1261, 90 pp.

Leopold, L.B., and Miller, J.P., 1956, Ephemeral streams—hydraulic factors and their relation to the drainage net: U.S. Geological Survey Professional Paper 282–A, 36 pp.

Leopold, L.B., Wolman, M.G., and Miller, J.P., 1964, Fluvial processes in geomorphology: San Francisco, W.H. Freeman, 522 pp.

Lewkowicz, A.G., 1988, Slope processes, in M.J. Clark (editor), Advances in periglacial geomorphology: New York, John Wiley, pp. 325–368.

Lézine, A-M., Casanova, J., and Hillaire-Marcel, C., 1990, Across an early Holocene humid phase in western Sahara: pollen and isotope stratigraphy: Geology, v. 18, pp. 264–267.

Long, A., Warnecke, L., Betancourt, J.L., and Thompson, R.S., 1990, Deuterium variations in plant cellulose from fossil packrat middens, in J.L. Betancourt, T.R. Van Devender, and P.S. Martin (editors), Packrat middens, the last 40,000 years of biotic change: Tucson, The University of Arizona Press, 469 pp.

Mabbutt, J.A., 1966, Mantle-controlled planation of pediments: American Journal of Science, v. 264, pp. 78–91.

Mabbutt, J.A., 1977, Desert landforms: Cambridge, Massachusetts, MIT Press, 340 pp.

Machette, M.N., 1985, Calcic soils of the southwestern United States, in D.L. Weide and M.L. Faber (editors), Quaternary soils and geomorphology of the American southwest: Geological Society of America Special Paper 203, pp. 1–21.

Mackay, D.A., 1981, Mt. Fyffe catchments detritus and stream survey: New Zealand Department of Lands and Survey, Marlborough Catchment and Regional Water Board Report, 28 pp.

Mackay, D.A., 1984, Kowhai management area detritus survey: New Zealand Department of Lands and Survey, Marlborough Catchment and Regional Water Board Report, 38 pp.

Magaritz, M., and Goodfriend, G.A., 1987, Movement of the desert boundary in the Levant from latest Pleistocene to early Holocene, in W.H. Berger and L.D. Labeyrie (editors), Abrupt climatic change: evidence and implications: Dordrecht, D. Reidel Publishing Company, pp. 173–183.

Marchand, D.E., and Allwardt, A., 1980, Late Cenozoic stratigraphic units, northeastern San Joaquin Valley, California: U.S. Geological Survey Bulletin 1470, 70 pp.

Marion, G.M., 1989, Correlation between long-term pedogenic CaCO$_3$ formation rate and modern precipitation in deserts of the American Southwest: Quaternary Research, v. 32, pp. 291–295.

Markgraf, V., Bradbery, J.P., Forester, R.M., McCoy, W., Singh, G., and Sternberg, R., 1983, Paleoenvironmental reassessment of the 1.6–million-year-old record from San Agustin Basin, New Mexico: New Mexico Geological Society Guidebook, 34th Field Conference, Socorro Region II, pp. 291–307.

Markgraf, V., and Scott, L., 1981, Lower timberline in central Colorado during the past 15,000 yr: Geology, v. 9, pp. 231–234.

Martinson, D.G., Pisias, N.G., Hays, J.D., Imbrie, J., Moore, T.C., and Shackleton, N.J., 1987, Age dating and the orbital theory of the ice ages; development of a high resolution 0 to 300,000 year chronostratigraphy: Quaternary Research, v. 27, pp. 1–29.

Matti, J.C., Tinsley, J.C., Morton, D.M., and McFadden, L.D., 1982, Holocene faulting history as recorded by alluvial history within the Cucamonga fault zone—a preliminary view, in J.C. Tinsley, J.C. Matti, and L.D. McFadden (editors), Late Quaternary pedogenesis and alluvial chronologies of the Los Angeles and San Gabriel Mountains areas, southern California, and Holocene faulting and alluvial stratigraphy within the Cucamonga fault zone; Field Trip 12, Cordilleran Section of the Geological Society of America, pp. 29–44.

Mayer, L., Gerson, R., and Bull, W.B., 1984, Alluvial gravel production and deposition—a useful indicator of Quaternary climatic changes in deserts, in A.P. Schick (editor), Channels processes—water, sediment, catchment controls: Catena Supplement 5, pp. 137–151.

Mayer, L., McFadden, L.D., and Harden, J. W., 1988, Distribution of calcium carbonate in soils; a model: Geology, v. 16, pp. 303–306.

McClure, H.A., 1978, Ar Rub' Al Khali, in J.G. Zotl and S.S. Sayari (editors), Quaternary in Saudi Arabia: Vienna, Springer-Verlag, pp. 252–263.

McFadden, L.D., 1982, The impacts of temporal and spatial climatic changes on alluvial soils genesis in southern California: Tucson, University of Arizona, Ph.D. dissertation, 430 pp.

McFadden, L.D., 1988, Climatic influences on rates and processes of soil development in Quaternary deposits of southern California, in J. Reinhardt and W.R. Sigleo (editors), Paleosols and weathering through geologic time: principles and applications: Geological Society of America Special Paper 206, pp. 153–177.

McFadden, L.D., and Hendricks, D.M., 1985, Changes in the content and composition of pedogenic iron oxyhydroxides in a chronosequence of soils in southern California: Quaternary Research, v. 23, pp. 189–204.

McFadden, L.D., Ritter, J.B., and Wells, S.G., 1989, Use of multiparameter relative-age methods for age estimation and correlation of alluvial fan surfaces on a desert piedmont, eastern Mojave Desert, California: Quaternary Research, v. 32, p. 276–290.

McFadden, L.D., and Tinsley, J.C., 1982, Soil profile development in xeric climates, in J.C. Tinsley, J.C. Matti, and L.D. McFadden (editors), Late Quaternary pedogenesis and alluvial chronologies of the Los Angeles and San Gabriel Mountains areas, southern California, and Holocene faulting and alluvial stratigraphy within the Cucamonga fault zone; Field Trip 12, Cordilleran Section, Geological Society of America, pp. 15–20.

McFadden, L.D., and Tinsley, J.C., 1985, The rate and depth of accumulation of pedogenic carbonate accumulation in soils: Formation and testing of a compartment model, in D.W Weide and M.L. Faber (editors), Soils and Quaternary geology of the southwestern United States: Geological Society of America Special Paper 203, pp. 23–42.

McFadden, L.D., Tinsley, J.C., and Bull, W.B., 1982, Late Quaternary pedogenesis and alluvial chronologies of the Los Angeles basin and San Gabriel Mountains areas, southern California, in J.C. Tinsley, J.C. Matti, and L.D. McFadden (editors), Late Quaternary pedogenesis and alluvial chronologies of the Los Angeles and San Gabriel Mountains areas, southern California, and Holocene faulting and alluvial stratigraphy within the Cucamonga fault zone; Field Trip 12, Cordilleran Section of the Geological Society of America, pp. 1–13.

McFadden, L.D., and Weldon, R.J., 1987, Rates and processes of soil development in Quaternary terraces in Cajon Pass, southern California: Geological Society of America Bulletin, v. 98, pp. 280–293.

McFadden, L.D., and Wells, S.G., 1989, Soilgeomorphic studies in the Cima volcanic field, eastern Mojave Desert, in T.J. Rice, Jr. (editor), Soils geomorphology relationships in the Mojave Desert, California, Nevada: Field Tour Guidebook for the 1989 Soil Science Society of America Annual Meeting Pre-Meeting Tour, October 12–24, 1989, pp. 40–59.

McFadden, L.D., Wells, S.G., and Dohrenwend, J.C., 1986, Influences of Quaternary climatic changes on processes of soil development on desert loess deposits of the Cima volcanic field, California: Catena, v. 13, pp. 361–389.

McFadden, L.D., Wells, S.G., Dohrenwend, J.C., and Turrin, B.D., 1984, Cumulic soils formed in eolian parent materials on flows of the late Cenozoic Cima volcanic field, California, in J.C. Dohrenwend (editor), Surficial geology of the eastern Mojave Desert, California, Field Trip 14: 97th Annual Meeting of the Geological Society of America, pp. 134–149.

McFadden, L.D., Wells, S.G., and Jercinovich, M.J., 1987, Influences of eolian and pedogenic processes on the origin and evolution of desert pavements: Geology, v. 15, pp. 504–508.

McGee, W.J., 1897, Sheetflood erosion: Geological Society of America Bulletin, v. 8, pp. 87–112.

McGlone, M.S., 1980, Late Quaternary vegetation history of the central North Island, New Zealand: University of Canterbury Ph.D. dissertation.

McGlone, M.S., 1983, Polynesian deforestation of New Zealand: a preliminary synthesis: Archaeology in Oceania, v. 18, pp. 11–25.

McGlone, M.S., 1988, New Zealand, in B. Huntley and T. Webb (editors), Vegetation history: Dordrecht, Boston, Kluwer Academic Publishers, pp. 557–599.

McGlone, M.S., and Bathgate, J.L., 1983, Vegetation and climate history of the Longwood Range, South Island, New Zealand, 12,000 b.p. to present: New Zealand Journal of Botany, v. 21, pp. 293–315.

McGlone, M.S., Nelson, C.S., and Hume, T.M., 1978, Palynology, age and environmental significance of some peat beds in the upper Pleistocene Hin-

uera Formation, South Auckland, New Zealand: Journal of the Royal Society of New Zealand, v. 8, pp. 385–393.

McGlone, M.S., and Topping, W.W., 1983, Late Quaternary vegetation, Tongariro region, central North Island, New Zealand: New Zealand Journal of Botany, v. 21, pp. 53–76.

McHargue, L.E., 1981, Late Quaternary deposition and pedogenesis on the Aquila Mountains piedmont, southeastern Arizona: Tucson, University of Arizona, M.S. thesis, 132 pp.

McKeague, J.A. and Day, J.H., 1966, Dithionite and oxalate extractable Fe and Al as aids in differentiating various classes of soils: Canadian Journal of Soil Science, v. 46, pp. 13–22.

McKeague, J.A., Brydon, J.E., and Miles, N.M., 1971, Differentiation of forms of extractable iron and aluminum in soils: Soil Science Society of America Proceedings, v. 35, pp. 33–38.

McKee, E.D., Hamblin, W.K., and Damon, P.E., 1968, K/Ar age of Lava Dam in Grand Canyon: Geological Society of America Bulletin, v. 79, pp. 133–136.

McKee, E.D., and Schenk, E.T., 1942, The lower canyon lavas and related features of Toroweap in Grand Canyon: Journal of Geomorphology, v. 5, pp. 243–273.

Mehringer, P.J., 1982, Early Holocene climate and vegetation in the eastern Sahara: the evidence from Selima Oasis, Sudan: Geological Society of America Abstracts with Programs, v. 14, p. 564.

Mehringer, P.J., and Ferguson, C.W., 1969, Fluvial occurrence of bristlecone pine (Pinus aristata) in the Mojave Desert mountain range: Journal of Arizona Academy of Science, v. 5, pp. 284–292.

Meisling, K.E., and Weldon, R.J., 1989, Late Cenozoic tectonics of the northwestern San Bernardino Mountains, southern California: Geological Society of America Bulletin, v. 101, pp. 106–128.

Melton, M.A., 1965a, Debris-covered hillslopes of the southern Arizona desert—consideration of their stability and sediment contribution: Journal of Geology, v. 73, pp. 715–729.

Melton, M.A., 1965b, The geomorphic and paleoclimatic significance of alluvial deposits in southern Arizona: Journal of Geology, v. 73, pp. 1–38.

Merriam, R., and Bischoff, J.L., 1975, Bishop ash: A widespread volcanic ash extended to southern California: Journal of Sedimentary Petrology, v. 45, pp. 207–211.

Merritts, D.J., Chadwick, O.A., and Hendricks, D.M., in press, Rates and processes of soil evolution on uplifted marine terraces, northern California: Geoderma.

Metzger, D.G., 1968, The Bouse Formation (Pliocene) of the Parker-Blythe-Cibola area, Arizona and California, in Geological Survey Research 1968: U.S. Geological Survey Professional Paper 600–D, pp. 126–136.

Metzger, D.G., Loeltz, O.J., and Irelna, B., 1973, Geohydrology of the Parker-Blythe-Cibola area, Arizona and California: U.S. Geological Survey Professional Paper 486–G, 130 pp.

Mew, G., Hunt, J.L., Froggatt, P.C., Eden, D.N., and Jackson, R.J., 1986, An occurrence of Kawakawa tephra from the Grey River valley, South Island, New Zealand: New Zealand Journal of Geology and Geophysics, v. 29, pp. 315–322.

Milankovitch, M.M., 1941, Canon of insolation and the ice-age problem: Beograd; Königlich Serbische Akademie, English translation by the Israel Program for Scientific Translations, published for the U.S. Department of Commerce and National Science Foundation, Washington, D.C., 1969, 633 pp.

Mitchell, D.L., 1976, The regionalization of climate in the western United States: Journal of Applied Meteorology, v. 15, pp. 920–927.

Mitchell, J.M., 1963, On the world-wide pattern of secular temperature change, in Changes of climate: UNESCO, Monograph 20, Proceedings of the Rome Symposium, Paris, pp. 161–181.

Moar, N.T., 1966, Plant fragments from Kettlehole Bog, Cass: New Zealand Journal of Botany, v. 4, pp. 596–598.

Moar, N.T., 1980, Late Otiran and early Aranuian grassland in central South Island: New Zealand Journal of Ecology, v. 3, pp. 4–12.

Morrison, R.B., 1964, Lake Lahontan; geology of the southern Carson Desert, Nevada: U.S. Geological Survey Professional Paper 401, 156 pp.

Morrison, R.B., and Davis, J.O., 1984, Quaternary stratigraphy and archaeology of the Lake Lahontan area; a reassessment, in J. Lintz (editor),

Western geological excursions, 1984 annual meeting of the Geological Society of America, v. 1, pp. 252–281, and 50 supplementary pages: University of Nevada Mackay School of Mines, Reno, pp. 252–281.

Morton, D.M., and Miller, F.K., 1975, Geology of the San Andreas fault zone north of San Bernardino between Cajon Creek and Santa Ana Wash, in T.C. Crowell (editor), San Andreas fault in southern California: A guide to San Andreas fault from Mexico to Carrizo Plain: California Division of Mines and Geology Special Report 118, pp. 136–146.

Moss, J. H. and Kochel R. C., 1978, Unexpected geomorphic effects of the Hurricane Agnes storm and flood, Conestoga drainage basin, southeastern Pennsylvania: Journal of Geology, v. 86, pp. 1–11.

Muhs, D.R., 1982, A soil chronosequence on Quaternary marine terraces, San Clemente Island, California: Geoderma, v. 28, pp. 257–283.

Muhs, D.R., 1983, Quaternary sea-level events on northern San Clemente Island, California: Quaternary Research, v. 20, pp. 322–341.

Muhs, D.R., 1985, Amino acid age estimates of marine terraces and sea levels on San Nicolas Island, California: Geology, v. 13, pp. 58–61.

Muhs, D.R., and Szabo, B.J., 1982, Uranium-series age of the Eel Point terrace, San Clemente Island, California: Geology, v. 10, pp. 23–26.

Musick, H.B., 1975, Barrenness of desert pavements in Yuma County, Arizona: Journal of Arizona Academy of Science, v. 10, pp. 24–28.

Neev, D., and Emery, K.O., 1967, The Dead Sea, depositional processes and environments of evaporites: Geological Survey of Israel Bulletin 41, 147 pp.

Neev, D., and Hall, J.K., 1977, Climatic fluctuations during the Holocene as reflected by the Dead Sea levels: in D.C. Greer (editor), Terminal lakes: Proceedings of the International Conference on Desertic Terminal Lakes, Ogden, Utah, pp. 53–60.

Nettleton, W.D., and Peterson, F.F., 1983, Aridisols, in L.B. Wilding, N.E. Smeck, and G.F. Hall (editors), Pedogenesis and soils taxonomy, v. 2, The soil orders: Amsterdam, Elsevier-Verlag, pp. 165–215.

Nettleton, W.D., Witty, J.E., Nelson, R.E., and Holly, J.W., 1975, Genesis of argillic horizons in soils of desert areas of the southwestern United States: Soil Science Society of America Proceedings, v. 39, pp. 919–926.

New Zealand Meteorological Service, 1985, 1951–80 climate maps, 1:2,000,000: Miscellaneous Publication 175, parts 4,6.

Nir, D., 1970, Les lacs quaternaires dans la region de Feiran (Sinai, central): Revue de Geographic Physique et de Geologie Dynamique, v. 12, p. 335–346.

North American Commission on Stratigraphic Nomenclature, 1983, North American stratigraphic code: American Association of Petroleum Geologists Bulletin, v. 67, pp. 841–875.

O'Loughlin, C.L., and Pearce, A.J., 1982, Erosion processes in the mountains, in J.M. Soons, and M.J. Selby (editors), Landforms of New Zealand: Auckland, Longman Paul, pp. 67–79.

Olmsted, F.H., Loeltz, O.J., and Irelna, B., 1973, Geohydrology of the Yuma area, Arizona and California: U.S. Geological Survey Professional Paper 486–H.

Olsen, P.E., 1986, A 40 million year lake record of early Mesozoic orbital climatic forcing: Science, v. 234. pp. 842–848.

Pachur, H.-J., and Braun, G., 1980, The palaeoclimate of the central Sahara, Libya, and the Libyan Desert, in M. Sarntheim, E. Seibold, and P. Rognon (editors), Palaeoecology of Africa, v. 12: Rotterdam, Balkema, pp. 351–363.

Pachur, H.-J., and Kropelin, S., 1987, Waid Howar: Paleoclimatic evidence from an extinct river system in the southeastern Sahara: Science, v. 237, pp. 298–300.

Pachur, H.-J., and Roper, H.-P., 1984, The Libyan (Western) Desert and northern Sudan during the late Pleistocene and Holocene: Berliner Geowissenschaftliche Abhandlungen, Reihe A, v. 50, pp. 249–234.

Parker, R. S., 1977, Experimental study of basin evolution and its hydrologic implications: Colorado State University Ph.D. dissertation, 331 pp.

Patton, P.C., and Schumm, S.A., 1975, Gully erosion, northwestern Colorado: a threshold phenomenon: Geology, v. 3, pp. 88–90.

Patton, P.C., and Schumm, S.A., 1981, Ephemeral-stream processes: implications for studies of Quaternary valley fills: Quaternary Research, v. 15, pp. 24–43.

Pavich, M.J., 1987, Application of mass spectrometric measurement of $^{10}$Be, $^{26}$Al, $^{3}$He to surficial geology, in A.J. Crone and E.M. Omdahl (editors), Proceedings of conference XXXIX, U.S. Geological Survey National Earthquakes-Hazards Reduction Program: U.S. Geological Survey Open-File Report 87–673, pp. 39–41

Pavich, M.J., Brown, L., Harden, J.W., Klein, J., and Middleton, R., 1986, $^{10}$Be distribution in soils from Merced River terraces, California: Geochimica et Cosmochimica Acta, v. 50, pp. 1727–1735.

Pearce, A. J., and Watson, A., 1983, Medium term effects of two landsliding episodes on channel storage of sediment: Earth Surface Processes and Landforms, v. 8, pp. 29–39.

Peltier, L., 1950, The geographical cycle in periglacial regions as it is related to climatic geomorphology: Association of American Geographers Annals, v. 40, pp. 214–236.

Peltier, W.R., and Tushingham, A.M., 1989, Global sea level rise and the greenhouse effect: Might they be connected?: Science, v. 244, pp. 806–810.

Penck, W., 1953, Morphological analysis of landforms: Translation of 1924 German book by H. Czech and K.C. Boxwell, London, MacMillan Co., 429 pp.

Perry, R.S., and Adams, J., 1978, Desert varnish: Evidence of cyclic deposition of manganese: Nature, v. 276, pp. 489–491.

Petit, J.R., Briat, M., and Royer, A., 1981, Ice-age aerosol content from east Antarctica ice core samples and past wind strength: Nature, v. 293, p. 391–394.

Péwé, T.L., 1983, The periglacial environment in North America during Wisconsin time, in S.C. Porter (editor), The late Pleistocene; late-Quaternary environments of the United States: Minneapolis, University of Minnesota Press, pp. 157–189.

Phillips, F.M., Leavy, B.D., Jannick, N.O., Elmore, D., and Kubik, P.W., 1986, The accumulation of cosmogenic chlorine-36 in rocks: a method for surface exposure dating: Science, v. 231, pp. 41–43.

Phillips, F.M., Zreda, M.G., Smith, S.S., Elmore, D., Kubik, P.W., and Sharma, P., 1990, Cosmogenic chlorine-36 chronology for glacial deposits at Bloody Canyon, eastern Sierra Nevada: Science, v. 248, pp. 1529–1532.

Pisias, N., 1978, Paleoceanography of the Santa Barbara Basin during the past 8000 years: Quaternary Research, v. 10, pp. 366–384.

Pisias, N.G., Martinson D.G., Moore, T.C., Shackleton, N.J., Prell, W., Hays, J., and Boden, G., 1984, High resolution stratigraphic correlation of benthic oxygen isotope record spanning the last 300,000 years: Marine Geology, v. 56, pp.119–136.

Porter, S.C., 1975a, Weathering rinds as a relative age criterion; application to subdivision of glacial deposits in the Cascade Range: Geology, v. 3, pp. 101–104.

Porter, S.C., 1975b, Equilibrium-line altitudes of late Quaternary glaciers in the Southern Alps, New Zealand: Quaternary Research, v. 5, pp. 27–47.

Porter, S.C., 1981, Glaciological evidence of Holocene climatic change, in T.M.L. Wigley, M.J. Ingram, and G. Farmer (editors), Climate and history: Cambridge, Cambridge University Press, pp. 82–110.

Porter, S.C., 1989, Some geological implications of average Quaternary glacial conditions: Quaternary Research, v. 32, p. 245–261.

Potter, R.M., and Rossman, G.R., 1977, Desert varnish; the importance of clay minerals: Science, v. 196, pp. 1446–1448.

Powell, J.W., 1875, Exploration of the Colorado River of the west (1869–72): Smithsonian Institute, Washington, D.C.

Prell, W.L., 1984, Monsoon climate of the Arabian Sea during the late Quaternary; a response to changing solar radiation, in A. Berger, J. Imbrie, J. Hays, G. Kukla, and B. Saltzman (editors), Milankovitch and climate, part 1: Reidel, Dordrecht, pp. 349–366.

Quade, J., Cerling, T.E., and Bowman, J.R., 1989, Systematic variations in the carbon and oxygen isotopic composition of pedogenic carbonate along elevation transects in the southern Great Basin, United States: Geological Society of America Bulletin, v. 101, pp. 464–475.

Reheis, M.C., 1987a, Climatic implications of alternating clay and carbonate formation in semiarid soils of south-central Montana: Quaternary Research, v. 27, pp. 270–282.

Reheis, M.C., 1987b, Gypsic soils on the Kane alluvial fans, Big Horn County, Wyoming: U.S. Geological Survey Bulletin 1590–C, 39 pp.

Reheis, M.C., 1990, Influence of climate and eolian dust on the major-element chemistry and clay mineralogy of soils in the northern Bighorn Basin, U.S.A.: Catena, v. 17, pp. 219–248.

Reheis, M.C., Harden, J.W., McFadden, L.D., and Shroba, R.R., 1989, Development rates of late Quaternary soils, Silver Lake Playa, California: Soil Science Society of America Journal, v. 53, pp. 1127–1140.

Ritchie, J.C., Eyles, C.H., and Haynes, C.V., Jr., 1985, Sediment and pollen evidence for an early to mid-Holocene humid period in the eastern Sahara: Nature, v. 324, pp. 352–355.

Ritter, D. F., 1967, Terrace development along the front of the Beartooth Mountains, southern Montana: Geological Society of America Bulletin, v. 78, pp. 467–484.

Ritter, D.F., 1972, The significance of stream capture in the evolution of a piedmont region, southern Montana: Zeitschrift für Geomorphologie, v. 16, pp. 83–92.

Ritter, D.F., 1986, Process geomorphology: Dubuque, Iowa, Wm. C. Brown, 603 pp.

Robertson-Rintoul, M.S.E., 1986, A quantitative soil-stratigraphic approach to the correlation and dating of post-glacial river terraces in Glen Feshie, western Cairngorms: Earth Science Processes, v. 11, pp. 605–611.

Rockwell, T.K., Johnson, D.L., Keller, E.A., and Dembroff, G.R., 1985, A late Pleistocene–Holocene soils chronosequence in the Ventura Basin, southern California, USA, in K.S. Richards, R.R. Arnett, and S. Ellis (editors), Geomorphology and soils: London, Allen and Unwin, pp. 309–326.

Rockwell, T.K., Keller, E.A., Johnson, D.L., and Dembroff, G.R., 1983, Soil chronosequences as instruments for dating Holocene and late Pleistocene faulting, western Transverse Ranges, California: Final Technical Report, U.S. Geological Survey Contract No. 14–08–0001–19781, 81 pp.

Rodgers, R.J., 1980, A numerical model for simulating pedogensis in semiarid regions: University of Utah Ph.D. thesis, 285 pp.

Rosholt, J.N., 1980, Uranium-trend dating of Quaternary sediments: U.S. Geological Survey Open-File Report, 41 pp.

Rosholt, J.N., 1985, Uranium-trend systematics for dating Quaternary sediments: U.S. Geological Survey Open-File Report 85–298, 34 pp.

Rosholt, J.N., Bush, C.A., Carr, W.J., Hoover, D.L., Swadley, W.C., and Dooley, J.R., Jr., 1985, Uranium-trend dating of Quaternary deposits in the Nevada Test Site area, Nevada and California: U.S. Geological Survey Open-File Report 85–540, 37 pp.

Rowell, D.L., 1981, Oxidation and reduction, in D.J. Greenland and M.H.B. Hayes (editors), The chemistry of soil processes: London, John Wiley & Sons, Ltd., pp. 401–461.

Ruhe, R.V., 1952, Topographic discontinuities of the Des Moines lobe: American Journal of Science, v. 250, pp. 46–56.

Ruhe, R.V., 1956, Geomorphic surfaces and the nature of soils: Soil Science, v. 82, pp. 441–455.

Salmon, O., and Schick, A.P., 1980, Infiltration tests, in A.P. Schick (editor), Arid zone geosystems: Research Report DA-JA-DAERO-78G-111, U.S. Army European Research Office, London, Department of Physical Geography, The Hebrew University of Jerusalem, pp. 55–115.

Schenker, A.R., 1977, Particle-size distribution of late Cenozoic gravels on an arid region piedmont, Gila Mountains, Arizona: Tucson, University of Arizona, M.S. thesis in geosciences, 118 pp.

Schick, A.P., 1970, Desert floods—interim results of observations in the Nahal Yael research watershed, southern Israel, 1965–70: Symposium on the Results of Research on Representative and Experimental Basins, Wellington, IAHS Publication 96, pp. 478–493.

Schick, A.P., 1974, Formation and obliteration of desert stream terraces—a conceptual analysis: Zeitschrift für Geomorphologie, Supp. 21, pp. 88–105.

Schick, A.P., 1977, A tentative sediment budget for an extremely arid watershed in the southern Negev, in D.O. Doehring (editor), Geomorphology in arid regions: Binghamton, State University of New York Publications in Geomorphology, pp. 139–163.

Schick, A.P. (editor), 1980, Arid zone geosystems: Research Report DA-JA-DAERO-78G-111, U.S. Army European Research Office, London, De-partment of Physical Geography, The Hebrew University of Jerusalem, 170 pp.

Schick, A.P., and Sharon, D. (editors), 1974, Geomorphology and climatology of arid watersheds: Project Report DAJA-72C-3874, U.S. Army European Research Office, Department of Geography, The Hebrew University of Jerusalem, Israel.

Schlesinger, W.H., 1985, The formation of caliche in soils of the Mojave Desert, California: Geochimica et Cosmochimica Acta, v. 49, pp. 57–66.

Schumm, S.A., 1963, Disparity between present rates of denudation and orogeny: U.S. Geological Survey Professional Paper 454–H, 13 pp.

Schumm, S.A., 1965, Quaternary paleohydrology, in H.E. Wright, Jr. and D.G. Frey (editors), The Quaternary of the United States, pp. 783–794.

Schumm, S.A., 1973, Geomorphic thresholds and complex response of drainage systems, in M. Morisawa (editor), Fluvial geomorphology: Binghamton, State University of New York Publications in Geomorphology, 4th Annual Meeting, pp. 299–310.

Schumm, S.A., 1977, The fluvial system: New York, John Wiley and Sons.

Schumm, S.A., 1985, Explanation and extrapolation in geomorphology; seven reasons for geologic uncertainty: Transactions of Japanese Geomorphological Union, v. 6, pp. 1–18.

Schumm, S.A., and Khan, H.R., 1972, Experimental study of channel patterns: Geological Society of America Bulletin, v. 83, pp. 1755–1770.

Schumm, S.A., and Lichty, R.W., 1965, Time, space and causality in geomorphology: American Journal of Science, v. 263, pp. 110–119.

Schumm, S.A., Mosley, M.P., and Weaver, W.E., 1987, Experimental fluvial geomorphology: New York, John Wiley, 413 pp.

Schumm, S.A., and Parker, R.S., 1973, Implications of complex response of drainage systems for Quaternary alluvial stratigraphy: Nature, v. 243, pp. 99–100.

Schwertmann, J., Murad, E., and Schulze, D.G., 1982, Is there Holocene reddening (Hematite Formation) in soils of axeric temperature areas?: Geoderma, v. 27, pp. 209–223.

Schwertmann, V., 1973, Use of oxalate for Fe extraction from soils: Canadian Journal of Soil Science, v. 53, pp. 244–246.

Schwertmann, V., and Taylor, R.M., 1977, Iron oxides, *in* J.B. Dixon and S.B. Weed (editors), Minerals in soil environments: Madison, Wisconsin, Soil Science of America, pp. 145–180.

Scott, K.M., 1971, Origin and sedimentology of 1969 debris flows near Glendora, California: U.S. Geological Survey Professional Paper 750–C, pp. 242–247.

Selby, M.J., 1982, Controls on the stability and inclinations of hillslopes on hard rock: Earth Surface Processes and Landforms, v. 7, pp. 449–467.

Sellers, W.D., 1983, A quasi-three-dimensional climate model: Journal of Climate and Applied Meteorology, v. 22, pp. 1557–1574.

Sellers, W.D., 1984, The response of a climate model to orbital variations, *in* A. Berger, J. Imbrie, J. Hays, G. Kukla, and B. Saltzman (editors), Milankovitch and climate, understanding the response to astronomical forcing: Boston, D. Reidel, pp. 765–788.

Sellers, W.D., and Hill, R.H. (editors), 1974, Arizona climate 1931–1972: Tucson, University of Arizona Press, 616 pp.

Shafiqullah, M., Damon, P.E., Lynch, D.J., Reynolds, S.J., Rehrig, W.A., and Raymond, R.H., 1980, K-Ar geochronology and geologic history of southwestern Arizona and adjacent areas: Arizona Geological Society Digest, v. 12, pp. 201–260.

Sharon, D., 1972, The spottiness of rainfall in a desert area: Journal of Hydrology, v. 17, pp. 161–175.

Sharp, L.P., and Nobles, L.H., 1953, Mudflow of 1941 at Wrightwood, southern California: Geological Society of America Bulletin, v. 64, pp. 547–560.

Shih, E.H., 1982, Multi-spectral reflectance and image textural signatures of arid alluvial geomorphic surfaces in the Castle Dome Mountains and piedmont, southwestern Arizona: Tucson, University of Arizona, M.S. thesis in geosciences, 287 pp.

Shih, E.H., and Schowengerdt, R.A., 1983, Classification of arid geomorphic surfaces using landsat spectral and textural features: Photogrammetric Engineering and Remote Sensing, v. 49, pp. 337–347.

Shimron, A., 1974, Geology of the Nahal Yael watershed, *in* A.P. Schick and D. Sharon (editors), Geomorphology and climatology of arid watersheds: Project Report DAJA-72C-3874, U.S. Army European Research Office, Department of Geography, The Hebrew University of Jerusalem, Israel.

Shlemon, R.J., 1978, Quaternary soil-geomorphic relationships, southeastern Mojave Desert, California and Arizona, *in* W.C. Mahaney (editor), Quaternary soils: University of East Anglia, Norwich, England, Geo Abstracts Ltd., pp. 187–207.

Sigalove, J.J., 1969, Carbon-14 content and origin of caliche: Tucson, University of Arizona, M.S. thesis in geosciences, 72 pp.

Silbering, N.J., Nichols, K.M., Bradshaw, J.D., and Blome, C.D., 1988, Limestone and chert in tectonic blocks from the Esk Head subterrane, South Island, New Zealand: Geological Society of America Bulletin, v. 99, pp. 1213–1223.

Singer, A., 1980, The paleoclimatic interpretation of clay minerals in soils and weathering profiles: Earth Science Reviews, v. 15, pp. 303–326.

Slate, J.L., 1985, Soil-carbonate genesis in the Pinacate volcanic field, northwestern Sonora, Mexico: University of Arizona, Geosciences Department, M.S. prepublication manuscript, 85 pp.

Slate, J.L., Bull, W.B., Ku, T-L., Shafiqullah, M., Lynch, D.J., and Huang, Y-P., in press, Soil-carbonate genesis in the Pinacate volcanic field, northwestern Sonora, Mexico: Quaternary Research.

Smith, G.I., 1979, Subsurface stratigraphy and geochemistry of late Quaternary evaporites, Searles Lake, California: U.S. Geological Survey Professional Paper 1043.

Smith, G.I., and Street-Perrott, F.A., 1983, Pluvial lakes of the western United States, *in* H.E. Wright, Jr. (editor), Late Quaternary environments of the United States: Minneapolis, University of Minnesota Press, pp. 190–212.

Smith, H.T.U., 1967, Past versus present wind action in the Mojave Desert region, California: U.S. Air Force Research Office Contract Report No. AF19(628)2486, 26 pp.

Snyder, C.T., and Langbein, L.B., 1962, The Pleistocene lake in Spring Valley, Nevada, and its climatic implications: Journal of Geophysical Research, v. 67, pp. 2385–2394.

Soil Survey Staff, 1975, Soil taxonomy: A basic system of soil classification for making and interpreting soil survey: U.S. Department of Agriculture,

Soil Conservation Service, Agriculture Handbook No. 436, Washington, D.C., U.S. Government Printing Office, 754 pp.

Soil Survey Staff, 1981, Replacement chapter to Handbook 18, Soil Survey Manual, released May 1981: U.S. Department of Agriculture, Soil Conservation Service.

Solomon, S.L., Beran, M., and Hogg, W., 1987, The influence of climatic change and climatic variability on the hydrologic regime and water resources: IAHS Publication No. 168, 640 pp.

Soons, J.M., 1979, Late Quaternary environments in the central South Island of New Zealand: New Zealand Geographer, v. 35, pp. 16–23.

Soons, J.M., 1980, Comments on R.J. Wasson on relict slope mantles: Australian Quaternary Newsletter, v. 14, p. 29.

Soons, J.M., and Burrows, C.J., 1978, Dates for Otiran deposits, including plant microfossils and macrofossils from Rakaia Valley: New Zealand Journal of Geology and Geophysics, v. 21, pp. 607–615.

Sowers, J.M., Amundson, R.G., Chadwick, O.A., Harden, J.W., Jull, A.J.T., Ku, T.L., McFadden, L.D., Reheis, M.C., Taylor, E.M., and Szabo, B.J., 1989, Geomorphology and pedology on the Kyle Canyon alluvial fan, southern Nevada, *in* T.J. Rice, Jr. (editor), Soils geomorphology relationships in the Mojave Desert, California, Nevada: Field Tour Guidebook for the 1989 Soil Science Society of America Annual Meeting Pre-Meeting Tour, October 12–14, 1989, pp. 93–112.

Spaulding, W.G., 1985, Vegetation and climates of the last 45,000 years in the vicinity of Nevada Test Site, south-central Nevada: U.S. Geological Survey Professional Paper 1329, 83 pp.

Spaulding, W.G., 1990, Vegetation dynamics during the last deglaciation, southeastern Great Basin, U.S.A.: Quaternary Research, v. 33, pp. 188–203.

Spaulding, W.G., and Graumlich, L.J., 1986, The last pluvial climatic episodes in the deserts of southwestern North America: Nature, v. 320, pp. 441–444.

Spaulding, W.G., Leopold, E.B., and Van Devender, T.R., 1983, Late Wisconsin paleoecology of the American Southwest, *in* H.E. Wright, Jr. (editor), Late Quaternary environments of the United States, v. 1 (The Late Pleistocene, S.C. Porter, editor): Minneapolis, University of Minnesota, pp. 259–293.

Springer, M.E., 1958, Desert pavement and vesicular layer of some desert soils in the desert of the Lahontan Basin, Nevada: Soil Science Society of America Proceedings, v. 22, pp. 63–66.

Stevens, G.R., 1957, Solifluction phenomena in the Lower Hutt area: New Zealand Journal of Science and Technology, v. 38B, pp. 279–296.

Stevens, G.R., 1974, Rugged landscape, the geology of central New Zealand: Wellington, A.W. Reed, 286 pp.

Stevens, P.R., and Walker, T.W., 1970, The chronosequence concept and soil formation: Quarterly Review of Biology, v. 45, pp. 333–350.

Stiller, M., 1979, The influence of Lake Hula drainage on the radiocarbon record and on the sedimentation rate in Lake Kinneret: Israel Journal of Earth Sciences, v. 28, pp. 1–12.

Stoddart, D., 1969, Climatic geomorphology, *in* R. Chorley (editor), Introduction to fluvial processes: London, Methuen and Co., pp. 189–201.

Strahler, A.N., 1952, Hypsometric (area-altitude) analysis of erosional topography: Geological Society of America Bulletin, v. 63, pp. 1117–1142.

Strahler, A.N., 1956, Quantitative slope analysis: Geological Society of America Bulletin, v. 67, pp. 571–596.

Strakhov, N.M., 1967, Principles of lithogenesis, Volume 1 (S.P. Fitsimmons, S.I. Tomkieff, and J.E. Hemingway, translators): Edinburgh, Scotland, Oliver and Boyd Ltd.

Street, A.F., and Grove, A.T., 1979, Global maps of lake-level fluctuations since 30,000 years B.P.: Quaternary Research, v. 12, pp. 83–118.

Stuiver, M., 1982, A high precision calibration of the ad radiocarbon time scale: Radiocarbon, v. 24, pp. 1–26.

Stuiver, M., and Reimer, P.J., 1986, A computer program for radiocarbon age calibration: Radiocarbon, v. 28, pp. 1022–1030.

Suggate, R.P., 1965, Late Pleistocene geology of the northern part of the South Island, New Zealand: New Zealand Geological Survey Bulletin 77, 91 pp.

Switzer, P., Harden, J.W., and Mark, R.K., 1988, A statistical method for estimating rates of soil development and ages of geological deposits: a design for soil chronological studies: Mathematical Geology, v. 20, pp. 49–61.

Taylor, B.D., 1981, Annual sediment yields, *in* Sediment management for southern California mountains, coastal plains, and shoreline; Part B, Inland sediment movements by natural processes: California Institute of Technology Environmental Quality Laboratory, EQL Report 17–B, pp. 31–81.

Taylor, B.R., Brown, W.M., and Brownlie, W.R., 1978, Sediment management for southern California coastal mountains, coastal plains, and shoreline: Progress Report No. 4, California Institute of Technology Environmental Quality Laboratory, EQL Open-File Report 78–1, 17 pp.

Teeri, J.A., 1979, The climatology of the $C_4$ photosynthetic pathway, *in* O.T. Solbrig et al. (editors), Topics in plant population biology: New York, Columbia University Press, pp. 356–374.

Thompson, L.G., Mosley-Thompson, E., Davis, M.E., Bolzan, J.F., Dai, J., Yao, T., Gundestrup, N., Wu, X., Klein, L., and Xie, Z., 1989, Holocene–late Pleistocene climatic ice core records form Qinghai-Tibetan Plateau: Science, v. 246, pp. 474–477.

Thomson, P.A., and MacArthur, R.S., 1969, Major river control, drainage and erosion control scheme for Kaikoura: Marlborough Catchment Board Report.

Thornes, J.B., and Brunsden, D., 1977, Geomorphology and time: New York, John Wiley, 208 pp.

Tonkin, P.J., and Basher, L.R., 1990, Soil stratigraphic techniques in the study of soil and landform evolution across the Southern Alps, New Zealand, *in* P.L.K. Knuepfer and L.D. McFadden (editors), Soils and Landscape Evolution, 1990 Binghamton Geomorphology Symposium: Geomorphology, v. 3, pp. 547–565.

Tricart, J., 1974, Structural Geomorphology: English translation by S. H. Beaver and E. Derbyshire of 1968 French edition, London, Longman Group, 305 pp.

Tricart, J., and Cailleux, A., 1972, Introduction to climatic geomorphology: London, Longman.

Tucker, W.C., Jr., 1979, The geology of the Aguila Mountains quadrangle, Yuma and Maricopa Counties, Arizona: University of Arizona, Geosciences Department, M.S. prepublication manuscript, 29 pp.

Turner, R.M., and Brown, D.E., 1982, Sonoran desertscrub, *in* D.E. Brown (editor), Biotic Communities of the American Southwest—United States and Mexico: Tucson, University of Arizona Press, pp. 181–221.

U.S. Department of Commerce, 1975, Monthly climatic data for the world: National Oceanic and Atmospheric Administration, v. 28.

U.S. Department of Commerce, 1985, Climatological data for each state. National Oceanic and Atmospheric Administration.

U.S. Department of Commerce, Environment Data Service, 1972, Climates of the world: Washington, D.C., Environmental Sciences Services Administration.

Van Campo, E., 1986, Monsoon fluctuations in two 20,000 yr B.P. oxygen-isotope pollen records off southwest India: Quaternary Research, v. 26, pp. 376–388.

Van Devender, T.R., 1973, Late Pleistocene plants and animals of the Sonoran Desert: A survey of ancient packrat middens in southwestern Arizona: Tucson, University of Arizona, Ph.D. dissertation, 179 pp.

Van Devender, T.R., 1977, Holocene woodlands in the southwestern deserts: Science, v. 198, pp. 189–192.

Van Devender, T.R., 1987, Holocene vegetation and climate in the Puerto Blanco Mountains, southwestern Arizona: Quaternary Research, v. 27, pp. 51–72.

Van Devender, T.R., and Spaulding, W.G., 1979, Development of vegetation and climate in the southwestern United States: Science, v. 204, pp. 701–710.

Van Devender, T.R., Thompson, R.S., and Betancourt, J.L., 1987, Vegetation history of the deserts of southwestern North America; the nature and timing of the late Wisconsin-Holocene transition, *in* W.F. Ruddiman and H.E. Wright, Jr. (editors), North America and adjacent oceans during the last glaciation: Boulder, Colorado, Geological Society of America, The Geology of North America, Chapter 15 in v. K-3, pp. 323–352.

Van Dissen, R.J., 1989, Late Quaternary faulting in the Kaikoura region, southeastern Marlborough, New

Zealand: Oregon State University, M.S. thesis in Geology, 72 pp.

Vogel, J.C., Fuls, A., and Danin, A., 1986, Geographical and environmental distribution of $C_3$ and $C_4$ grasses in the Sinai, Negev, and Judean deserts: Oecologia, v. 70, pp. 258–265.

Vogel, J.G., and Waterbolk, H.T., 1972, Groningen radiocarbon dates, geological samples, Dead Sea series, Israel, Lisan: Radiocarbon, v. 14, pp. 46–47.

Wahrhaftig, C., 1965, Stepped topography of the southern Sierra Nevada: Geological Society of America Bulletin, v. 76, pp. 1165–1190.

Walcott, R.I., 1979, Plate motion and shear strain rates in the vicinity of the Southern Alps, in R.I. Walcott and M.N. Creswell (editors), The origin of the Southern Alps: Royal Society of New Zealand Bulletin 18, pp. 5–12.

Walcott, R.I., 1984, The major structural elements of New Zealand, in R.I. Walcott (compiler), An introduction to the recent crustal movements of New Zealand: Royal Society of New Zealand Miscellaneous Series, v. 7, pp. 1–6.

Walsh, J.E., 1984, Snow cover and atmospheric variability: American Scientist, v. 72, pp. 51–57.

Wardle, P., 1985, New Zealand timberlines. 3. A synthesis: New Zealand Journal of Botany, v. 23, pp. 263–271.

Washburn, A.L. 1980, Geocryology, a survey of periglacial processes and environments: New York, Wiley, 406 pp.

Wasson, R. J., 1979, The identification of relict periglacial slope mantles: Australian Quaternary Newsletter: v. 13, pp. 26–34.

Weldon, R.J., 1986, Late Cenozoic geology of Cajon Pass; implications for tectonics and sedimentation along the San Andreas fault: California Institute of Technology, Ph.D. thesis, 400 pp.

Weldon, R.J., 1989, Origin of fill terraces in the central Transverse Ranges, California: Transactions, American Geophysical Union, v. 70, no. 43, p. 1125.

Weldon, R.J., and Humphreys, E., 1986, A kinematic model of southern California: Tectonics, v. 5, pp. 33–48.

Weldon, R.J., and Sieh, K.E., 1985, Holocene rate of slip and tentative recurrence interval for large earthquakes on the San Andreas fault in Cajon Pass, southern California: Geological Society of America Bulletin, v. 96, pp. 793–812.

Wells, P.V., 1976, Macrofossil analysis of woodrat (Neotoma) middens as a key to Quaternary vegetational history of arid America: Quaternary Research, v. 6, pp. 223–248.

Wells, P.V., 1979, An equitable glaciopluvial in the west; pleniglacial evidence of increased precipitation on a gradient from the Great Basin to the Sonoran and Chihuahuan Deserts: Quaternary Research, v. 12, pp. 311–325.

Wells, P.V., and Berger, R., 1967, Late Pleistocene history of coniferous woodlands in the Mojave Desert: Science, v. 155, pp. 1640–1647.

Wells, P.V., and Woodcock, D., 1985, Full-glacial vegetation of Death Valley, California; juniper woodland opening to yucca semidesert: Madrono, v. 32, pp. 11–23.

Wells, S.G., Anderson, R.Y., McFadden, L.D., Brown, W.J., Enzel, Y., and Miossec, J.L., 1989, Late Quaternary paleohydrology of the eastern Mojave River drainage, southern California; quantitative assessment of the late Quaternary hydrologic cycle in large arid watersheds: New Mexico Water Resources Research Institute Technical Completion Report 14–08–0001–G1312, Geology Department, University of New Mexico at Albuquerque, 253 pp.

Wells, S.G., and Dohrenwend, J.C., 1985, Relic sheet-flood bedforms on late Quaternary alluvial-fan surfaces in the southwestern United States: Geology, v. 13, pp. 512–516.

Wells, S.G., Dohrenwend, J.C., McFadden, L.D., Turrin, B.D., and Mahrer, K.D., 1985, Late Cenozoic landscape evolution of lava flow surfaces of the Cima volcanic field, Mojave Desert, California: Geological Society of America Bulletin, v. 96, pp. 1518–1529.

Wells, S.G., and McFadden, L.D., 1987, Comment on "Isotopic evidence for climatic influence on alluvial-fan development in Death Valley, California": Geology, v.15, pp. 1178–1180.

Wells, S.G., McFadden, L.D., and Dohrenwend, J.C., 1987, Influence of late Quaternary climatic changes on geomorphic and pedogenic processes on a desert piedmont, eastern Mojave Desert, California: Quaternary Research, v. 27, pp. 130–146.

Wells, S.G., McFadden, L.D., Dohrenwend, J.C., Bullard, T.F., Feilberg, B.F., Ford, R.L., Grimm, F.P., Miller, F.P., Orbock, S.M., and Pickle, J.D., 1984, Lake Quaternary geomorphic history of Silver Lake, eastern Mojave Desert, California; an example of the influence of climatic change in desert piedmonts, *in* J.C. Dohrenwend (editor), Surficial geology of the eastern Mojave Desert, California: 1984 Annual Meeting Field Trip 14 Guidebook, Geological Society of America, Reno, Nevada, pp. 69–87.

Wells, W.G., 1981, Some effects of brush fires on erosion processes in coastal southern California, *in* Erosion and sediment transport in Pacific Rim steeplands: Christchurch, New Zealand, International Association of Hydrological Sciences Publication 132, pp. 305–342.

Wells, W.G., and Brown, W.M., 1982, Effects of fire on sedimentation processes, *in* Sediment management for southern California mountains, coastal plains, and shoreline; Part D, Inland sediment movements by natural processes: California Institute of Technology Environmental Quality Laboratory, EQL Report 17–D, pp. 83–121.

Whitehouse, I.E., 1979, Stream aggradation following recent glacier retreat: New Zealand Journal of Hydrology, v. 18, pp. 49–51.

Whitehouse, I.E., 1983, Distribution of large rock avalanche deposits in the central Southern Alps, New Zealand: New Zealand Journal of Geology and Geophysics, v. 26, pp. 271–279.

Whitehouse, I.E., 1985, The frequency of high-intensity rainfalls in the central Southern Alps: Journal of the Royal Society of New Zealand, v. 15, pp. 213–226.

Whitehouse, I.E., and McSaveney, M.J., 1983, Diachronous talus surfaces in the Southern Alps, New Zealand, and their implications to talus accumulations: Arctic and Alpine Research, v. 15, pp. 53–64.

Whitehouse, I.E., McSaveney, M.J., and Chinn, T.J., 1980, Dating your scree: Journal of the Tussock Grasslands and Mountainlands Institute, Review 39, pp. 15–24.

Whitehouse, I.E., McSaveney, M.J., Knuepfer, P.L.K., and Chinn, T.J., 1986, Growth of weathering rinds on torlesse sandstone, Southern Alps, New Zealand, *in* S.M. Colman and D.P. Dethier

(editors), Rates of chemical weathering of rocks and minerals: Academic Press, pp. 419–435.

Whitley, D.S., and Dorn, R.I., 1987, Rock art chronology in eastern California: World Archaeology, v. 19, pp. 150–164.

Whitley, D.S., and Dorn, R.I., 1988, Cation-ratio dating of petroglyphs using PIXE, *in* Nuclear Instruments and Methods in Physics Research B35: Amsterdam, Elsevier, pp. 410–414.

Wickens, G.E., 1975, Changes in the climate and vegetation of the Sudan since 20,000 b.p.: Boissiera, v. 24, pp. 43–65.

Wickens, G. E., 1982, Paleobotanical speculations and Quaternary environments in the Sudan, *in* M.A.J. Williams and D.A. Adamson (editors), A land between two Niles: Rotterdam, Balkema, pp. 23–50.

Wilson, C.J.N., Switsur, R.V., and Ward, A.P., 1988, A new radiocarbon age for the Oruanui (Wairakei) eruption, New Zealand: Geological Magazine, v. 125, pp. 297–300.

Wilson, L., 1968, Morphogenetic classification, *in* R.W. Fairbridge (editor), Encyclopedia of geomorphology: New York, Reinhold Book Corp., pp. 717–728.

Winkler, E.M., 1975, Stone; properties, durability in man's environment: 2nd edition, New York, Springer-Verlag, 232 pp.

Winograd, I.J., Szabo, B.J., Coplen, T.B., and Riggs, A.C., 1988, A 250,000 year climatic record from Great Basin vein calcite; implications for Milankovitch theory: Science, v. 242, p. 1275.

Wolman, M.G., and Gerson, R., 1978, Relative scales of time and effectiveness of climate in watershed geomorphology: Earth Surficial Processes and Landforms, v. 3, pp. 189–208.

Womack, W.R., and Schumm, S.A., 1977, Terraces of Douglas Creek, northwestern Colorado: Geology, v. 5, pp. 72–76.

Wright, H.E., Jr., 1984, Sensitivity in response time of natural systems to climatic change in the late Quaternary: Quaternary Science Reviews, v. 3, pp. 91–131.

Yaalon, D.H., 1971, Soil forming processes in time and space, *in* Paleopedology–origin, nature, and dating of paleosols: ISSS and Israel University Press, Jerusalem, pp. 29–39.

Yaalon, D.H., 1975, Conceptual models in pedogenesis; can soil-forming functions be solved?: Geo-derma, v. 14, pp. 189–205.

Yaalon, D.H., and Ganor, E., 1973, The influence of dust on soils in the Quaternary: Soil Science, v. 116, pp. 146–155.

Yair, A., and Lavee, H., 1976, Runoff-generative pro-cess and runoff yield from arid talus-mantled slopes: Earth Surface Processes, v. 1, p. 235.

Yair, A., and Lavee, H., 1982, Factors affecting spatial variability of runoff generation over arid hill-slopes, southern Israel: Israel Journal of Earth Sciences, v. 31, pp. 133–143.

Yatsu, E., 1966, Rock control in geomorphology: To-kyo, Sozosha Publishing Company, 135 pp.

Zeeman, E.C., 1976, Catastrophe theory: Scientific American, v. 234, pp. 65–83.

Zuener, F.E., 1959, The Pleistocene period: London, Hutchinson and Company, 447 pp.

# Glossary

**Accuracy** The sum of precision (instrumental error) and interpretational uncertainties of a measured value, generally indicated as a ± range about the most likely value.

**Active channel** The bed, banks, and floodplain of a stream subject to modifications by streamflows.

**Active layer** Materials above perennially frozen ground that are subject to annual freeze-thaw and related periglacial processes.

**Aggradation** A disequilibrium mode of operation that raises the altitude of a reach of an active stream channel by depositing bedload.

**Aggradation event** Alluviation of a valley floor, commonly in response to climatic change.

**Aggradation surface** A fill-terrace tread that marks the end of an aggradation event. It is the fundamental climatic stream-terrace landform where aggradation is the result of geomorphic responses to climatic change.

**Alfisol** The classification order of soils typically formed under forest or savanna vegetation with strongly seasonal precipitation. They commonly have gray-brown A and reddish brown B horizons. Examples include **haploxeralf** (simple alfisol formed under a xeric–strongly seasonal–moisture regime), and **palexeralf** (alfisol formed

under a xeric moisture regime, and which is sufficiently old to have argillic horizons thicker than 50 cm and with more than 35 percent clay).

**Allometric change** The tendency to orderly adjustment among interdependent materials, processes, and landforms in a geomorphic open system.

**Alluvial fan** A fluvial deposit whose surface forms the segment of a cone that radiates downslope from the point where the stream leaves a source area.

**Alluvial fill** A stratigraphic unit of fluvial deposits in a stream valley or on an alluvial fan deposited by one or more aggradation events.

**Alluvial geomorphic surface** Fluvial landforms such as alluvial fans and stream terraces that commonly are formed at climatically induced times of deposition or erosion.

**Alluvial slope** A fluvial piedmont surface of either depositional or erosional origin.

**Antecedent stream** A stream that was present before the formation of a fold, fault scarp, or other tectonic landform, and which maintains its course despite subsequent earth deformation.

**Argillic horizon** A soil-profile horizon where clay minerals have accumulated sufficiently to contrast with the overlying source horizon and to coat and bridge grains with oriented clay.

**Aridisol** The classification order of soils typically formed under sparse vegetation in dry climate. They commonly have argillic, calcic, gypsum or halite-rich, or siliceous horizons. Examples include **haplargids** (simple clayey soils of arid regions); **calciorthids** (calcareous soils of arid regions with a calcic horizon less than 1 m below the surface but no argillic horizon); and **camborthids** (soils of arid regions with only a cambic horizon).

**Astronomical clock** A correlation dating method based on uniform, predictable variations of the earth's orbital parameters, which cause past, present, or future climatic changes.

**Base level** The altitude below which a reach of a stream cannot degrade its bed. Sea level is the ultimate base level.

**Base-level change** Increase or decrease of the altitude of a reach of stream as a result of tectonic or fluvial processes.

**Base-level fall** Lowering of a reach of a stream by tectonic or erosional processes, relative to the adjacent upstream reach.

**Base level of erosion** A concept that integrates the system, equilibrium, and base-level concepts by referring to the longitudinal profile below which a stream cannot degrade. See **static equilibrium; type 1** and **type 2 dynamic equilibrium.**

**Base-level processes** Fluvial aggradation or degradation, or tectonic uplift that changes the altitude of a reach of a stream.

**Base-level rise** Elevation of a reach of a stream by tectonic or depositional processes, relative to the adjacent upstream reach.

**Bedrock** Rock that underlies soil, alluvium, or colluvium.

**Boulder acoustic wave speeds** Acoustic velocities of boulders whose weathering-induced variations can be used as a sensitive tool for dating gravelly deposits.

**Bounding fault** A fault zone that coincides with the mountain–piedmont junction; it generally should be considered active.

**Calcic horizon** A soil-profile horizon where calcium and other carbonate salts have accumulated sufficiently to contrast with the parent material.

**Calciorthid** See **aridisols.**

**Calibrated age** Temporal determination based on rates of tectonic, geomorphic, or pedogenic processes that have been dated.

**Cambic horizon** Commonly a sandy soil-profile horizon where initial stages of pedogenesis provide incipient visible changes of structure, color, or carbonate and clay content when compared with the parent material

**Camborthid** See **aridisols.**

**Catena** A topographic sequence of soils of about the same age and occurring under similar climatic conditions. A study of a soil catena examines the spatial variation of soil-profile characteristics as related to position on and gradient of a slope.

**Cation-ratio dating** A calibrated dating method based on differential leaching rates of mobile to immobile cations in rock varnish.

**Climatic perturbation** A climatic change of sufficient strength to cause aggradation or degradation in a fluvial system.

**Closed system** Interactions between variables that are characterized by irreversibility, such as some chemical reactions.

**Cobble weathering rind** Light-colored zones of oxidation whose inner boundaries parallel the surfaces of boulders and cobbles of rocks such as basalt and graywacke.

**Cobble-weathering stage** Relative disaggregation of gravel clasts resulting from chemical weathering.

**Colluvium** Loose detritus deposited on hillslopes by fluvial and mass-movement processes.

**Complex response** Progression of morphologic or stratigraphic changes that occur within a fluvial system after a single (external) perturbation changes the bedload transport rate.

**Complex-response terraces** Fill-cut, strath, or fill terraces, formed during a degradation event, that result from internal adjustments as dependent variables interact in a reach of a fluvial system that has been elevated by aggradation or uplift.

**Consequent stream** A stream whose course is dependent on the form of preexisting landforms.

**Correlated age** Temporal determination based on demonstration of equivalence to events or deposits that have known numerical ages. Examples are dated volcanic ash layers, global magnetic polarity reversals, and variations in the earth's orbital parameters (the astronomical clock).

**Coseismic mass movement** An earthquake-triggered landslide, rock avalanche, rock fall, or debris flow.

**Creep** Slow downslope mass movement of colluvial materials under gravitational stresses.

**Debris flow** A mass movement of rocky detritus that is intermediate between a landslide and a water flood; debris flows may move down hillslopes and then flow down stream channels.

**Debris slide** Mass movement consisting of downslope movement of thin surficial layer above bedrock or permanently frozen materials.

**Degradation** A disequilibrium mode of operation that lowers the altitude of a reach of a stream.

**Degradation event** Stream-channel downcutting into Quaternary alluvium and bedrock between the times of formation of an aggradation surface and attainment of the lowest possible base level of erosion for the stream. This would be a type 1 dynamic equilibrium condition in tectonically active reaches.

**Degradation terrace** Fill-cut and strath complex-response terraces formed during a degradation event.

**Denudation rate** Overall erosional lowering of the drainage basin surface (in mm/ky).

**Dependent variable** A parameter of a geomorphic system that is influenced by other dependent variables as well as by the independent variables of the system. An example is vegetation.

**Desert pavement** A closely packed surficial layer of rock fragments, commonly protecting underlying gravelly alluvium from erosion.

**Diachronous stream terrace** A terrace whose tread formed over a period of time. Differentiating diachron-

ous from synchronous terraces is in part a function of terrace age and resolution of dating methodologies.

**Diamict** A nongenetic term for a poorly sorted bouldery sheet or lens of massive to thick-bedded silty to clayey sedimentary deposit that may be formed by a variety of glacial, periglacial, fluvial, and mass movement processes.

**Disequilibrium** A reach of a stream that is aggrading, or is downcutting, and has yet to attain the base level of erosion.

**Displacement** Horizontal offset or vertical throw of a landform by relative movement of blocks cut by a fault.

**Drainage basin** The watershed whose divides direct flow of water derived from precipitation down hillslopes into channels of a drainage net that conveys streamflow to the drainage basin mouth.

**Dynamic allometric change** The interrelations of measurements made of a landform or process at different times; it is time dependent. An example is the changes that occur as a rising stream fills its channel.

**Dynamic equilibrium** An adjustment between streamflow variables that occurs when a reach of a stream has downcut to the minimum gradient needed to transport the imposed sediment load. See **base level of erosion; type 1** and **type 2 dynamic equilibrium.**

**Earthquake recurrence interval** The mean time span between characteristic surface-rupture events for a segment of a fault.

**Effective precipitation** The amount of water available to support plants, weather rocks and form soil profiles, and supply runoff from hillslopes. It is a function of temperature, precipitation, and windiness.

**Elapsed time** The time span since the last surface rupture along a segment of a fault.

**Entic haploxeroll** See **mollisols.**

**Entisols** The classification order of soils typically formed on very young or unstable surfaces (undergoing erosion or deposition). They have minimal horizonation. Examples include **torriorthents** and **xerorthents** formed under torric and xeric moisture regimes.

**Ephemeral stream** A stream that flows only briefly in direct response to rainfall or snowmelt. Its streambed remains above the groundwater table.

**Equifinality** A landform or end point of a geomorphic process caused by one of several indeterminate interactions among variables.

**Equilibrium** A condition of balance between processes operating in a fluvial system.

**Equivalent potential temperature** Calculated from monthly values of maximum temperature, relative humidity, and barometric pressure. This multivariable parameter reduces the influence of altitude on airmass characteristics at the land interface.

**Escarpment** A steeply sloping mountain front formed by uplift and fluvial erosion that rises above a gently sloping lowland.

**Fault** A single fracture, or a zone of fractures, along which there has been tectonic displacement of one side relative to the other.

**Fault displacement** Relative horizontal or vertical tectonic dislocation of a landform or subsurface feature that crosses a fault.

**Fault segmentation** Subdivision of a fault zone based on styles, timing, and amounts of late Quaternary displacements, and on spatial changes in geologic structures and landform assemblages.

**Fault trace** The line described by the intersection of a fault plane and the land surface.

**Fill-cut terrace** Surface formed by lateral stream erosion into recently deposited alluvium during brief periods of static equilibrium followed by renewed stream-channel downcutting.

**Fill terrace** A terrace formed by valley-floor aggradation and subsequent stream-channel incision into the alluvium that leaves remnants of the former active channel as the tread of a paired terrace.

**Flexural isostasy model** A major delay of isostatic induced uplift that occurs where, after plutonic emplacement, a batholithic crustal root is held down in an elastic manner. Uplift begins when extensional stresses break the elastic plate; the mountain range then can rise rapidly, and commonly is tilted. Erosional removal of mass adds another component of isostatic adjustment (Chase and Wallace,1986,1988).

**Floodplain** A planar lowland bordering a stream channel that is covered with water at high stages of streamflow.

**Fluvial system** The interacting hills and streams of a drainage basin. Hillslope and adjacent stream subsystems comprise the watershed, or input reaches, and piedmont depositional subsystems comprise the output reaches of fluvial systems.

**Fragipan** A hard and brittle soil layer that is sufficiently dense to inhibit root penetration.

**Gelifluction** Soil flow under periglacial conditions.

**Geomorphic surface** A mappable landscape element formed during a discrete time period.

**Geomorphic system** A group of interacting landforms and processes.

**Geomorphic threshold** A transition that separates different reversible processes that tend to change part of an open system.

**Geomorphology** The science of landscape description, genesis, and classification.

**Gley soil** A mottled, gray soil formed under reducing conditions that reduce iron; gleying generally occurs where aeration is minimal because of poor drainage.

**Gradational threshold** A change between different modes of operation of a geomorphic system that may result during interaction of variables with ranges of values, instead of an abrupt transition defined by critical values of controlling variables.

**Graded stream** See **base level of erosion, dynamic equilibrium,** and **static equilibrium** .

**Grus** Monomineralic detritus resulting from decay of granitic rocks.

**Haplargid** See **aridisols.**

**Hillslope sediment reservoir** Accumulated colluvium and soils on hillslopes that are sources of sediment yielded to streams. Net reservoir gain or loss is sensitive to climatic change.

**Illuviation** Translocation of soil materials such as clay minerals, iron oxyhydroxides, and calcium carbonate; generally to deeper horizons in a soil profile.

**Independent variable** A fluvial system parameter that is not influenced by other variables within the drainage basin. Examples include tectonic uplift and climatic change.

**Inselberg** An erosional bedrock outlier that rises above a pediment; commonly it is a remnant of a former spur ridge.

**Intermittent stream** A stream that flows during part of the year. Its water is derived from rainfall, snowmelt, or occasional groundwater discharge into the stream channel.

**Iron oxyhydroxides** Minerals resulting from oxidation and hydrolytic weathering of rocks; they include ferrihydrite, goethite, and hematite. Hematite is the most stable pedogenic iron oxyhydroxide.

**Isopleth** A line on a map that defines the areal pattern of a constant value of a variable.

**Leading edge** Landform assemblages in a reach downstream from an active strike-slip fault that are being tectonically translocated into the path of the stream where it crosses the fault zone.

**Lithology** The physical character of igneous, metamorphic and sedimentary rocks based on color, mineralogical composition, grain size, and structural fabric.

**Local base levels** Resistant outcrops, lakes, and dams at the downstream ends of reaches of streams, which remain at about the same altitude.

**Longitudinal profile** A topographic profile along a valley or stream.

**Maximum-horizon index** Sum of normalized numerical values of pedogenic properties of the soil-profile horizon that differs most from the parent material. The value commonly is multiplied by the thickness of the horizon.

**Mode of operation** The behavior of a reach of stream where interactions among variables cause one of three conditions: aggradation, degradation, or equilibrium.

**Mollisols** The classification order of soils typically formed under grass in the presence of abundant calcium. They commonly have highly organic, nearly black surface horizons and argillic horizons. Examples include **haploxerolls** (simple mollisols formed under a xeric moisture regime), and **argixerolls** (have argillic horizons).

**Mountain–piedmont junction** The planimetric junction between a mountainous escarpment and an adjacent piedmont. It tends to be straight in tectonically active settings and sinuous in inactive settings.

**Nickpoint** An abrupt increase in stream gradient represented by a waterfall that migrates upstream by the process of headcutting.

**Nickzone** A short reach of anomalously steep stream gradient. It commonly is a degraded nickpoint that has migrated upstream.

**Numerical age** Temporal determination based on known rates of change described by ratios. Examples include radioactive decay of carbon-14 and radiogenic disruption of crystal lattices as measured by thermoluminescence.

**Open systems** The interaction of mass and energy in a physical environment within a defined space through time.

**Orbital parameters** Three features of the earth's orbit may cause climatic changes: (1)– variations of eccentricity (the slightly elliptical path of the earth's orbit around the sun), (2)– obliquity of axial tilt of the earth to the plane of the orbit, and (3)– precession of the seasonal equinoxes (progressive changes in seasonal timing of the earth's orbital perihelion and aphelion that result from a wobble in the earth's axis of rotation).

**Packrat** Commonly the genus *Neotoma*. It has the habit of collecting bits of vegetation and other objects and storing them in rocky crevices and small caves; these homes are termed **middens.**

**Paired terrace** Remnants of a former valley floor along one or both sides of the present stream that formerly were continuous.

**Paleoseismology** The study of prehistorical earthquakes based on geomorphology, structural geology, stratigraphy, and geochronology.

**Paleosol** A buried soil profile that is overlain by a younger soil profile.

**Parent material** Rock or unconsolidated sedimentary deposits with minimal weathering characteristics, and on which pedogenic processes act to make a soil profile.

**Partial area contribution** That portion of a watershed that supplies runoff of water and sediment load to a stream during a rainfall event.

**Pediment** An erosion surface formed by retreat of an escarpment and by continued planation and degradation of rocks and basin fill downslope from the escarpment.

**Pedon** A single soil profile in a landscape. It is the smallest soil descriptive unit that includes all the genetic soil horizons.

**Perennial stream** A stream that flows continuously as a result of surface runoff and influx of groundwater.

**Periglacial** Geomorphic processes and conditions where frozen ground and frost action are important.

**Persistence time** The time span of the new equilibrium condition after a stream has adjusted to a perturbation.

**Perturbation** A change in one of the variables of a system such as displacement along a fault or increased rainfall.

**Phi scale** Particle diameters of the Wentworth size classes used for analyses of sedimentary particle-size distribution; expressed as negative logarithms to the base 2.

**Piedmont** A plain that slopes away from the base of a mountainous escarpment.

**Piedmont foreland** A belt of former piedmont terrain that commonly is between bounding and internal thrust faults.

**Plant-species threshold** The ratio between factors that if increased favor growth and reproduction of resident (numerator) and invading (denominator) species in a plant community.

**Polygenetic system** Alternation between modes of totally different landscape-shaping processes. Examples include periglacial and fluvial erosion on alpine hillslopes, and eolian and fluvial processes in the Sahara Desert.

**Precision** The instrumental error of a measured value, generally indicated as a plus or minus ( ± ) range about the mean value.

**Process-response model** A hypothesis regarding the likely interactions between variables that define self-enhancing and self-arresting feedback mechanisms and crossings of thresholds and attainment of equilibrium conditions, which result in certain rates and magnitudes of geomorphic processes and landscape change.

**Reach** That part of a stream and its adjacent hillslopes with characteristics that differ from the adjacent upstream and downstream parts of a valley.

**Reaction time** The interval between the time of initiation of a perturbation to a fluvial system and the time that a change in system behavior is first observed.

**Refugia** A place where a plant or animal community can survive a climatic change.

**Relative age** Temporal position in a sequence of landscape or stratigraphic units.

**Relaxation time** The time needed to establish a new equilibrium or threshold condition after a fluvial system has reacted to a perturbation (Thornes and Brunsden, 1977).

**Relict soil profile** A soil that has remained at the land surface. Its polygenetic properties commonly reflect a combination of initial and recent soil-forming processes.

**Resisting power** The power needed to entrain and transport the bedload supplied to the reach of a stream. It consists of those variables (such as hydraulic roughness and the amount and size of bedload) that if increased, favor deposition of bedload and valley-floor aggradation.

**Response time** The sum of reaction and relaxation times.

**Rock mass strength** The total resistance of a rock type to the applied stresses of flowing water, taking into account cohesive strength and presence of fractures, joints, shears, and groundwater head differentials (Selby, 1982).

**Rock varnish** A brownish-black coating of iron oxyhydroxides and manganese oxides, clay, and minor organic matter that accretes on rock surfaces.

**Scarp** A linear cliff or slope of any height that separates surfaces.

**Sediment yield** Volume of earth materials yielded per square kilometer of drainage basin per year.

**Seismology** Instrumental measurements and analyses of historical earthquakes.

**Self-arresting feedback mechanism** Interactions between variables that tend to eliminate changes that are occurring in geomorphic processes or landforms.

**Self-enhancing feedback mechanism** Interactions between variables that tend to perpetuate changes in geomorphic processes or landforms.

**Semilogarithmic profile** A graph of altitude (ordinate) versus logarithm of length (abscissa) of a valley floor in the downstream direction. It is dimensionless where both values are expressed in terms of percent.

**Slope distance** The distance along the land surface between two points on a hillslope.

**Slope fall** The vertical distance between two points on a hillslope.

**Slope length** The horizontal distance between two points on a hillslope.

**Soil-development index** Sum of normalized numerical values of pedogenic properties of all soil-profile horizons compared with the parent material. Value commonly is multiplied by thickness of the soil profile.

**Soil-leaching index** A comparison of the effectiveness of available moisture in soil-profile formation based on precipitation and evapotranspiration.

**Soil profile** A vertical sequence of soil horizons above the parent material.

**Soils chronosequence** A temporal sequence of soils formed on alluvial geomorphic surfaces of different ages under similar topographic and climatic conditions (Stevens & Walker, 1970).

**Solifluction** Slow flow of soil and colluvial materials that is saturated because of underlying frozen ground.

**Static allometric change** Refers to the interrelations of measurements made of a landform or process at a single time; it is time independent. An example, is the changes that occur in the downstream direction for a stream at bankfull stage.

**Static equilibrium** An adjustment between all the interacting streamflow variables of a stream system so as to transport the sediment load with neither aggradation nor degradation of the streambed (Leopold and Bull, 1979).

**Steady state** A landscape condition where every slope and every form is adjusted to every other.

**Strath** A beveled bedrock surface formed by streamflow.

**Strath terrace** A stream terrace formed by channel incision into a strath. Strath terraces represent pauses in downcutting through bedrock during attainment of equi-

librium conditions; they may be genetically the same as fill-cut terraces.

**Stratigraphic superposition** Progressively younger overlapping alluvial fills that occur as a sequence above each other.

**Stream-channel dissection** Formation of a drainage net on an alluvial geomorphic surface by locally derived runoff.

**Stream-channel incisement** Fluvial lowering of a valley floor by trunk streamflows derived from watershed runoff.

**Stream-gradient index** (SL index) The product of the slope of a reach of a stream times the length from the headwaters divide to the midpoint of the reach (Hack, 1973). The stream-gradient index is a function of tectonically induced steepening, streamflow characteristics, resistance of earth materials to erosion, and size of streambed gravel clasts.

**Stream order** The relative position of stream segments in a channel network where stream orders increase systematically downstream.

**Stream power** The power available to transport bedload. It consists of those variables (such as discharge and gradient) that if increased, favor bedload transport and fluvial degradation.

**Stream terrace** Former level of a broad valley floor that was created by aggradation or by lateral erosion, but now is above the present valley floor because of incision by the stream. A terrace consists of a tread and a riser, which separates the tread from the stream or a lower terrace. Paired stream terraces have continuity along a valley but unpaired terraces occur only in one reach.

**Summer monsoon** Seasonal influx of rain drawn into a large continental area by atmospheric low pressure caused by prolonged intense solar heating.

**Synchronous terrace** A terrace that appears to have a single time of tread formation within the constraints of the dating methods used.

**Tafoni** Cavernous weathering of massive rocks, such as crystalline plutonic rocks and sandstone, during subaerial exposure that results from variations in case hardening, salt weathering, and chemical weathering.

**Tectonically induced aggradation** Fluvial deposition that occurs in the adjacent reach downstream from a zone of tectonic base-level fall. It is caused by stream-channel headcutting that increases sediment yields as it migrates upstream.

**Tectonically induced downcutting** Stream-channel downcutting that occurs upstream from a zone of tectonic base-level fall. It begins a nickpoint or nickzone that migrates upstream.

**Tectonic geomorphology** The study of the influences of vertical and horizontal earth deformation on geomorphic processes and resulting landforms, and evaluation of tectonic history as recorded by landscapes.

**Tectonic landform** A landscape element initially created by the processes of vertical or horizontal earth deformation. Examples are fault scarps, anticlines, and subduction trenches.

**Tectonic perturbation** A tectonic increase in relief, or horizontal disruption of fluvial landforms, of sufficient strength to affect rates of aggradation or degradation in a fluvial system.

**Tectonics** The study of deformation and evolution of faults and folds and the crustal blocks between them.

**Tectonic stream terrace** The strath beveled while a stream is at type 1 dynamic equilibrium that, if subject to stream-channel incisement, becomes a strath terrace. If buried by the deposits of an aggradation event, the tectonic strath remains a buried feature until exposed by tectonically induced downcutting.

**Tephra** Pyroclastic material ejected from a volcano.

**Terrace flight** A sequence of stream or marine terraces that descends from oldest to youngest.

**Terrace riser** A fluvial scarp above the tread of a stream terrace; both are formed at the same time. Heights of terrace risers determine the vertical separations between the treads of a stream-terrace flight.

**Terrace tread** A remnant of a former valley floor that has been abandoned by the stream as a result of stream-channel incision.

**Threshold intersection point** A point on a longitudinal profile that separates aggrading and degrading reaches of a stream.

**Threshold of critical power** A transition between the modes of net deposition and net erosion by streams, or vice versa. Opposing tendencies for aggradation and degradation are described by a ratio whose numerator consists of variables that if increased favor degradation and whose denominator consists of variables that if increased favor aggradation.

**Topographic inversion** Landscape change where valley floors become ridgecrests through processes of differential denudation of terrains underlain by rocks with greatly different weathering and erosion properties.

**Topographic segmentation** The effects of tectonism that result in segmented mountain fronts, fault scarps, and longitudinal profiles of streams and alluvial fans.

**Topographic succession** Progressively younger stream terraces that are inset into a valley and whose treads occur as a sequence below each other.

**Torriorthent** See **entisols.**

**Trailing edge** Landform assemblages in a reach downstream from an active strike-slip fault that are being tectonically translocated away from the path of the stream where it crosses the fault zone.

**Transport-limited slope** A scarp or other hillslope that lacks a cliffy slope element. It tends to change morphology by the processes of slope decline at rates that are largely controlled by the rates at which erosional processes remove weakly cohesive earth materials.

**Triangular facet** A triangular or rectilinear surface of a truncated end of a spur ridge associated with normal faults, anticlines, thrust faults, and erosional base-level falls. The degree of fluvial dissection increases with increase of the ratio of consequent valley downcutting rate/uplift rate.

**Triangular talus facets** Triangular facets that form on talus-covered slopes where resistant cliff-forming strata are underlain by weak slope-forming strata. These lithologically controlled hillslope landforms commonly record alternating accumulation and dissection of talus that reflect major late Quaternary climatic changes.

**Trunk stream** The principal through-flowing stream-course that connects the different reaches of a fluvial system. It commonly heads in a mountainous area and increases in size by addition of tributary streamflows.

**Type 1 dynamic equilibrium** A condition of adjustment between processes operating in a reach of a stream that occurs when rates of tectonically induced downcutting equal rates of uplift, thus allowing the longitudinal profile of a stream to attain a succession of base levels of erosion in a tectonically active environment. Flights of strath terraces, broad straths, and longitudinal profiles of valley floors that plot as straight lines on semilogarithmic graphs are common in such reaches.

**Type 2 dynamic equilibrium** Present in streams that have a strong tendency toward, but lack of clear attainment of, the base level of erosion. Diagnostic landforms include valley floors that are narrow and lack strath terraces, but whose longitudinal profiles plot as straight lines on semilogarithmic graphs.

**Typic argixeroll** See **mollisols**.

**Typic haploxeralf** See **alfisols**.

**Typic palexeralf** See **alfisols**.

**Typic xerorthent** See **entisols**.

**Unpaired terraces** Isolated flats occurring at different heights above a stream that never formed a synchronous single tread; they commonly form during continuous downcutting and episodic point-bar formation.

**Variable** An object or an attribute that varies in time or space.

**Vertisols** The classification order of soils with swelling clays that typically form under strongly seasonal precipitation. They commonly have wide, deep shrinkage cracks when dry.

**Weathering-limited slope** A cliffy slope element on a scarp or other hillslope. It tends to change morphology by the processes of parallel erosional retreat, whose rates are largely controlled by the rates at which materials are made available for geomorphic processes to remove.

# Index

Topics and place names are indexed using page numbers set in four type styles. A plain arabic number denotes a text discussion. The italic letter *f* following an arabic numeral denotes a figure. An italic letter *t* following an italic numeral designates a table. Definitions are in boldface numerals (for example, **46**). Use Glossary (pp. 313–321) as principal source of complete definitions.

**Classification of climates.**

| Precipitation | | Temperature | |
|---|---|---|---|
| *Class* | *Mean Annual (mm)* | *Class* | *Mean Annual (°C)* |
| Extremely arid | <50 | Pergelic | >0 |
| Arid | 50–250 | Frigid | 0–8 |
| Semiarid | 250–500 | Mesic | 8–15 |
| Subhumid | 500–1000 | Thermic | 15–22 |
| Humid | 1000–2000 | Hyperthermic | >22 |
| Extremely humid | >2000 | | |
| *Class* | *Seasonality Index (Sp)[a]* | *Class* | *Seasonality Index (St)[b] (°C)* |
| Nonseasonal | 1–1.6 | Nonseasonal | <2 |
| Weakly seasonal | 1.6–2.5 | Weakly seasonal | 2–5 |
| Moderately seasonal | 2.5–10 | Moderately seasonal | 5–15 |
| Strongly seasonal | >10 | Strongly seasonal | >15 |

[a]Precipitation seasonality index (Sp) is the ratio of average total precipitation for the three wettest consecutive months (Pw) divided by average total precipitation for the three driest consecutive months (Pd).

$$Sp = \frac{Pw}{Pd}$$

[b]Temperature seasonality index (St) is mean temperature of the hottest month (Th) minus mean temperature of the coldest month (Tc), in °C.

$$St = Th - Tc$$